# Ultraprecision Machining and Metrology

This comprehensive introduction to ultraprecision machining and metrology provides an essential foundation for students and engineers, offering an in-depth analysis of key methods, technologies, and practical applications.

Ultraprecision machining is a critical enabling technology for producing high-value mechanical, optical, optoelectronic, and biomedical components with complex geometries and extreme precision. This book delivers a structured exploration of ultraprecision machining and metrology, covering essential topics such as system configuration, tooling, machining mechanism modeling, and surface characterization. Dedicated chapters on surface and subsurface metrology, as well as the machinability of both ductile and brittle materials, make it an indispensable resource for understanding machine design, manufacturing processes, and related materials science.

Ideal for students and researchers, this book serves as a vital reference for those working in precision machining, MEMS, advanced manufacturing, and precision metrology.

# Ultraprecision Machining and Metrology

Jiwang Yan

CRC Press
Taylor & Francis Group
Boca Raton London New York

CRC Press is an imprint of the
Taylor & Francis Group, an **informa** business

Designed cover image: Getty Images

First edition published 2026
by CRC Press
2385 NW Executive Center Drive, Suite 320, Boca Raton FL 33431

and by CRC Press
4 Park Square, Milton Park, Abingdon, Oxon, OX14 4RN

*CRC Press is an imprint of Taylor & Francis Group, LLC*



ISBN: 978-1-032-27357-0 (hbk)
ISBN: 978-1-032-27362-4 (pbk)
ISBN: 978-1-003-29238-8 (ebk)

DOI: 10.1201/9781003292388

Typeset in Times
by SPi Technologies India Pvt Ltd (Straive)

# Contents

Author ........ x
Preface ........ xi
Acknowledgments ........ xiii

**Chapter 1** Introduction ........ 1

1.1 Trends of Manufacturing ........ 1
1.2 Applications of Ultraprecision Machining ........ 1
1.2.1 Optical Applications ........ 1
1.2.2 Electronic Applications ........ 15
1.2.3 Biomedical and Biochemical Applications ........ 18
1.3 Future Perspectives of Ultraprecision Machining ........ 20
1.3.1 Precision and Accuracy ........ 20
1.3.2 Surface Integrity ........ 21
1.3.3 Shape Flexibility ........ 21
1.3.4 Productivity ........ 21
1.3.5 Sustainability ........ 22
1.4 Importance of Ultraprecision Metrology ........ 22
1.5 Outline of this Book ........ 23
References ........ 23

**Chapter 2** Ultraprecision Machining Methods ........ 25

2.1 Introduction ........ 25
2.1.1 Definition of "Precision" and "Ultraprecision" ........ 25
2.1.2 Methodologies for Manufacturing ........ 25
2.2 Cutting ........ 25
2.2.1 Turning ........ 26
2.2.2 Fly Cutting ........ 29
2.2.3 Milling ........ 31
2.2.4 Shaping/Planning ........ 33
2.3 Grinding ........ 34
2.4 Lapping ........ 36
2.5 Polishing ........ 37
2.5.1 Mechanical Polishing ........ 38
2.5.2 Chemomechanical Polishing ........ 38
2.5.3 Float Polishing ........ 40
2.5.4 Elastic Emission Machining ........ 40
2.6 Electrical Machining ........ 41
2.6.1 Electrical Discharge Machining ........ 41
2.6.2 Electrical Chemical Machining ........ 42

2.7 Laser Processing ........ 45
2.7.1 Femtosecond Pulsed Laser Machining ........ 45
2.7.2 Laser Polishing ........ 45
2.7.3 Laser Recovery ........ 46
2.8 Electron Beam Polishing ........ 47
2.9 Ion Beam Figuring ........ 49
2.10 Plastic Injection Molding/Imprinting ........ 50
2.11 Glass Molding ........ 51
2.12 Hybrid Processes ........ 53
2.12.1 EDM-Grinding Hybrid Process ........ 53
2.12.2 Cutting and Photolithography Hybrid Process ........ 54
References ........ 56

**Chapter 3** Ultraprecision Machine Tool System ........ 59

3.1 Introduction ........ 59
3.2 Machine Structures ........ 60
3.3 Slides, Guideways, and Spindles ........ 61
3.3.1 Sliding Bearing ........ 62
3.3.2 Rolling Bearing ........ 63
3.3.3 Hydrostatic Bearing ........ 63
3.3.4 Pneumatic Bearing ........ 64
3.3.5 Magnetic Bearing ........ 65
3.4 Axis Drives ........ 65
3.4.1 Servo Motor ........ 65
3.4.2 Linear Motor ........ 66
3.5 Positioning and Feedback Control ........ 67
3.5.1 Laser Interferometer ........ 68
3.5.2 Grating Scale ........ 68
3.6 Chucking and Clamping ........ 71
3.6.1 Vacuum Chucking System ........ 71
3.6.2 Mechanical Clamping System ........ 72
3.7 Cooling and Lubricating ........ 73
3.8 Environmental Controlling Technology ........ 75
3.8.1 Thermal Control ........ 75
3.8.2 Vibration Control ........ 76
References ........ 78

**Chapter 4** Tooling Technologies ........ 80

4.1 Introduction ........ 80
4.2 Tool Materials ........ 80
4.2.1 Single-Crystal Diamond ........ 81
4.2.2 Poly-Crystal Diamond ........ 84
4.2.3 Nano-Polycrystal Diamond ........ 84

4.2.4 Cubic Boron Nitride ... 86
4.2.5 Tungsten Carbide ... 88
4.3 Tool Design ... 89
4.3.1 General Shape Design ... 90
4.3.2 Edge Angle Design ... 93
4.3.3 Crystal Orientation (of Diamond) ... 93
4.3.4 Edge Sharpness ... 96
4.4 Tool Fabrication ... 96
4.4.1 Mechanical Polishing ... 96
4.4.2 Thermo-Chemical Polishing ... 97
4.4.3 Ion Milling ... 98
4.4.4 Laser Machining ... 99
4.4.5 Electro-Discharge Machining ... 101
References ... 101

**Chapter 5** Machining Mechanism and Machinability of Ductile Materials ... 105

5.1 Introduction ... 105
5.2 Machining Mechanism ... 106
5.2.1 Chip Formation ... 106
5.2.2 Cutting Force and Energy ... 109
5.2.3 Surface Roughness ... 111
5.2.4 Surface Formation ... 118
5.3 Machinability of Metal Materials ... 120
5.3.1 Copper and Its Alloys ... 120
5.3.2 Aluminum and Its Alloys ... 122
5.3.3 Nickel Alloy ... 125
5.3.4 Gold and Other Rare Metals ... 128
5.3.5 Ferrous Metals ... 130
5.3.6 Titanium/Magnesium Alloys ... 134
5.3.7 High-Entropy Alloys ... 135
5.4 Machinability of Polymetric Materials ... 137
5.4.1 Photoresist ... 138
5.4.2 PMMA ... 140
5.4.3 Cellulose Nanofiber-Reinforced Plastics ... 143
References ... 147

**Chapter 6** Machining Mechanism and Machinability of Brittle Materials ... 151

6.1 Introduction ... 151
6.2 Machining Mechanism ... 152
6.2.1 Brittle–Ductile Transition ... 152
6.2.2 Ductile Cutting Methods ... 156
6.2.3 Effect of Tool Edge Geometry ... 159
6.2.4 Effects of Crystal Orientation ... 165

6.3 Machinability of Crystalline Brittle Materials ................ 167
6.3.1 Silicon ................ 167
6.3.2 Germanium ................ 170
6.3.3 Calcium Fluoride ................ 173
6.3.4 Indium Phosphide ................ 177
6.3.5 Gallium Arsenide ................ 182
6.3.6 Zinc Selenide ................ 185
6.3.7 KDP ................ 189
6.3.8 Silicon Carbide ................ 192
6.4 Machinability of Glass ................ 195
6.4.1 Borosilicate Glass ................ 195
6.4.2 Soda Lime Glass ................ 197
6.4.3 BK7 Glass ................ 200
6.4.4 Fluoride Glass ................ 203
6.5 Machinability of Ceramics ................ 204
6.5.1 Reaction-Bonded Silicon Carbide ................ 204
6.5.2 Tungsten Carbide ................ 210
6.5.3 Zirconia ................ 213
References ................ 216

**Chapter 7** Surface Metrology ................ 222

7.1 Introduction ................ 222
7.2 Contact Measurement ................ 223
7.2.1 Stylus Profilometry ................ 223
7.2.2 Scanning Probe Microscopy ................ 225
7.2.3 Ultrahigh Accuracy 3D Profilometer ................ 226
7.3 Noncontact Measurement ................ 228
7.3.1 Scanning Electron Microscopy ................ 228
7.3.2 Confocal Laser Scanning Microscope ................ 231
7.3.3 White Light Interferometer ................ 233
7.3.4 Laser Probe Profilometer ................ 235
7.4 On-Machine Measurement of Workpiece ................ 236
7.4.1 On-Machine Stylus Profiling ................ 238
7.4.2 On-Machine Interferometer Profiling ................ 239
7.4.3 On-Machine Laser Probe Profiling ................ 240
7.5 On-Machine Measurement of Tools ................ 241
7.5.1 Tool-Workpiece Transcription Method ................ 242
7.5.2 Stylus Profiling of Tools ................ 243
7.5.3 Noncontact Tool Measurement ................ 244
References ................ 249

**Chapter 8** Subsurface Metrology ........ 252

8.1 Introduction ........ 252
8.2 Subsurface Damage Classification ........ 252
8.2.1 Microcrack ........ 252
8.2.2 Grain Refinement ........ 253
8.2.3 Phase Transformation ........ 254
8.2.4 Dislocation ........ 256
8.2.5 Residual Stress ........ 256
8.3 Subsurface Damage Evaluation Methods ........ 257
8.3.1 Slanted Polishing ........ 257
8.3.2 Small Tool Polishing ........ 258
8.3.3 Transmission Electron Microscope ........ 259
8.3.4 Nanoindentation Tests ........ 261
8.3.5 Chemical Etching ........ 264
8.3.6 X-Ray Diffraction ........ 269
8.4 Laser Micro-Raman Spectroscopy ........ 271
8.4.1 Principle of Raman Scattering ........ 271
8.4.2 Raman Spectroscopy of Subsurface Damage ........ 272
8.4.3 Quantitative Evaluation of Subsurface Damage Depth ........ 274
8.4.4 Raman Mapping Measurement ........ 278
References ........ 279

**Index** ........ 283

# Author

**Jiwang Yan** earned a PhD at Tohoku University, Japan in 2000 and is currently a Professor of Mechanical Engineering at Keio University, Japan (2012–present), leading the Precision Machining and Nano Processing Laboratory. As adjunct appointments, he has served as a Specially Appointed Professor at Tokyo Institute of Technology (2017–2024) and Leverhulme Visiting Professor at the University of Manchester (2018–2019). His research areas include ultraprecision machining, micro/nano manufacturing, laser processing, and manufacturing sustainability. As a principal investigator, he has led 20+ nationally funded projects and 70+ industrial projects. He has authored/co-authored 300+ peer-reviewed international journal papers, given 150+ keynote/invited talks, and received 40+ awards. He is the author of the book *Micro Nano Scale Laser Processing of Hard Brittle Materials* (Elsevier 2019) and the editor of the book *Micro and Nano Fabrication Technology* (Springer 2018). He is a member of the Japan Society of Mechanical Engineers (JSME), the Japan Society for Precision Engineering (JSPE), the Japan Society for Abrasive Technology (JSAT), the Japan Society for Applied Physics (JSAP), the American Society for Precision Engineering (ASPE), and the European Society for Precision Engineering and Nanotechnology (euspen). Since 2022, he has been serving as the executive director of JSPE and the Chairman of the Publication Sector and a board member of the Japan Society of Laser Technology (JSLT). He also serves on the editorial boards of several journals, including *International Journal of Machine Tools and Manufacturer (IJMTM)*, *Journal of Material Processing Technology (JMPT)*, and *International Journal of Extreme Manufacturing (IJEM)*. He was elected as a Fellow of the International Academy for Production Engineering (CIRP) and a Fellow (foreign) of the Engineering Academy of Japan (EAJ) in 2024 and a Fellow of the Society of Manufacturing Engineers (SME) in 2025.

# Preface

Ultraprecision machining and metrology represent two critical pillars of modern manufacturing, playing a fundamental role in industries ranging from optics and semiconductors to biomedical devices, and beyond. As we continue to push the boundaries of performance, functionality, and stability, the demand for parts and components with nanoscale tolerances and extreme surface qualities is rapidly increasing. This book aims to provide graduate students, engineers, and researchers with an in-depth understanding of the principles, phenomena, techniques, and applications associated with ultraprecision machining and metrology required to ensure the quality of products.

The field of ultraprecision machining, although deeply rooted in traditional methods of material removal has evolved significantly with the advent of new technologies, new materials, and new applications. It involves the precise shaping of materials to achieve surface finishes and dimensional accuracies that were once thought impossible. However, achieving and maintaining such precision presents a complex set of challenges that go far beyond conventional machining practices. The scientific understanding of the interactions between tools and materials at the micro/nanometer scale, as well as the mechanisms behind the formation of ultraprecision surfaces, has become extremely important.

Similarly, metrology, the science of measurement, plays a vital role in ensuring the accuracy and repeatability of ultraprecision machining processes. As tolerances shrink, measurement systems must evolve to accommodate these extreme requirements. The field of ultraprecision metrology encompasses a wide range of instruments and techniques. Accurate measurement is essential not only for verifying part dimensions but also for monitoring and controlling the machining process itself to achieve the desired outcomes consistently.

While most of the previous books have focused separately on machining and metrology, this book brings these two closely related fields together, much like two wheels of a cart. It provides a comprehensive overview of both fields, focusing on the latest advancements in methodologies, tools, and systems, covering both theoretical foundations and practical applications. In particular, this text is designed to support the educational needs of graduate students who wish to specialize in advanced manufacturing technologies. It is also intended as a resource for practicing engineers and researchers who seek to deepen their understanding of ultraprecision machining and metrology, whether in academia, industry, or research institutions.

As the mechanism of ultra-precision machining varies significantly depending on the material properties of the workpiece and the machining conditions, this book provides a systematic discussion of this in considerable detail using a lot of illustrations and photographs to assist in understanding the invisible microscopic processes. By combining theoretical and practical insights with application examples, this book will equip readers with the knowledge needed to further explore this exciting and rapidly evolving field.

I hope this book serves as an informative guide for all those who wish to contribute to the ultraprecision manufacturing of tomorrow.

**Jiwang Yan**
Yokohama, Japan
March 2025

# Acknowledgments

In completing the manuscript of this book, I would like to express my sincere gratitude to Dr. Weihai Huang, who is a JSPS postdoctoral research fellow in my lab. Dr. Huang has assisted me in various aspects of the preparation of this book, including the collection and compilation of references, creation of some figures and ALT text for the electronic version, as well as proofreading of the manuscript. I am deeply thankful for his invaluable support. I also extend my thanks to Ms. Asako Yanagi, the secretary of my lab, for contacting many publishers, companies, and research institutions to obtain permission to reproduce published content.

Much of the content of this book is based on research results from my own lab, and I would like to extend my heartfelt appreciation to the students, researchers, and collaborators who have worked alongside me throughout this journey. I am also grateful to the companies with which I have engaged in joint research, and I express my sincere thanks to all involved. Without their efforts and contributions, this book would not have been possible.

Finally, I wish to acknowledge my family members for their understanding and support during the writing of this manuscript. The many weekends and holidays spent working meant that I could not always devote time to family activities, and I am truly grateful for their patience and cooperation during this time.

# 1 Introduction

## 1.1 TRENDS OF MANUFACTURING

Manufacturing is the use of machines, tools, and labor to produce goods for use or sale [1]. Manufacturing has been the most fundamental activity of human beings since ancient times. Based on the efforts of infinite number of people for thousands of years, nowadays, manufacturing has become an extremely broad and complicated integration of modern science and technology and the strongest driving force for the social economy.

Manufacturing technology should always be a technology that increases the value, quality, and functionality of products, and meanwhile reduces cost, waste, energy/resource consumption, and pollution. Nowadays, manufacturing is advancing in the directions summarized in Table 1.1. According to their similarity and correlation, some trends are grouped together in the table.

Among the above-mentioned trends, high value-added manufacturing through precision/ultraprecision machining and micro/nanomanufacturing is especially important as it determines the quality, stability, reliability, and functionality of various daily used high-end products. In other words, these manufacturing technologies are closely related to the quality and safety of our lives.

## 1.2 APPLICATIONS OF ULTRAPRECISION MACHINING

Ultraprecision machining is a highly advanced manufacturing process that achieves extremely tight tolerances and surface finishes. Typical applications of ultraprecision include optical elements, biomedical parts, semiconductor substrates, IT devices, as well as aerospace components.

### 1.2.1 Optical Applications

Optical component manufacturing is the biggest field of ultraprecision machining applications. Optical components require extremely tight tolerances and precise geometries to ensure optimal performance. Even minor deviations can lead to significant optical aberrations, affecting image quality. The quality of the surface finish is also critical. The requirements of both high accuracy and high surface quality can be satisfied by ultraprecision machining. Ultraprecision machining is widely used in the production of lenses, mirrors, and other optical components for cameras, telescopes, and microscopes, where precision is critical for performance.

For example, a list of optical applications (not all inclusive) of ultraprecision cutting, a major ultraprecision machining method, is shown in Table 1.2.

DOI: 10.1201/9781003292388-1

**TABLE 1.1**
**Recent Trends in Manufacturing**

| Trends | Objectives |
|---|---|
| – Intelligent manufacturing<br>– Smart manufacturing<br>– Virtual manufacturing | • To realize data collection, analytics, and communication among machines and controllers by using smart/intelligent equipment and sensors.<br>• To improve productivity, accessibility, flexibility, and safety by using the internet of things (IOT).<br>• To realize remote monitoring, servicing, operation and maintenance by using digital twins, machine learning, AI (artificial intelligence), AR (augmented reality), VR (virtual reality). |
| – Flexible manufacturing<br>– Agile manufacturing<br>– Additive manufacturing | • To respond swiftly/flexibly to market changes while maintaining quality/cost control.<br>• To allow for rapid prototyping and customization of products<br>• To reduce production cost and lead time |
| – Clean manufacturing<br>– Green manufacturing<br>– Sustainable manufacturing | • To reduce energy consumption and protect the environment<br>• To protect employees' safety and health<br>• To realize carbon neutrality |
| – High-value manufacturing<br>– Precision manufacturing<br>– Micro/nano manufacturing | • To create/add value for products<br>• To improve product quality/stability/reliability<br>• To create new functionality<br>• To enable miniaturization, integration, and mobility<br>• To realize energy/resource-saving production |

#### 1.2.1.1 Polygon Mirror

An example of flat surface machining is polygon mirror machining. Polygon mirrors are widely used in laser printers, scanners, barcode readers, and other inspection systems, as shown in Figure 1.1. They are also used in laser machining systems for scanning the laser beam. Polygon mirrors are made of aluminum with high purity. A polygon motor consists of multiple flat mirror faces uniformly distributed along a rotation axis. The shape and number of faces depend on the required scanning/printing speed, dot density, motor rotation speed, etc. The normal requirement of polygon mirror machining accuracy is very high. Namely, flatness is $\lambda/4$–$\lambda/6$ ($\lambda$ is the light wavelength), indexing error <30", angular error <90", and reflectivity >85%.

The machining method of polygon mirrors is fly cutting using single-crystal diamond tools. Usually, a number of polygon mirror workpieces are stacked together on an indexing rotary table and machined at one time. The cutting tools are fixed on the flange of a rotating spindle. Balance weights are necessary to counteract the unbalance of the spindle due to the diamond tool. The reflection coefficient of aluminum after diamond cutting is 90–92% to He–Ne laser with a

**TABLE 1.2**
**Applications of Ultraprecision Cutting in Optical Field**

| Category | Name | Shape | Material | Application |
|---|---|---|---|---|
| **Mirror** | Plane mirror | Flat | Oxygen-free copper | $CO_2$ laser system |
| | Spherical mirror | Spherical (convex or concave) | Oxygen-free copper | $CO_2$ laser system<br>Laser resonator |
| | Paraboloidal mirror | Paraboloidal | Oxygen-free copper | $CO_2$ laser system<br>Condenser |
| | Off-axis paraboloidal mirror | Paraboloidal | Oxygen-free copper<br>Nickel plating | Optical system<br>Laser nuclear fusion |
| | Ellipsoidal mirror | Ellipsoidal | Nickel plating | Condenser<br>X-ray microscope |
| | Hyperbolic mirror | Hyperbolic | Nickel plating | X-ray microscope |
| | Conical mirror | Conical (outer/inner) | Oxygen-free copper | Laser machine |
| | Polygon mirror | Multi-plane | Aluminum | Scanner<br>Laser printer |
| **Lens** | Laser resonating element | Plane | Gadolinium Gallium Garnet (GGG) | Laser resonator |
| | Aspherical lens | Aspherical | Optical crystal (Si, Ge, ZnSe, ZnS) | Infrared optical system |
| | Fresnel lens | Grooved | Optical crystal (Si, Ge, ZnSe, ZnS) | Infrared optical system |
| **Mold** | Lens array mold | Dimpled | Nickel plating | Image sensor<br>Optical fiber communication |
| | Light guiding plate mold | Grooved | Nickel plating | Flat panel display |
| | Fresnel lens mold | Grooved | Oxygen-free copper<br>Nickel plating | Projector<br>Solar condenser |
| | Aspherical lens mold | Aspherical | Nickel plating, Tungsten carbide, Silicon carbide | Camera<br>Smartphone<br>CD/DVD player/recorder |
| | Freeform optical mold | Sculptured | Oxygen-free copper<br>Steel | AR/VR optics |

wavelength of 633 nm. After machining, a thin protective coating is usually deposited on the aluminum surface to protect the surface, which will cause a slight decrease in the reflection coefficient.

#### 1.2.1.2 Laser Reflecting Mirror

$CO_2$ lasers (wavelength of 10.6 μm) are used in laser cutting, welding, and heat treatment of thick metal sheets. In low-power (~5 kW) laser systems, infrared lenses made of optical crystals such as ZnSe are used to concentrate the laser beam. However, for high-power (~ a few tens of kW) lasers, the lenses will be

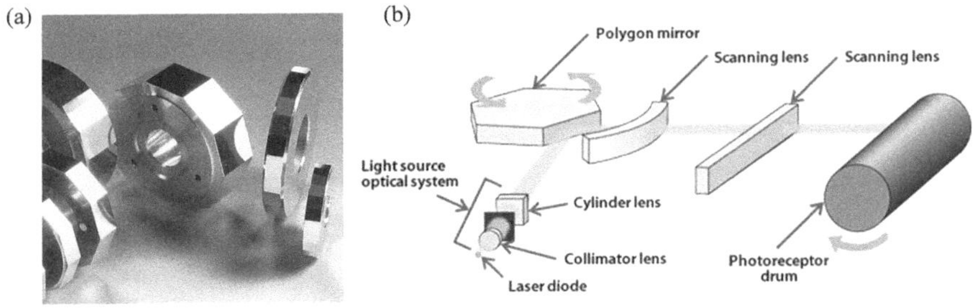


**FIGURE 1.1** Polygon mirrors and their application in laser printers.

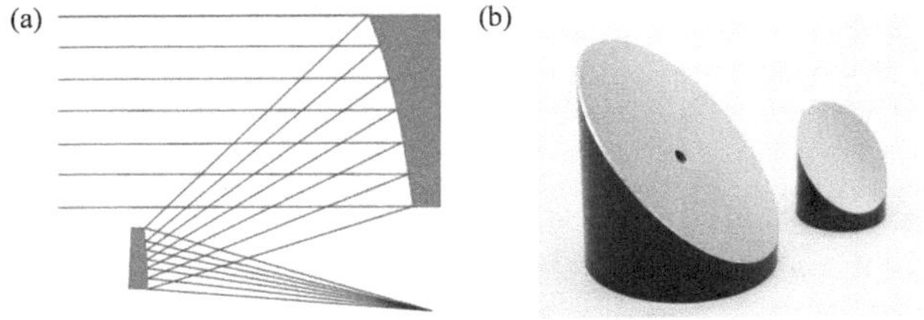


**FIGURE 1.2** Off-axis mirrors: (a) principle and (b) photograph. (Courtesy of Firebirds.)

thermally damaged. In this case, multiple reflecting mirrors, which are made of metal, such as oxygen-free copper, are used in the optical system of the $CO_2$ laser to concentrate the laser beam. The mirrors include planar mirrors, spherical (concave and convex) mirrors, parabolic mirrors, and off-axis mirrors. The mirror surface must have high form accuracy ($\lambda/5$–$\lambda/40$), low surface roughness (3–5 nm Ra), high reflection rate, and high threshold resistance to laser damage.

Figure 1.2 shows an optical path diagram and a photograph of off-axis mirrors [2]. The laser beam is directed to the mirror along the optical axis and then reflected and concentrated by the mirror to the focus. The mirrors are made of aluminum alloy or oxygen-free copper, and water cooling is performed on the back of the mirrors. To machine the mirrors, multiple workpieces are mounted to the circumference of a round jig fixed to the end of the machine spindle and are machined together. As the weight of the workpiece and the jig is large, to guarantee the machining accuracy, the stiffness and accuracy of the spindle are extremely important. The balancing of the jig and workpiece at a relatively high rotation speed is also important.

### 1.2.1.3 Aspherical Lens

An aspheric surface is expressed as the difference between a sphere and an asphere at different heights above the optic axis. An axisymmetric aspheric surface can be described by

$$z(x) = \frac{Cx^2}{1+\sqrt{1-(k+1)C^2x^2}} + \sum_{i=1}^{m} a_i x^i \quad (1.1)$$

where $C = 1/r$, in which $r$ is the radius of curvature of the spherical surface; $x$ is the distance from the optic axis ($Z$); $k$ is the conic constant, a parameter representing the eccentricity of the conic surface; $a_i$ for even $i$ are the aspheric deformation constants, and $a_i$ for odd $i$ are aspheric coefficients used to define other polynomial curves by setting $C = 0$.

Aspherical lenses can remarkably eliminate spherical aberrations of spherical lens. The function of a large group of spherical lenses can be replaced by a single aspherical lens, which enables miniaturization and low-cost production of optical systems. Most optical lenses used in the visible light wavelength are made of glass and plastics.

PMMA is the major plastic material for optical lenses such as camera lenses, objective lenses for CD pickup systems, lenses for VTR, and lenses for eyeglasses. In some of these applications, the shape of the lenses is aspherical. Plastic lenses are manufactured by replication methods, such as press molding and injection molding. To obtain high-accuracy plastic lenses, the form accuracy of the molding die, alternatively termed core or insert, is required to be very high, normally at the submicron level. The surface roughness of the molding die used for visible light is required to be Ry < 30 nm, Ra < 10 nm. Molds for plastic aspherical lenses are usually made of metals such as copper or stainless steel plated with nickel-phosphorus (NiP) alloy and machined by diamond turning. The machining and measurement technologies for ultraprecision molding dies for plastic lenses have been established since the 1980s.

A glass lens has many predominant advantages over the plastic counterpart in aspects of hardness, refractive index, light permeability, stability to environmental changes in terms of temperature and humidity, and so on. For this reason, glass lenses have been needed increasingly in the field of high-resolution digital cameras, mobile phone cameras, and CD/DVD players and recorders. Conventionally, glass aspherical lenses are fabricated by a series of material removal processes such as grinding, lapping and polishing, which requires a long production cycle and results in high production cost. As an alternative approach, glass molding press (GMP) process has been demonstrated to be an effective way to efficient production of glass aspherical lenses [4]. Figure 1.3 presents optical path diagrams of spherical and aspherical lenses, an example of lens design, and some glass aspherical lenses produced by glass molding technology.

A key technology for manufacturing aspherical lenses is mold fabrication. Molds for glass lenses are conventionally made of cemented carbide or silicon carbide, which require ultraprecision grinding and polishing to obtain high form accuracy and low surface roughness. The fabrication of glass molding dies is more difficult than that of plastic molding dies, thus is nowadays still a research focus for researchers and engineers in this area.

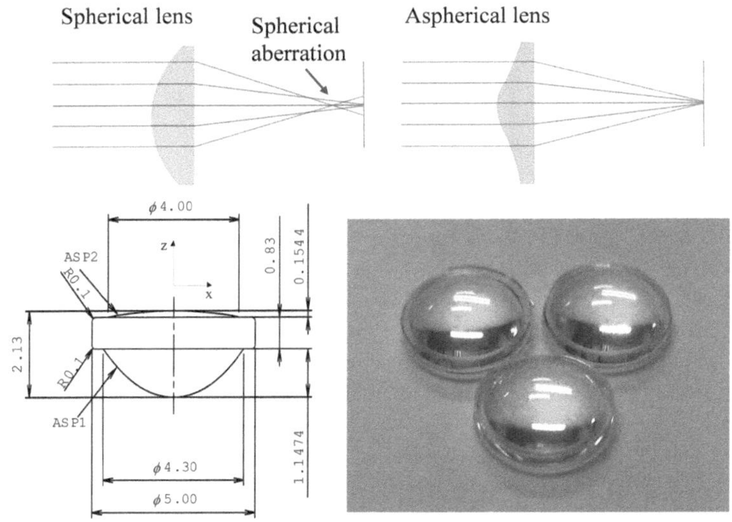


**FIGURE 1.3** Example of glass aspherical lenses produced by glass molding [3].

The light with a longer wavelength than that of the visible light is infrared light. Recently, infrared optical lenses have been required more and more in various fields such as night vision systems of vehicles, dark field sensing systems for security, and so on. Infrared optical lenses are usually made of brittle materials such as single crystalline germanium and silicon. Silicon is an optical material commonly used in infrared imaging systems as lenses, windows, mirrors, and output couplers for low-power lasers. Germanium has very high permeability and high refractive index in the wavelength range of 2–14 microns; thus, it has been an excellent substrate material for infrared lenses with extensive applications in thermal imaging systems, dark-field optical instruments, astronomical telescopes, night-vision systems, and so on. Recently, other new infrared transparent materials, such as chalcogenide glass, are being developed. Press molding technology is useable to fabricate infrared lenses from chalcogenide glass but the service life of the molding die is very short due to thermo-chemical damages. Currently, the fabrication of infrared lenses is still dependent on material removal processes.

Conventional infrared lenses are of spherical shape and are manufactured by curve-generator grinding and polishing. Recently, aspherical lenses are required more and more, and these complicated-shaped lenses are manufactured by diamond turning. Studies have revealed that although silicon and germanium are typically hard brittle materials and difficult to machine, they can be machined in a ductile mode without generating brittle fractures on the machined surface. This technology, usually called ductile (mode) machining, can be used to fabricate infrared aspherical lenses with nanometer level surface roughness without the necessity of subsequent polishing process. The details of ductile machining technology and the ductile machinability of various hard brittle materials will be given in Chapter 6 of this book.

#### 1.2.1.4 Fresnel Lens

A Fresnel lens is a plano-convex or plano-concave lens that is cut into concentric narrow rings, namely Fresnel zones, and flattened. It retains the optical characteristics of a plano-convex or plano-concave lens but is much smaller in thickness, and therefore has less absorption losses. There are numerous optical designs that can benefit from the application of Fresnel lenses. If the Fresnel zones are sufficiently narrow, the surface of each zone can be approximately made conical (with straight cross section) and not spherical (with arc cross-section) for ease of production. However, for recent high-precision optical systems, Fresnel lenses with spherical zones with arc cross sections are necessary.

Figure 1.4 shows a schematic of the geometry of a Fresnel lens and examples of aspherical lenses of single-crystal germanium produced by diamond turning. The shape of the Fresnel structure, $z(x)$, can be described by the modulus function (MOD) of the equivalent plano-concave or plano-convex lens shape $z_0(x)$ and the zone depth $d$ as:

$$z(x) = MOD\left[z_0(x), d\right] \quad (1.2)$$

where MOD is a function that returns the remainder when one number is divided by another. For example, MOD (4, 3) = 1.

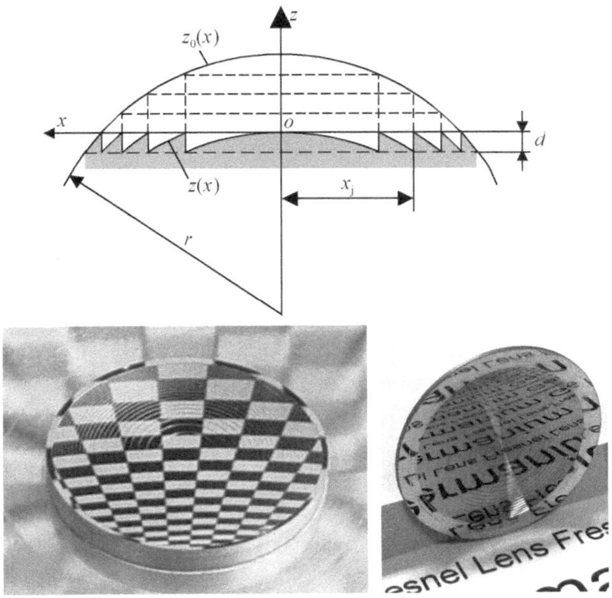


**FIGURE 1.4** Schematic of Fresnel lens and examples of Fresnel lenses of single crystal germanium produced by diamond turning.

The surface function of a spherical lens shape $z_0(x)$ can be simplified from that of an aspherical surface, the topology of which has been defined in Equation (1.1), as:

$$z_0(x) = \frac{Cx^2}{1+\sqrt{1-(k+1)C^2x^2}} + \sum_{i=1}^{m} a_i x^i \tag{1.3}$$

where the definitions of parameters are the same as those in Equation (1.1).

Therefore, for a spherical shape, we can just set $k = 0$ and $a_i = 0$ in Equation (1.3). Thus, the function of a spherical lens surface can be described as:

$$z_0(x) = \frac{Cx^2}{1+\sqrt{1-C^2x^2}}. \tag{1.4}$$

This equivalent lens surface is cut into concentric rings by cylindrical surfaces at the zone steps. The radial coordinate of each zone step can be calculated by:

$$x_j = \sqrt{j \cdot d\left(\frac{2}{C} - j \cdot d\right)} \tag{1.5}$$

where $j$ is the sequential number of the Fresnel zone counted from the optical axis $z$, and $d$ is the zone depth, as in Equation (1.2).

The fabrication of Fresnel lens molds by diamond turning on ductile metals like oxygen-free copper and electroless-plated nickel has been established for decades. However, the literature on the fabrication of Fresnel structures on brittle crystalline materials such as germanium is very little. Germanium is a nominally hard brittle material. At room temperature, it responds in an extremely brittle manner thus it is very difficult to machine. Micro grinding has also been reported as a fabrication method of Fresnel structures on hard brittle materials, but the groove profile distortions resulting from the roundness of the grinding wheel remain a critical problem. By using ductile machining technology, Yan et al succeeded in fabricating infrared Fresnel lenses on germanium by diamond turning as shown in Figure 1.4. The details of ductile machinability of germanium will be given in Chapter 6 of this book.

#### 1.2.1.5 Diffraction Optical Elements

A diffraction optical element (DOE) is an optical component with periodical grating structures, such as micro grooves, which split and diffract light into several beams traveling in different directions. DOE modifies the optical wavefront via subdivision and individual modification of the phase and/or amplitude of the incident light. According to their applications, DOEs can roughly be divided into three categories: diffraction gratings, holographic elements, and diffraction lenses. Diffraction gratings are commonly used in monochromators and spectrometers.

DOEs offer a high degree of freedom in their optical function that cannot be reached with refractive optics.

Diffraction gratings have linear segments with uniform diffraction angles. The directions of the light beams depend on the spacing of the grating and the wavelength of the light so that the grating acts as the dispersive element. The brazed angle, depth, and pitch of the gratings depend on the application of the element and the wavelength of the incident light. For visible light and ultraviolet light, the grating pitch is 1/600–1/1200 mm, and the blazed angle is 5–10°. For soft X-ray and vacuum ultraviolet light, the pitch is 1/1200–1/3600 mm, and the angle is 1–3°, respectively. In a few cases, an extremely fine pitch of the grating ~1/10000 mm is required.

Although various manufacturing methods, such as photolithography, laser direct writing, and electron beam lithography, have been developed for fabricating DOEs, ruling, a process of direct mechanical machining using a ruling engine or a diamond turning machine, is still the major method used for fabricating masters with fine diffractive gratings. To fabricate these kinds of extremely fine diffraction gratings, the depth of which is in the range of 0.1–0.01 μm order, ruling is the most suitable method.

In the ruling engine method, linear diffractive rulings are generated by a specially polished diamond tool by utilizing the burnishing effects. The diamond tool consists of two conical surfaces having different apex angles. The two conical surfaces are oriented in opposite directions but have the same base plane. The ruling engines have similar structures to the ultraprecision machine tools for machining flat surfaces, like shapers or planners. In a shaper machine, as will be introduced in Chapter 2, the tool is reciprocated at a high speed and the workpiece is stepped in a perpendicular direction to provide the feeding movement; while in a planner machine, the workpiece is reciprocated at a high speed and the tool is stepped in a perpendicular direction to provide the feeding direction. Unlike in diamond turning where a thin layer of material is removed from the workpiece surface thus the form accuracy of the machined surface depends on the machine accuracy, in ruling, however, material removal does not occur. The pressure between the burning tool and the workpiece is constant during ruling, thus the criterion for surface generation is the substrate itself rather than the ruling engine. From this meaning, the flatness and surface quality of the substrate material for diffraction gratings should be strictly guaranteed.

Normally, a thin layer of ductile metal materials, such as aluminum (Al), copper (Cu), silver (Ag), and gold (Au), is coated on a flat glass substrate and used as the material for grating master. The gratings are then replicated from the master mold to plastic materials. The plastic replica is adhered to a glass substrate and a thin layer of aluminum is coated on the glass substrate.

The ruling method is effective when generating extremely fine gratings, but it is difficult to generate gratings with large depth and width, such as diffraction lenses. In this case, diamond turning is necessary. The diamond-turned metal mold can be used for press or injection molding of plastic lenses. Specially prepared NiP-plated metal molds can also be used for the press molding of DOEs on glasses with low softening points.

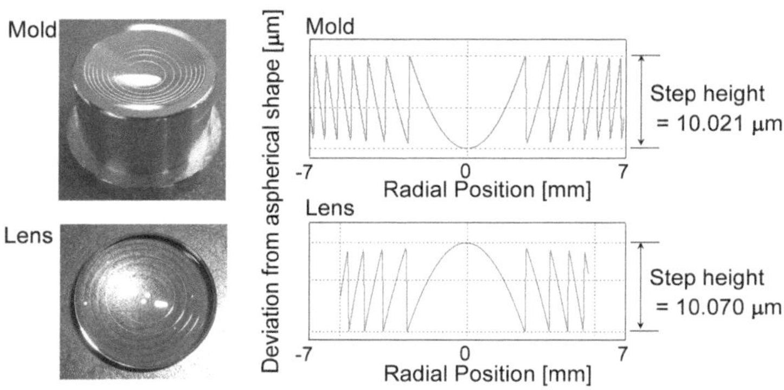


**FIGURE 1.5** A diamond-turned steel mold plated with NiP and a press-molded glass aspherical lens with diffractive gratings as well as their cross-sectional profiles [4].

An example is given in Figure 1.5 to show the diamond turning of a NiP-plated steel mold for molding glass aspherical lenses with diffractive gratings [4]. The steel substrate is plated with electroless Ni–10 wt%P (thickness 100 μm). To prevent deformation by heat during molding, the substrate is heated at a temperature higher than the molding temperature. The NiP layer is machined by diamond turning. After cutting the NiP layer, a release agent coating made of rare metals is applied on the mold surface. The mold was then used for molding the DOE onto K-PG375 glass. The molded glass DOE is also shown in the figure. The DOEs made of glass were molded for 500 shots with the same mold. From the cross-sectional profiles of the mold and the lens, it is demonstrated that the transfer error during the molding process is about 50 nm, indicating that the preheated NiP-plated steel mold is useable for press molding glass DOE.

#### 1.2.1.6 Micro Lens Array

A micro lens array (MLA) is an optical element in which multiple micro lenses are arranged and arrayed to achieve specific optical functions by adjusting their shapes, focal lengths, etc. They significantly improve the integration and performance of optical devices. Its main functions are light collection and diffusion, and it is used in a variety of fields and applications such as CCD/CMOS image sensors, optical communication equipment, optical interfaces, digital cameras, high-brightness LED, etc. For example, an MLA with a lens diameter of several tens micrometers can increase 15% of luminance in the backlighting module of liquid crystal displays. MLAs with lens diameter of several hundred micrometers are used in the Shack–Hartmann wavefront sensors to divide the wavefront surface into a number of beamlets.

MLAs for visible light are made of viscoelastic materials, i.e., plastics and glass, thus replication techniques such as plastic injection molding and glass molding press are used for their fabrication. In the replication processes, the fabrication of MLA molding dies is a crucial issue. Like aspherical lens fabrication mentioned in Section 1.2.1.3, copper and NiP-plated steels are popular mold materials for plastic MLAs, while tungsten carbide and silicon carbide are commonly used as die materials for molding glass MLAs.

A typical method for machining MLA molds is micro end milling, which has been widely used in machining various microstructured surfaces on metals [5]. The tool materials used for micro end milling include tungsten carbide, polycrystalline diamond (PCD), and single-crystalline diamond (SCD). All these tool materials can be used for machining metal molds, but for machining tungsten carbide or silicon carbide molds, PCD or SCD tools are necessary.

Figure 1.6 shows micrographs of micro lens arrays molds machined on copper by a single-crystal diamond end mill. The lens dimple has a diameter of 282 μm and a depth of 10 μm. The dimple surface is very smooth with a surface roughness of 6 nm Ra and a form error of 67 nm. It can be seen that the boundaries of the dimples are very sharp without burr formation. These MLA molds can be used as injection molds for manufacturing plastic micro lens arrays for industry.

For harder mold materials such as tungsten carbide and silicon carbide, polycrystalline diamond (PCD) end mills prepared using EDM were used for cutting micro dimples. Figure 1.7 shows SEM micrographs of the dimple array machined using a PCD endmill. No brittle fractures can be found thus the dimple surface has

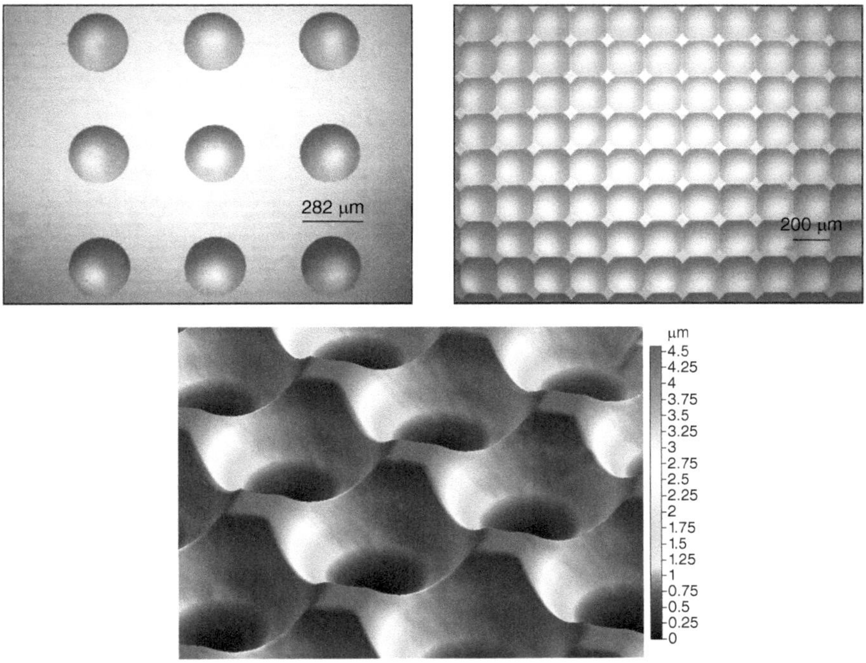


**FIGURE 1.6** Examples of micro lens array molds machined on copper by a single-crystal diamond end mill [6].

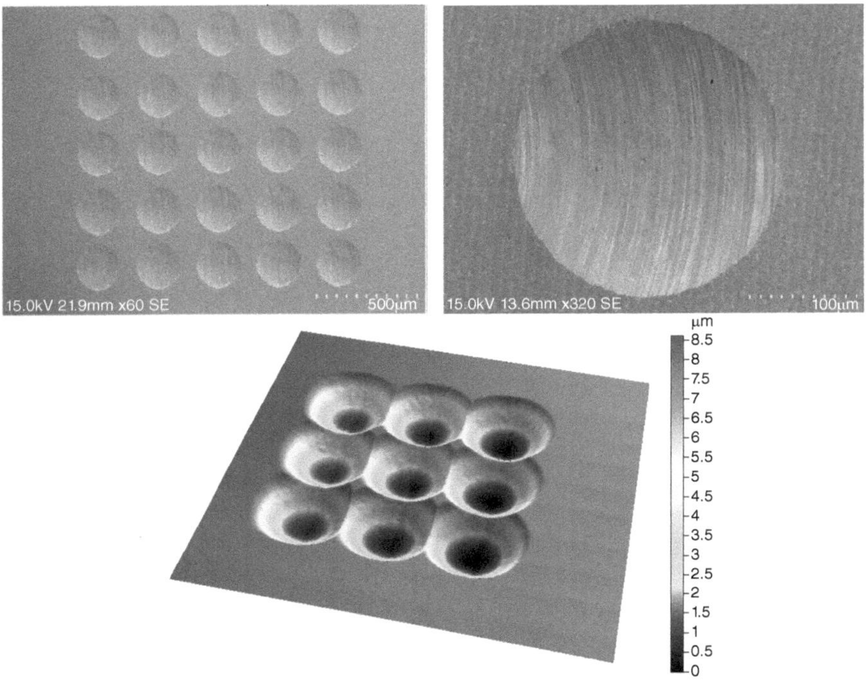


**FIGURE 1.7** Examples of micro lens array molds machined on silicon carbide by a polycrystalline diamond end mill [7].

been machined in a ductile mode. The boundary between the dimple surface and the uncut surface is very clear, and there are slight tool marks along the cutting direction. All the dimples are uniformly generated. It should be pointed out that the surface of the machined dimple using the PCD tool is quite rough. This is because the unevenness of the cutting edge was directly imprinted on the workpiece surface. Surface roughness can be improved by cross-feeding the tool. This effect can be achieved by either rotating the workpiece along the dimple center or swinging the tool surface vertically to the cutting direction.

#### 1.2.1.7 Freeform Optics

In AR/VR applications, to meet the stringent demands for size and performance, complex optical designs are increasingly used to reduce the light path and correct astigmatism in optical devices, enhancing performance. Freeform optics offer exceptional optical efficiency and reduce the need for multiple optical elements due to their flexible shape design. Freeform surface designs provide significant advantages over conventional optics, enabling more compact and simplified systems, expanding the field of view, and improving image resolution. These designs often involve complex surfaces described by equations, such as the Alvarez lens, which adjusts focal length through transverse shifts of two surfaces and is expressed by XY-polynomials, and the Zernike polynomial for describing wavefront aberrations, to control light paths and wavefronts more precisely and efficiently. Such innovations have led to the increased integration of freeform optics into head-mounted displays (HMDs) for VR and AR applications.

The molds for freeform optics are typically made by ultraprecision machining. The end milling method as mentioned in Section 1.2.1.6 is versatile for most optical designs; however, the main problem is that the machining cycle is time-consuming. Each lenslet is typically machined individually even if the number of lenslet arrays increases. As an alternative method, tool servo-based diamond turning is one of the manufacturing techniques that can fabricate complicated optics shapes, freeform optics, and structured surfaces with higher machining efficiency compared to milling.

Two kinds of tool servo turning methods have been developed: slow tool servo (STS) and fast tool servo (FTS). STS is realized by synchronization between the C-axis and other linear axes (XZ axes) of a machine tool, which enables the fabrication of non-rotationally symmetric surfaces including complex freeform shapes. Using the diamond turning method, freeform surface fabrication with STS is distinctly faster than end milling. Figure 1.8 shows some freeform optics machined by STS diamond turning.

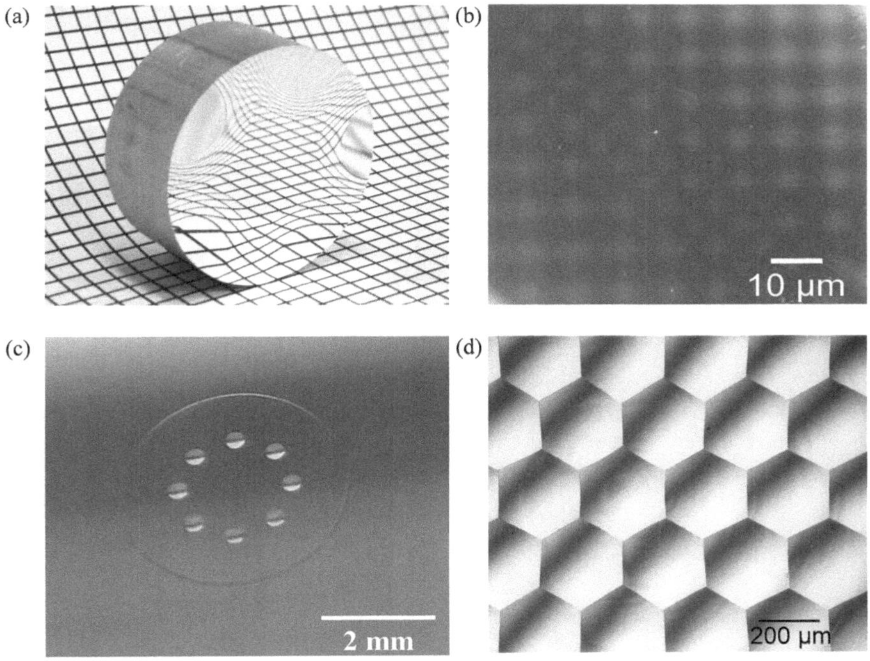


**FIGURE 1.8** Freeform optics machined by STS diamond turning: (a) freeform surface on aluminum, (b) two-dimensional sine wave [8], (c) spherical MLA [9], and (d) hexagonal MLA on single-crystal silicon [10].

Although STS enabled faster fabrication of freeform optics than milling, the following error induced by the machine table acceleration still remains a problem. The oscillation frequency of the tables is limited by their big mass, as a result, the spindle rotation rate must be controlled low enough to suppress the acceleration-induced following error. To solve this problem and realize much higher productivity of axis-symmetric freeform components, fast-tool servo (FTS) diamond turning was developed which uses a specially equipped small-mass unit capable of high-frequency movement. FTS systems are typically driven by piezo actuators, voice coil motors, or Maxwell's electromagnetic force methods. The mechanical stroke limitation of conventional FTS driven by piezoelectric actuators is commercially available up to 100 μm, which restricts its application in producing high-precision freeform optics with millimeter-scale sagitta. Recently, a new type of long-stroke FTS, driven by a voice coil motor and high-frequency servo control technology with high-resolution positioning, has become commercially available. This FTS system allows for the fabrication of a wide range of designs, from micrometer-scale structures to millimeter-scale sagittal freeform surfaces, including sine waves, dual sinusoidal patterns, and large freeform shapes, with increased productivity. This new type of FTS can be widely used for fabricating surface structures with a millimeter scale sagitta. Figure 1.9 presents two examples of freeform optics machined by FTS-based diamond turning in the author's laboratory.

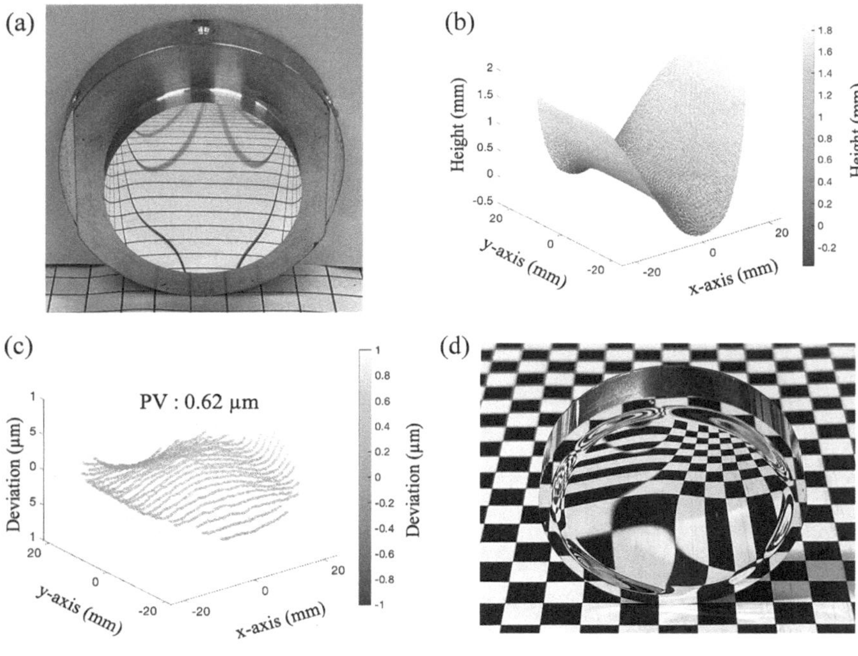


**FIGURE 1.9** (a) Optical image, (b) surface topography, (c) form error of a cylindrical surface machined by FTS-based diamond turning [11]. (d) Another freeform surface.

#### 1.2.1.8 Retro-Reflective Film

A retro-reflective film is a film that reflects light in a direction parallel to the incident light back to the light source by using micro glass bead balls or micro prisms. The micro prism type retro-reflective film has higher reflective efficiency than micro glass beads and the reflective lighting is back to a more concentrated area than glass beads. The beam of light stays tight and gathered together over a long distance. Owing to these features, micro prism type retro-reflective film is nowadays used for high visibility reflective highway signs and warning reflective tapes on vehicles and uniforms. In the case of highway signs, the drivers can see the highly bright road sign, but other drivers or persons on the road cannot see it.

Micro pyramids as shown in Figure 1.10 are typically used as micro prism type retro-reflective film. Especially, the triangular pyramids are popularly used. Micro triangular pyramid molds made of copper, aluminum, and nickel-plated steel are used for replicating the micro pyramids to the polymer sheets. The angular accuracy of the pyramid faces is extremely important for the final reflection ratio of the film thus the ultraprecision machining of the molds is an essential step. Among various manufacturing methods, fly cutting using a V-shaped diamond tool is the most widely used method to machine the micro pyramids. In this method, the workpiece is placed horizontally on the worktable while the diamond tool rotates at a high speed by an air-bearing spindle. The details of the manufacturing methods will be introduced in Chapter 2.

### 1.2.2 Electronic Applications

#### 1.2.2.1 Magnetic Disk

Until the 1990s, diamond turning has been a major manufacturing method for magnetic disks of computers, especially high-performance computers. The disk is made of aluminum alloys having a thickness of 1–2 mm, as shown in Figure 1.11. Before coating the recording media, the aluminum substrate surface must be flattened and smoothed to an extremely low surface roughness (~angstrom level),

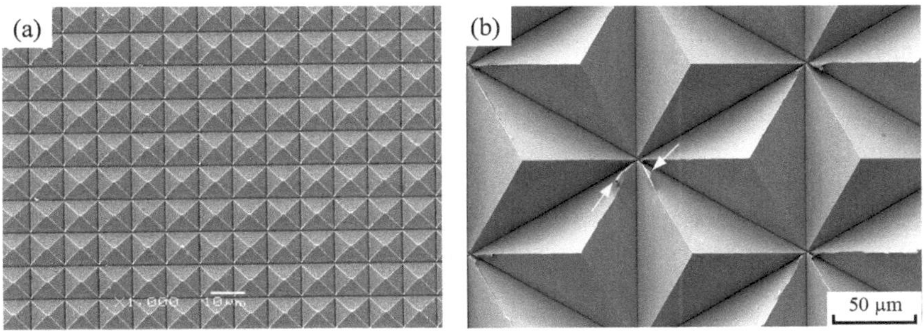


**FIGURE 1.10** Micro pyramids machined on metal molds for replicating micro prism type retro-reflective films: (a) square pyramids [12] and (b) triangular pyramids [13].

**FIGURE 1.11** A computer hard disk unit.

because the surface roughness determines the recordable density of information. For this purpose, it is important to optimize the geometry of the diamond tool and its setting angles to the disk surface. As a recent trend, instead of using a flat surface, nanometer-level textures are needed on the substrate surface in order to increase the recording density. In this case, the nanoscale texture generation might be also accomplished by diamond turning using extremely sharp tools.

It should be noted that in recent years, glass is used as a hard disk substrate material instead of aluminum, especially for small-diameter magnetic disks for notepad computers and other portable devices. For glass substrates, instead of diamond turning, polishing is mainly used. The fundamentals of polishing will be given in Chapter 2.

#### 1.2.2.2 Semiconductor Wafer

Commercially important semiconductor materials include silicon (Si), silicon carbide (SiC), gallium arsenic (GaAs), gallium nitride (GaN), indium phosphide (InP), and diamond. Figure 1.12 shows a photograph of silicon wafers, and ingots from which the wafers are produced. Most of these semiconductors are hard brittle crystalline materials, having very low fracture toughness and high hardness, thus they tend to behave in brittle manners during machining at room temperature. Currently, these materials are machined by a series of abrasive machining processes such as grinding and lapping followed by chemomechanical polishing (CMP). Abrasive machining involves multipoint contacts between abrasive grains with irregular shapes/sizes and the substrate, the kinetics and material removal mechanisms are very complicated and, thus in turn, are relatively difficult to control. The fundamentals of abrasive machining will be given in Chapter 2.

An alternate approach of wafer fabrication might be single-point diamond turning. Compared to abrasive machining processes, diamond turning has high controllability. Diamond turning has been demonstrated to be capable of fabricating extremely smooth surfaces on brittle materials by controlling the material

**FIGURE 1.12** Single-crystal silicon wafers. (Courtesy of Mitsubishi Materials.)

removal mechanism in a ductile mode, as will be introduced in Chapter 7. However, currently there are two main problems that act as drawbacks to using diamond turning for fabricating semiconductor wafers. One is the problem of tool wear. Tool wear during machining silicon and other semiconductors is very severe, which raises the production cost and degrades the surface quality of the wafers. The tool wear is difficult to suppress because it involves very complex mechanisms affected by thermal, chemical, mechanical and electrical factors. Another problem is the subsurface damage. As will be discussed in Chapter 7, even if the machining is performed in a ductile mode, considerable subsurface damage will occur to the surface layer of the substrate. The damage may include various microstructure changes such as dislocations and residual stresses as well as phase transformations. For this reason, to eliminate the subsurface damage resulting from ductile machining, a subsequent polishing process is usually necessary. In other words, diamond turning might be used as a pre-processing technology for semiconductor wafers before the CMP process.

#### 1.2.2.3 LSI Substrate

Three-dimensional implementation of large-scale integrated (LSI) chips is a key technology in the current rapidly advancing information-based society. LSI chips are usually stacked and pressure-bonded via micro gold bumps electroplated on silicon wafers. Due to the low strength of the LSI chips, pressure-bonding is performed under a low load to prevent damage to the chips. For this reason, the bonding surfaces of the bumps need to be precisely flattened. To avoid contamination, photoresist, which is used as a mask in the bump electroplating, needs to be flattened together with the bumps and removed after the planarization step. The conventional planarization method is CMP in which the material removal rate in the center region of the bump tends to be higher than in the fringe area, leading to

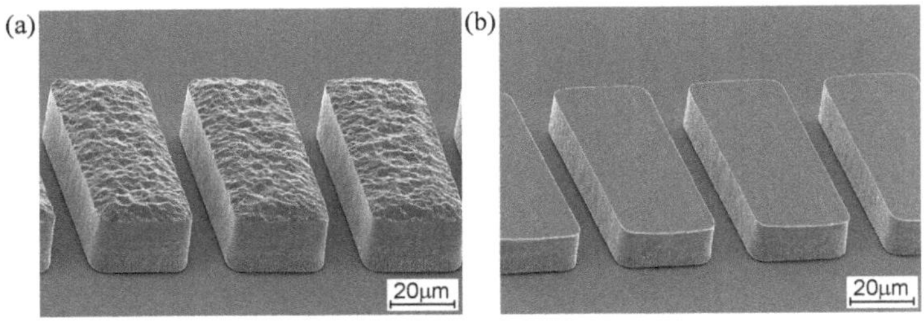


**FIGURE 1.13** SEM micrographs of gold bumps (a) before and (b) after diamond turning [14].

a concave bump surface. This phenomenon is sometimes called "dishing". Low efficiency and chemical pollution are also problems to be solved in CMP.

As an alternative method, ultraprecision diamond turning is used for the planarization of the LSI substrate. It involves cutting two different materials, gold and photoresist. As gold is a highly ductile material, burrs might be generated around the bump, which leads to short-circuiting of LSI chips. In contrast, the significant thermo-plasticity of the photoresist affects the cutting of the bumps. To solve this problem, cutting tests were performed on a photoresist/gold bump composite under various conditions and fundamental cutting characteristics were investigated. The feasibility of ultraprecise bump planarization by diamond turning was demonstrated. Figure 1.13 shows SEM micrographs of bumps before and after cutting. Before cutting, the bump surface was very rough, and after cutting, the surface became distinctly smoother. The surface roughness is 18 nm Ry and the flatness is 43 nm after cutting. The diamond-turned bumps can be directly used for low-pressure bonding in LSI implementation.

### 1.2.3 Biomedical and Biochemical Applications

#### 1.2.3.1 Artificial Joints

Hip joints of humankind involve articular cartilage slides between the acetabular and femoral head. Articular cartilage damage usually occurs at the hip joint. This is because the hip joint experiences high local pressure up to 5 times that of the average body weight. Nowadays artificial joints have been applied as medical treatments for joint diseases. In clinical applications, various biomaterials that have bioinert, corrosion-resistant, and strong properties were used as artificial joints, such as corrosion-resistant metals, polymers, ceramics, and composites which are applicable to the human body brought about new medical treatments using artificial joints. As shown in Figure 1.14, an artificial hip joint consists of three main components: femoral stem, femoral head, and hip socket. In addition to alumina ceramics, zirconia ceramics are used alongside for femoral head in total hip arthroplasty. This is because zirconia ceramics are superior to alumina ceramics in terms of mechanical strength and fracture toughness. The high-precision machining of these biomaterials is essential for preventing local

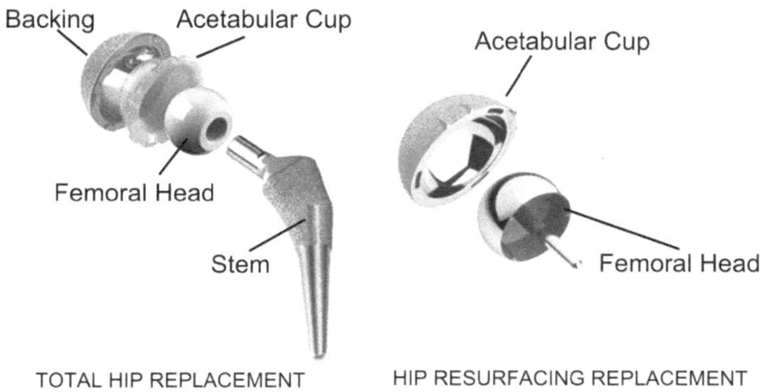


**FIGURE 1.14** Example of artificial hip joint consists of femoral stem, femoral head, and hip socket [15].

stress concentration and wear damage between the femoral head and hip socket and ensuring the stability and service life of the artificial joints. The machining process of artificial joints includes turning, milling, drilling, polishing, etc. for each step of which, the form accuracy and surface quality must be strictly controlled.

### 1.2.3.2 Microfluidics

Microfluidics are systems that can process small quantities of fluids by using micro channels having dimensions of tens to hundreds of micrometers. Microfluidic systems are widely used in biochemical and biomedical processes such as capillary electrophoresis, isoelectric focusing, immunoassays, flow cytometry, sample injection in mass spectrometry, PCR amplification, DNA analysis, separation and manipulation of cells, cell patterning, and so on. Figure 1.15 is an example of microfluidics, which was constructed as a "Lab on Chip" system.

**FIGURE 1.15** Example of a microfluidics-based lab on chip system [16].

Microfluidic systems can be fabricated on silicon, glass, and polymers. Silicon and glass microfluidics are manufactured by photolithography, which was originally developed for creating small features on circuits in the semiconductor industry. Photolithography needs masks to define the shape of microfluidics. Recently, microfluidics made of polymers such as polydimethylsiloxane (PDMS) have increased greatly. PDMS is a transparent elastic polymer permeable to oxygen and carbon dioxide and hence is useful for housing cells. Polymer microfluidics can be fabricated through replication processes such as injection molding and press molding. Molding processes are less time-consuming than photolithography and there is no need to use masks, thus contributing significantly to microfluidics production.

The metal molds used for molding polymer microfluidics are usually machined by micro end milling technologies. To guarantee the accuracy and surface quality of the molds in the micro end milling process, tooling technology is essential. The details of end milling and its tooling will be discussed in Chapter 2.

## 1.3 FUTURE PERSPECTIVES OF ULTRAPRECISION MACHINING

It is difficult to predict the future of manufacturing technology based on the past and the present. Manufacturing advances on the base of industrial and social needs as well as the mastered level of science and technology by humankind. For ultraprecision machining technology, the following aspects will become more and more important for the future R&D.

### 1.3.1 Precision and Accuracy

It should be pointed out that precision and accuracy are terms having different meanings. Precision refers to the repeatability of the machining process, either component to component or batch to batch, as well as across the same component, also defined surface uniformity, while accuracy is the ability of a process to generate a machined surface having the specified roughness and form accuracy. As shown by the famous Taniguchi chart [1], the accuracy and precision of machining processes change with time. In recent years, the accuracy/precision of machining is approaching the nanometer scale. There is no doubt that a much higher level of accuracy/precision, i.e., atomic scale level, will be reached in the near future. This is because a lot of methodologies will enable us to precisely control the amount of energy input to the machining region at a higher and higher spatial and temporal resolution, and in turn, the precision/accuracy of material removal, deformation, and addition will be higher and higher. The precision and accuracy of machining are especially important for short-wavelength optical applications. For example, extreme ultraviolet (EUV) lithography is a technology used in the semiconductor industry for manufacturing integrated circuits. It uses 13.5 nm EUV light to create intricate patterns on semiconductor substrates. As wavefront error requirements for diffraction-limited imaging scale with wavelength, and thus the evolution from deep ultraviolet (DUV) of 193 nm to EUV of 13.5 nm implies a factor of about 14 times smaller allowable wavefront error for the EUV system. This means the wavefront should have less than 1 nm rms deviation from a perfect sphere. In the context of a 4-mirror system, each surface should contribute no more than 0.5 nm rms of wavefront error. For mirrors, the allowable surface figure error is one-half of the wavefront error, i.e., the allowable average figure error for the 4-mirror system is 0.25 nm rms. Apart from machining, assembly is also an important step. Each component must be positioned within a nanometer accuracy to ensure the final image quality. This technology is the highest level of ultraprecision machining in the world at this moment.

### 1.3.2 Surface Integrity

The term surface integrity was coined by Michael Field and John F. Kahles in 1964. It includes two aspects: topography characteristics and surface layer characteristics. The topography is made up of surface roughness, waviness, errors of form, and flaws. The surface layer characteristics that can change through processing are plastic deformation, residual stress, subsurface cracks, hardness change, phase change, grain size change, and so on. As the accuracy and precision level of machining is higher, the effect of surface integrity on the performance of the product becomes more significant than before. In the future, the disorder of a single layer of atoms at the surface may be regarded as a problem, and such a problem will definitely be solved by new technologies.

### 1.3.3 Shape Flexibility

In optics field, the shapes of optical components have shifted from flat or spherical surfaces to aspherical surfaces, and further to freeform surfaces. With the increase in surface complexity, the degree of freedom of machines used for machining is becoming higher and higher thus the difficulty of machining process control and the surface metrology is increasing. Such needs for complicated surfaces will be increasing in the future also in other industrial areas. Thus, the development of machining and measurement technologies that are flexible to shape complexity is necessary. Although additive manufacturing has become an effective method for fabricating three-dimensional complicated surfaces, to date its attainable form accuracy and surface quality are far from the level required by ultraprecision machining.

### 1.3.4 Productivity

The initial cost for ultraprecision machining is high because of the investment in expensive machines, tools, and related facilities. The running cost is also high due to the tool and energy consumption we well as equipment maintenance. On the other hand, the productivity of ultraprecision machining processes is usually very low. Ultraprecision machining is time-consuming due to several reasons.

The small machining scale and the low material removal/deformation rate are the first. Another reason is that the speed of workpiece-tool movement is limited for kinematic or thermal problems or process stability reasons. Also, process repeatability is still a challenging issue for most ultraprecision processes, thus the machining-metrology-machining cycle is repeated several times before the final accuracy requirement is satisfied. From this meaning, productivity improvement based on machining methodological innovation is a future task for the manufacturing community.

### 1.3.5 Sustainability

Ultraprecision machining is a knowledge-intensive technological area. Some of the knowledge can be digitized and thus is transferable. However, in some cases, the success of machining technology is achieved by empirical technique or based strongly on the personal skill and experience of operators. This type of knowledge is hard to transfer to the future. To improve the sustainability of ultraprecision machining, deterministic processes are highly expected to reduce or eliminate the necessity of trial and error, or the personal skills of the operators. For this purpose, the application of artificial intelligence (AI) to ultraprecision machining and metrology may be an effective solution.

## 1.4 IMPORTANCE OF ULTRAPRECISION METROLOGY

Metrology, the science of measurement, has great significance in ensuring the quality of machining. Metrology for manufacturing can be roughly divided into two categories, surface metrology and subsurface metrology. Surface metrology involves the measurement of surface geometrical parameters, including but not limited to size, shape, form errors, and surface roughness, which are important aspects for evaluating the accuracy and precision of machining. On the other hand, subsurface metrology involves quantitatively characterizing the material microstructural changes, i.e., phase transformations, residual stresses, subsurface cracks, and other subsurface features to assess the stability of a machining process and the quality and performance of the resulting products. The combination of surface metrology and subsurface metrology can totally evaluate the quality of machining from geometrical, mechanical, physical, and chemical properties and even the functionality of a machined surface.

The importance of metrology rests on two key aspects. Firstly, metrology plays a crucial role in assessing whether a product meets the quality standards. Secondly, analyzing metrology results provides valuable information that is subsequently fed back into the manufacturing process. This iterative feedback loop serves to enhance and refine the manufacturing processes, ultimately guaranteeing the precision and accuracy of machining.

Machining and metrology are two closely related fields, mutually supported and interdependent, like two wheels of a cart. Neither one can develop or progress without the other. The mutual interdependence of machining and metrology will strengthen with the evolution of the required accuracy and precision of industrial products.

## 1.5 OUTLINE OF THIS BOOK

This book will focus on the fundamentals and applications of both Ultra-Precision Machining and Metrology (UPMM). Chapter 1 is an introduction to this topic by highlighting some recent applications of UPMM. Chapter 2 briefly introduces typical methodologies for ultraprecision machining, including mechanical, non-mechanical, and hybrid approaches. Chapters 3 and 4 describe machine tool systems and tooling technologies which are two important supporting bases for ultrapre-cision machining. Chapters 5 and 6 comprehensively describe the ultraprecision machining mechanisms for ductile materials and brittle materials, respectively. The understanding of physics governing the machining processes is essential to design and optimize the machining conditions. Chapter 7 deals with surface metrology for ultraprecision machining, including the measurement of surface roughness and form accuracy which is crucial for controlling and improving the product quality. Chapter 8 introduces another important topic of metrology, namely, measurement and characterization of subsurface damage of workpiece caused by machining processes.

As aforementioned, UPMM is an advanced enabling technological area that serves the manufacturing of high-value and high-end products for our society. This is a partially established and partially emerging research and development area receiving concentrated interests from multidiscipline students, researchers, and engineers. This book will provide a systematic integration of UPMM-related aspects including system construction, tooling, machining mechanism modeling, and surface/subsurface metrology. By reading the book, the readers will be endowed with both practical knowledge and a fundamental understanding of the underlying physical concepts of the UPMM methodologies and technologies. This book will be helpful for course study for graduate students in universities, scientific research in academic institutions, and process developments at the shop level of industry.

## REFERENCES


[1] Taniguchi, N. (1983) Current status in, and future trends of, ultraprecision machining and ultrafine materials processing. *CIRP Annals* 32 (2): 573–582.

[2] Hejduk, M., Heazlewood, B.R. (2019) Off-axis parabolic mirror relay microscope for experiments with ultra-cold matter. *Review of Scientific Instruments* 90 (12): 123701.

[3] Zhou, T., Yan, J., Yoshihara, N., Kuriyagawa, T. (2010) Study on nonisothermal glass molding press for aspherical lens. *Journal of Advanced Mechanical Design, Systems and Manufacturing* 4 (5): 806–815.

[4] Masuda, J. (2007) Application of the NiP-plated steel molds to glass lens molding. *Proceedings of the 10th International Symposium on advances in abrasive technology*, 2007: 123–130.

[5] Uhlmann, E., Piltz, S., Schauer, K. (2005) Micro milling of sintered tungsten–copper composite materials. *Journal of Materials Processing Technology* 167 (2): 402–407.

[6] Yan, J., Zhang, Z., Kuriyagawa, T., Gonda, H. (2010) Fabricating micro-structured surface by using single-crystalline diamond endmill. *The International Journal of Advanced Manufacturing Technology* 51 (9): 957–964.

[7] Zhang, Z., Peng, H., Yan, J. (2013) Micro-cutting characteristics of EDM fabricated high-precision polycrystalline diamond tools. *International Journal of Machine Tools and Manufacture* 65: 99–106.
[8] Nagayama, K., Yan, J. (2021) Deterministic error compensation for slow tool servo-driven diamond turning of freeform surface with nanometric form accuracy. *Journal of Manufacturing Processes* 64: 45–57.
[9] Mukaida, M., Yan, J. (2017) Ductile machining of single-crystal silicon for micro-lens arrays by ultraprecision diamond turning using a slow tool servo. *International Journal of Machine Tools and Manufacture* 115: 2–14.
[10] Mukaida, M., Yan, J. (2017) Fabrication of hexagonal microlens arrays on single-crystal silicon using the tool-servo driven segment turning method. *Micromachines* 8 (11): 323.
[11] Hashimoto, T., Yan, J. (2025) Time delay compensation in high-speed diamond turning of freeform surface using independent fast tool servo with a long stroke. *Precision Engineering* 93: 515–527.
[12] Yan, J., Oowada, T., Zhou, T., Kuriyagawa, T. (2009) Precision machining of micro-structures on electroless-plated NiP surface for molding glass components. *Journal of Materials Processing Technology* 209 (10): 4802–4808.
[13] Gao, J., Xu, Z., Lei, Y., Huang, S. (2024) Fly-cutting processing of micro-triangular pyramid arrays and synchronous micro-scrap removal. *Micromachines* 15 (5): 655.
[14] Yan, J. (2011) Ultraprecision cutting of photoresist/gold composite microstructures. *CIRP Annals* 60 (1): 133–136.
[15] Mattei, L., Di Puccio, F., Piccigallo, B., Ciulli, E. (2011) Lubrication and wear modelling of artificial hip joints: A review. *Tribology International* 44 (5): 532–549.
[16] Scientific Reports (2023). Microfluidics for Biomedical Research. https://www.nature.com/collections/eebbeabjjg (accessed Feb. 2025).

# 2 Ultraprecision Machining Methods

## 2.1 INTRODUCTION

### 2.1.1 Definition of "Precision" and "Ultraprecision"

It is not easy to exactly define and discriminate what is "precision machining" and what is "ultraprecision machining". The reason for this is that the concept of "precision" is changing with time and technical area. According to the prediction of Taniguchi [1], the achievable machining accuracy levels of normal machining, precision machining, and ultraprecision machining in the 2000s are 1 μm, 0.01 μm, and 0.001 μm (1 nm), respectively. Unfortunately, to the knowledge of the author of this book, the practical precision level of machining in the manufacturing industry at the time point of 2025 is still somewhat lower than this prediction. Thus, in this book, "ultraprecision machining" is defined as the machining processes/technologies where the 0.001 μm (1 nm) level surface roughness and submicron level (0.1 μm) form accuracy can be achieved, and "precision machining" represents the 10 nm (0.01μm) level surface roughness and micron (1 μm) level form accuracy. This definition reflects and matches with the current technological level of precision manufacturing in the global industry.

### 2.1.2 Methodologies for Manufacturing

There are numerous methods for manufacturing. Generally, they can be classified into three major categories according to material volume change, namely, material subtraction (volume decrease), addition (volume increase), and deformation (no volume change) manufacturing processes. The material subtraction, addition, or deformation may be realized by different kinds of methodologies such as mechanical, electrical, thermal, chemical, optical, and hybrid methods with various types of energy inputs. However, not all methods can be used for precision and ultraprecision manufacturing. Currently, most methods for precision and ultraprecision manufacturing are material subtractive (removal) processes. The major methods used for precision and ultraprecision manufacturing are listed in Table 2.1. The following sections will introduce the fundamentals of each method.

## 2.2 CUTTING

Cutting is the most important machining technology for fabricating complex shapes on metal materials. According to the relationship of tool-workpiece relative

DOI: 10.1201/9781003292388-2

**TABLE 2.1**
**Methods for Precision and Ultraprecision Manufacturing**

| | | | |
|---|---|---|---|
| Material subtraction | Cutting | Turning | |
| | | Fly cutting | |
| | | Milling | |
| | | Shaping/planning | |
| | Abrasive machining | Grinding | |
| | | Lapping | |
| | | Polishing | Mechanical polishing |
| | | | Chemomechanical polishing |
| | | | Float polishing |
| | | | Elastic emission machining |
| | Non-traditional machining | Electro discharge machining | |
| | | Electro chemical machining | |
| | | Chemical machining | |
| | | Energy beam machining | Laser machining |
| | | | Electron beam machining |
| | | | Ion beam machining |
| Material addition | 3D printing | Two-photon polymerization | |
| | | Projection micro stereolithography | |
| Material deformation | Plastic injection molding/imprinting | | |
| | Glass molding press | | |

movement, cutting can be divided into several subcategories: turning, fly cutting, milling, planning and shaping, etc.

### 2.2.1 Turning

Among a number of methods, diamond turning is the most popular one to realize ultraprecision cutting. It is a process of mechanical fabrication for precision elements using numerically controlled lathes equipped with natural or synthetic diamond tools. This technology emerged in the 1960s with the manufacturing demands in the USA. Due to the fact that the tool-workpiece contact area is extremely small compared to conventional machining, it is also termed single-point diamond turning (SPDT). Nowadays, SPDT is a well-established technology for the fabrication of non-ferrous metal mirrors for optical applications.

Some of the features that enable the SPDT technology are extremely sharpened diamond cutting tools, air-bearing spindles, pneumatic/hydrostatic slides, feedback control, vibration isolation, temperature control, etc. The lathe used for

SPDT usually rests on a high-quality granite base with fine surface quality. The granite base is placed on air suspensions on a specially constructed solid foundation, keeping its working surface strictly horizontal. The machine tables are placed on the top of the granite base and can be moved with high degree of accuracy using high-pressure aerostatic or hydraulic suspensions with high stiffness. The machine elements are usually driven by servomotors via hydrostatic screws or directly driven by linear motors. The movement of the machine elements is measured and monitored by extremely high-resolution displacement sensors such as laser hologram scales. The workpiece is usually attached to a vacuum chuck using negative air pressure after being strictly centered and then rotated at a high speed by an aerostatic or hydraulic bearing spindle. A detailed introduction to ultraprecision machine tool systems will be given in Chapter 3 of this book.

A schematic of diamond turning operation is shown in Figure 2.1, and a photograph of the main section of a diamond turning machine in Figure 2.2. The material removal in SPDT is realized by the relative movement between the

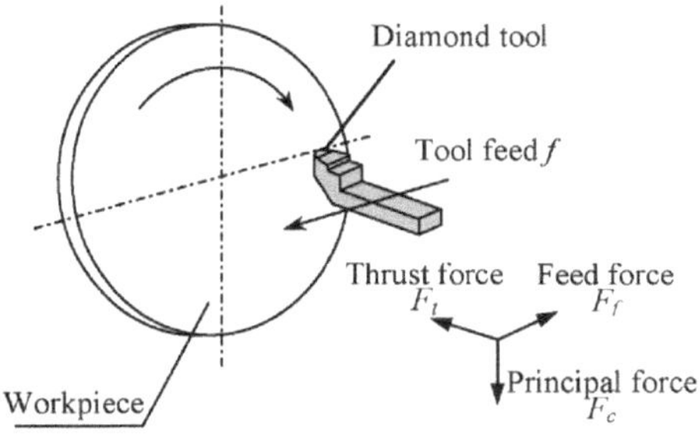


**FIGURE 2.1** Schematic presentation of diamond turning [2].

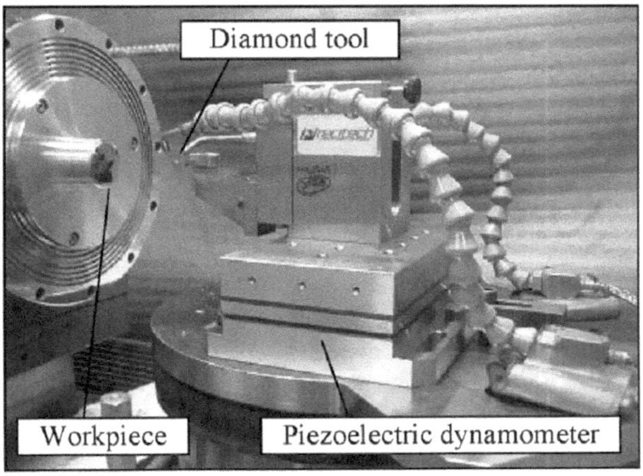


**FIGURE 2.2** Photograph of the main section of a diamond turning machine [2].

workpiece and the cutting tool, that is, the linear movement of the machine tables and the rotation of the workpiece. The change of the machining scale, i.e., the undeformed chip thickness, is realized by changing the tool feed rate and depth of cut. In recent years, ultraprecision machine tools have been developed with multi-axis numerical control at a resolution better than one nanometer per step to fabricate nano-precision components with complicated shapes.

Diamond turning is a major machining method for laser reflecting mirrors, molding dies for plastic injection molding of aspherical lenses, Fresnel lenses, DOEs, and other optical elements as introduced in Chapter 1. Most nonferrous metals and polymers are suitable for diamond turning to achieve nanometer-level surface roughness, the machinability of which will be detailed in Chapter 5. Some brittle materials, such as single crystal silicon, germanium, calcium fluoride, and some types of glasses, etc., are also diamond machinable, the machinability of which will be detailed in Chapter 6.

Two types of diamond turning machines are used, 2-axis (XZ) machines and 3-axis (XZB) machines. In the ductile machining of silicon and germanium lenses, a small radius round-nosed diamond tool with high form accuracy can be used under 2-axis numerical control, as shown in Figure.2.3(a). The tool radius is usually about 1 mm or much smaller. In this case, a truly ductile response only occurs along the apex of the tool tip where the undeformed chip thickness is smaller than a critical value (critical depth $d_c$), while the upper material is fractured. To obtain a crack-free surface, tool feed $f$ must be extremely small (~1 μm/rev). As a result, ductile machining is performed at a very low material removal rate. When machining a large-diameter component, tool wear will become a critical problem. In order to realize ductile machining of brittle materials at a high tool feed rate, and in turn a high material removal rate, three-axis aspherical surface machining using a straight-nosed tool is preferable, as shown in Figure 2.3(b).

In Figure 2.3(b), diamond turning of brittle materials is performed by using a straight-nosed diamond tool to fabricate convex surfaces under 3-axis numerical control. This method was termed the straight-line enveloping method. The tool is moved in the X and Z directions and simultaneously rotated about a B-axis that is perpendicular to the X–Z plane. The objective aspheric surface is then enveloped by the straight edge of the tool. The undeformed chip thickness $h$ is uniform within the entire cutting region and is determined by tool feed $f$ and cutting edge angle $\kappa$. Therefore, by using a sufficiently small $\kappa$, an extremely small undeformed chip thickness $h$ can be obtained even at a large tool feed $f$. Ductile machining at a large tool feed improves both machining efficiency and tool life.

However, for the straight-nosed tools, it is difficult to generate concave surfaces because of the interference between the tool and the workpiece. To machine concave surfaces by high-efficiency ductile machining, using a large radius round-nosed diamond tool, the nose radius of which ranges from a few millimeters to a few tens of millimeters, is effective. It has been demonstrated that this method can achieve both high production efficiency and high surface quality.

Three-axis (XZB) diamond turning is also effective for fabricating a Fresnel lens (see Section 1.2.1.4). As shown in Figure 2.4, the Fresnel structure is generated on a flat substrate using a diamond tool which has a sharply pointed

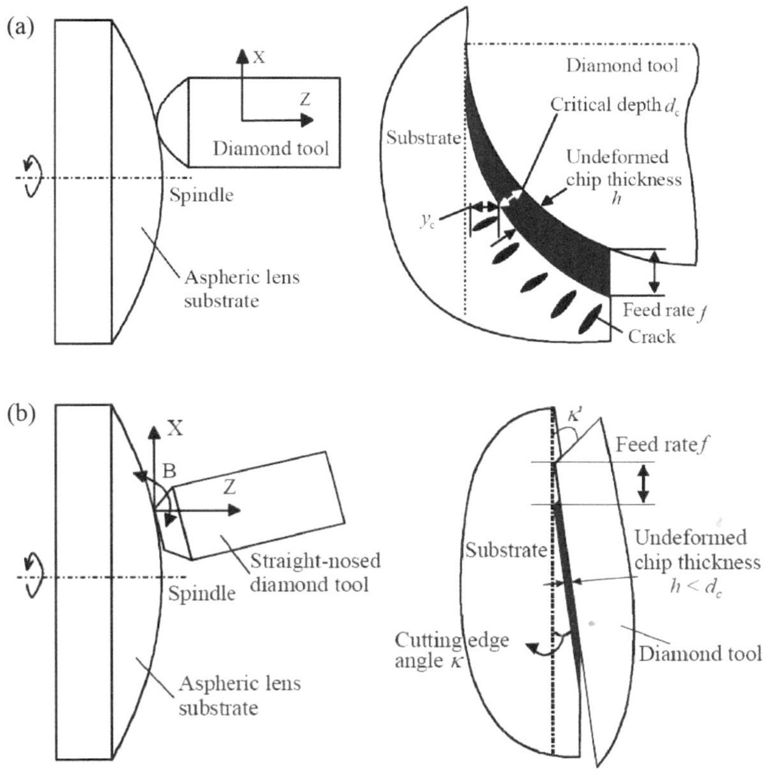


**FIGURE 2.3** Typical methods for diamond turning: (a) 2-axis control, (b) 3-axis control [3].

V-shaped tip. The machining operation for one micro groove consists of two steps. First, the tool moves along the Z axis, namely the vertical axis, to generate the cylindrical surface at the zone step. Next, the tool moves and at the same time rotates under X–Z–B three-axis simultaneous control to generate the spherical surface. By using a small cutting edge angle$\kappa$, the undeformed chip thickness $h$ can be decreased without reducing the tool feed $f$. Therefore, if $h$ is thinned to be smaller than the critical undeformed chip thickness, ductile-cut surface can be obtained. This method enables ductile machining of brittle materials under a high tool feed; thus, the tool wear can be reduced. In addition, the precutting process may cause subsurface damage such as potential microcracks and dislocations in the surface layer. By off-setting the lens apexes a few microns lower than the substrate surface, the precut-induced damage layer will be removed during the finishing cut, without damaging the Fresnel structures.

## 2.2.2 Fly Cutting

The word "fly cutting" originally means cutting a fixed workpiece using a rotating tool and is synonymous with milling. In the ultraprecision machining field, it refers to ultraprecision milling that uses only one cutting edge which is fixed at the edge of a tool head driven by the main spindle of the machine, as shown in Figure 2.5; while in milling, as will be discussed in the next section, the tool usually has a small diameter with multiple edges and driven by a tool spindle. Fly cutting is usually used for cutting flat surfaces such as polygon mirrors as introduced in Section 1.2.1.1 and LSI substrates in Section 1.2.2.3. Fly cutting has also been combined with shaping/planning (see Section 4.2.4) for microgrooving on flat surfaces to make stampers for replicating light guiding plates for liquid crystal displays.

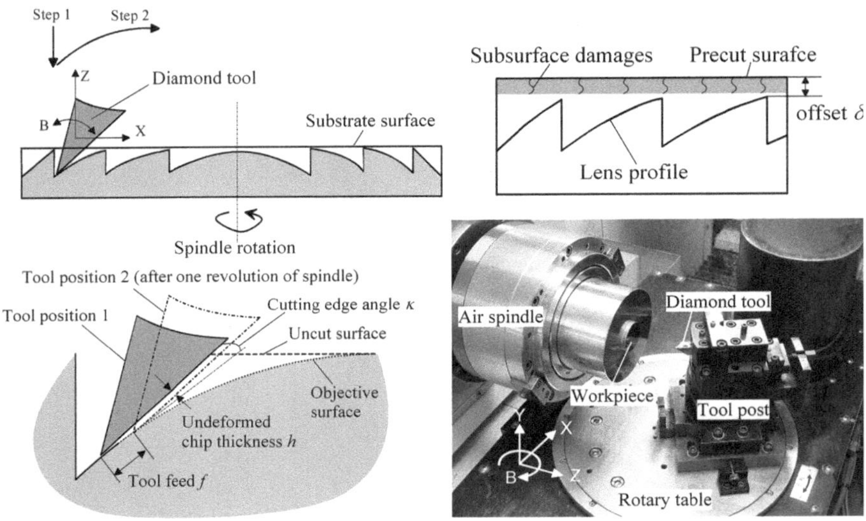


**FIGURE 2.4** Schematic of 3-axis diamond turning of single-crystal germanium for Fresnel lens [4], [5].

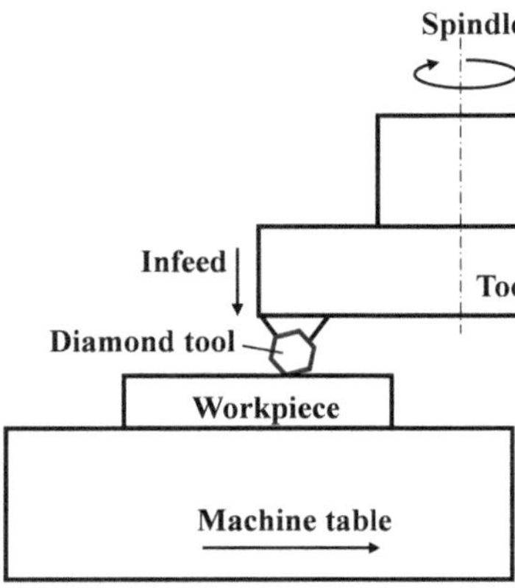


**FIGURE 2.5** Schematic of fly cutting.

### 2.2.3 Milling

Milling is a machining process of using a rotary cutter to remove material from a workpiece. The cutter which has multiple edges makes many individual cuts on the workpiece in a single run, making milling an efficient method for fabricating three-dimensional complicated shapes and surface structures.

Milling can be roughly classified into three types, namely, side milling, face milling, and end milling, as shown in Figure 2.6, according to the shapes of the milling cutters used. Side milling is mainly used to create grooves or flat surfaces using a cutter that has edges on its peripheral surface. In side milling (Figure 2.6(a)), the cutter axis is parallel to the workpiece surface and perpendicular to the cutter feeding direction. Face milling is used to create large flat surfaces using a milling cutter that cuts the workpiece with edges on its face (Figure.2.6(b)). In face milling, the cutter axis is perpendicular to both the workpiece surface and the tool feeding direction. End milling is a mix of side and face milling, where the edges on the side as well as on the end of the cutter are engaged in cutting simultaneously. End milling cutters include flat-bottomed ones (Figure.2.6(c)) and rounded ones (Figure.2.6(d)). The latter is also referred to as ball end mills and is used for curved surface generation. For all types of milling operations, the edges of a milling cutter remove workpiece material at regular intervals as the cutter rotates. As a result, the surface after milling contains regular cutter marks which are formed by cutter revolution and tool/workpiece feed. Then the height of these regular cutter marks is the main cause of the surface roughness of the finished workpiece.

As mentioned in Chapter 1, end milling is widely used for machining microstructured surfaces such as micro lens arrays (Section 1.2.1.6) and micro fluidics (Section 1.2.3.2), where the machining scale, namely, depth of cut, is in or below the micron

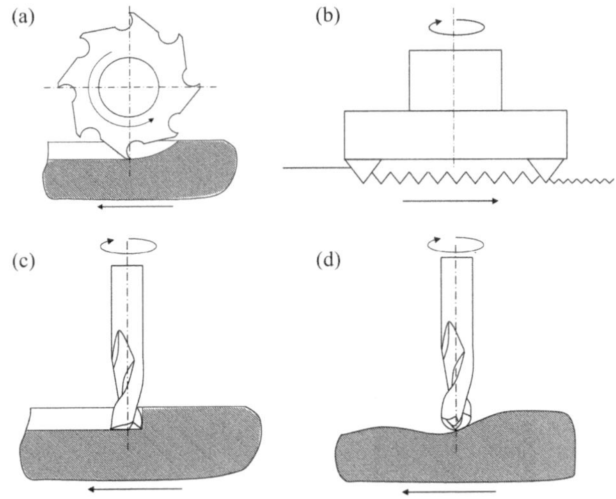


**FIGURE 2.6** Typical milling processes: (a) side milling, (b) face milling, (c) flat end milling, and (d) ball end milling.

level, and the machined feature size is smaller than 1 mm. Most of these micro lens arrays are made of plastics or glasses by using replication technologies such as plastic injection molding and glass molding press. In these replication processes, the fabrication of precision molding dies is a key step where micro end milling is useful.

It should be pointed out that in micro end milling, the undeformed chip thickness is extremely small and comparable to the edge radius. This means that the effective tool rake angle becomes a highly negative one. This situation is distinctly different from macro milling where the cutter can be assumed sharp. The tool-workpiece interference under highly negative rake angle causes very complicated material deformation in the cutting region. The bluntness of the cutter edge causes an increase in cutting force and subsurface damage. Edge bluntness will also cause burr formation when cutting metals, which deteriorates severely the profile accuracy and surface quality of machined parts. Therefore, burr formation is a critical problem limiting the product quality of micro milling. On the other hand, the edge bluntness-induced negative effective rake angle is sometimes beneficial for cutting hard and brittle materials because it can prevent from brittle fractures by generating compressive stress in the cutting region.

It should be pointed out that in micro end milling of dimples or grooves, the axis of the end mill is inclined to the workpiece surface by a certain angle (typically 45°), as shown in Figure 2.7, rather than using a perpendicular axis. This is to avoid difficulty in material removal at the center of the ball end mill where the cutting speed is zero. It is also noteworthy that when feeding the end mill in X direction or Y direction, though the same arc cross-section micro grooves can be generated, different tool feed directions lead to different surface formation mechanisms, and in turn, different surface roughness.

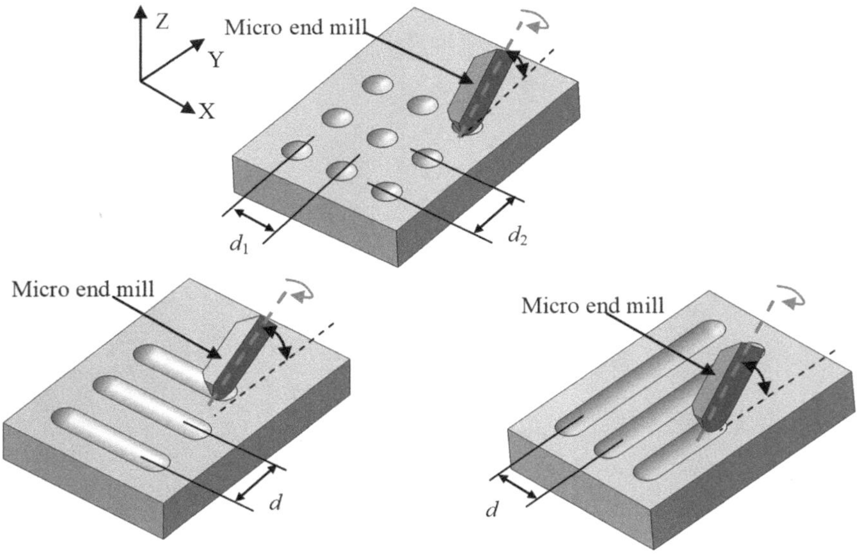


**FIGURE 2.7** Micro end milling for round dimples and grooves by using an inclined tool axis [6].

Tooling is also an important issue for micro end milling. Three kinds of tool materials are usually used for micro end milling: tungsten carbide (WC), polycrystalline diamond (PCD), and single crystalline diamond (SCD). The tooling of micro endmills will be detailed in Chapter 4.

### 2.2.4 Shaping/Planning

Shaping and planning are popular precision cutting methods especially for large flat workpieces with microgrooves. A shaper operates by moving a cutting tool backward and forward across the workpiece. The workpiece mounts on a rigid square table which can traverse sideways underneath the reciprocating tool which is mounted on a ram. The table motion is usually controlled by a precise feed mechanism. The ram slides back and forth above the workpiece and the tool can be positioned to cut the flat surface on the top of the workpiece. A planer is a type of machine tool analogous to a shaper but larger and with the entire workpiece moving beneath the cutter instead of the cutter moving above a stationary workpiece. A schematic presentation of shaping/planning is shown in Figure 2.8(a).

Planers and shapers have been used generally for two types of work: generating large accurate flat surfaces and cutting straight microgrooves. For example, planers are used for producing precision stamping dies and plastic injection molds for light-guiding plates of liquid crystal displays, large-scale linear Fresnel lenses,

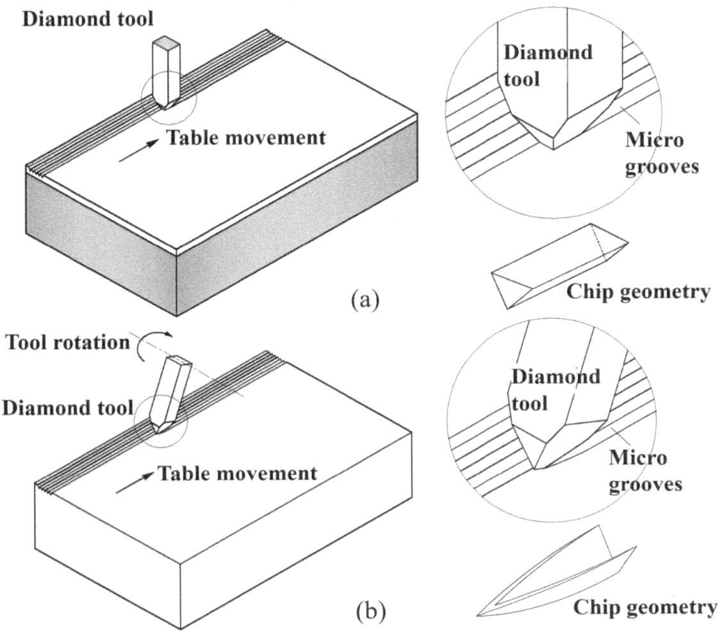


**FIGURE 2.8** Schematics of (a) shaping/planning, (b) shaping/planning with a fly cutter.

and so on. Recently developed shaper machines have also enabled the fabrication of curved microgrooves and three-dimensional microstructures by controlling the orientation angle and moving direction of the cutting tool and the depth of cuts.

However, the cutting speed in shaping and planning is very low due to the moving speed of the machine table is limited. In addition, the cross-sectional geometry of the chip in shaping and planning is in a triangular shape, leading to material stagnation in front of the tool tip. This chip formation mechanism causes significant burr formation. To solve this problem, fly cutting is combined with planning/shaping, as shown in Figure 2.8(b), where the fly cutter is rotated with a spindle. This method enables interrupted cutting at a higher speed and the cross-sectional shape of the chip is in a V shape, enabling fluent chip flow without material stagnation [6]. This chip geometry prevents burr formation in the grooving process. This combined microgrooving process has been used in machining stampers for replicating microprism-type retro-reflective films introduced in Section 1.2.1.8, and other kinds of functional surfaces.

## 2.3 GRINDING

Grinding, lapping, and polishing are three main methods of abrasive machining which uses abrasive grits to remove material from the workpiece. Grinding is a fixed abrasive machining method, whereas lapping and polishing are loose abrasive machine methods. This section focuses on grinding, and lapping and polishing will be introduced in Sections 2.4 and 2.5.

Ultraprecision grinding has been a major method for ultraprecision machining especially for materials with high hardness and strength as such materials will cause severe tool wear in cutting. Brinksmeier et al. [7] made a definition for ultraprecision grinding, which is a material removal process:

1. with fixed abrasives (in order to distinguish it from lapping and polishing),
2. which are in interrupted contact with the workpiece surface (in order to distinguish it from honing), and
3. where the abrasive surface is characterized by at least one statistical distributed parameter (in order to exclude milling but to include grinding with specially designed or "engineered" wheels).
   An ideal ultraprecision grinding process should yield functional surfaces with a surface finish adequate for optical applications (but not exclusively optical applications), i.e.
4. a root-mean-squares figure accuracy $< \lambda/10$ with $\lambda < 1$ μm,
5. a surface roughness Sq $< \lambda/100$, and
6. a damage-free sub-surface zone for avoiding light scatter (in the case of optically transparent materials) and/or crack formation (under load, or even without).
   Moreover, it is an implicit but important requirement that an ultraprecision grinding process (like other ultraprecision machining processes) should be deterministic, i.e.

7. form, roughness, and damage tolerances should be reached in one single machining step (in order to avoid iterative or additional machining and testing).

Ultraprecision grinding is mostly used to produce high-quality surfaces made from hard brittle materials. Figure 2.9(a) shows a typical ultraprecision grinding machine tool setup, where the workpiece is fixed on the work spindle, rotating at a low rotation rate (~100 rpm), and the grinding wheel is fixed on the grinding spindle, rotating at a high rotation rate (~20,000 rpm). To grind the entire surface, a feed motion is given to the grinding wheel along the x-axis and a depth of cut is et along the z-axis direction of the machine tool. Through the multipoint cutting actions of ultrafine abrasive grits on the grinding wheel, as schematically illustrated in Figure 2.9(b), ultraprecision grinding can fabricate parts with high surface quality at reduced tool wear compared to single-point diamond turning.

Ultraprecision grinding has been widely used in various industries such as optics, semiconductors, and optical communications. It is usually employed as a pre-finish machining step (finish is performed by polishing) in the production of glass lenses and other optical elements, and for the manufacturing of optical mold inserts for large volume production of aspherical lenses [7]. To create aspherical surfaces, advanced multi-axis CNC machines equipped with high-stiffness structures, vibration isolation systems, and nanometer-level positioning control that accommodate the complex geometries of aspherical profiles are necessary. By changing the machine axes arrangement, various grinding methods have been proposed. For example, "cross grinding" is a method where a grinding wheel with a V-shaped cross-section fixed on a spindle 45° tilted with respect to the horizontal plane is used to machine micro aspherical lenses [8], while "parallel grinding" is a method where a partially spherical grinding wheel fixed on a horizontal spindle is used for grinding aspherical surfaces, as shown in Figure 2.10 [9, 11].

(a)

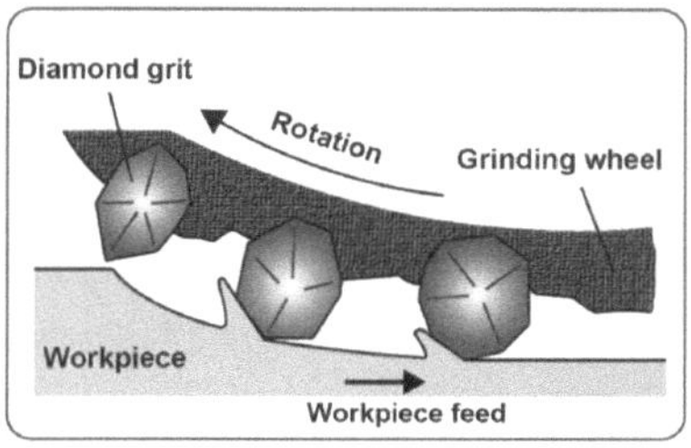

(b)

**FIGURE 2.9** (a) Photograph of the main section of a grinding machine, and (b) schematic of the grinding principle.

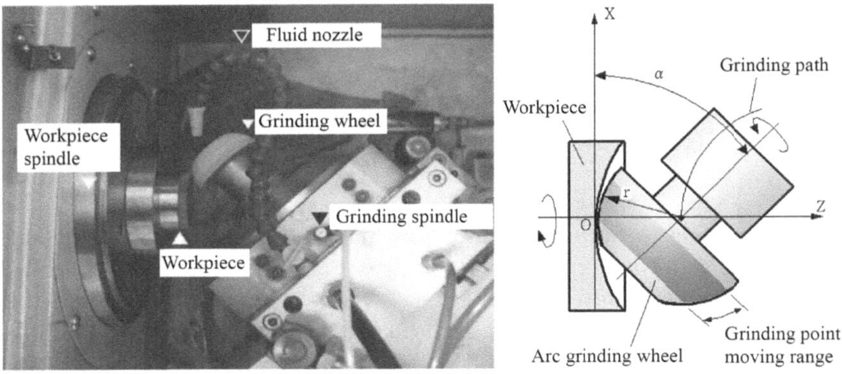


**FIGURE 2.10** Grinding process of aspherical surfaces [9].

As a grinding wheel contains numerous randomly distributed abrasive grains with different shapes, the mechanism of material removal via multi-edge cutting is more complicated than that of single-edge cutting methods. Due to the abrasive grains being fixed by the bond material and the protrusion heights of their edges being different, the interaction mechanism between the abrasive grains and workpiece may involve sliding, plowing, and cutting. After grinding, there will be clear grinding marks on the ground surface which are caused by the ploughing and scratching effects of abrasive grains. These grinding marks increase the roughness of the ground surface, which makes it difficult to use grinding as the final finishing process. To reduce surface roughness, the optimization of grinding parameters and path planning, the selection of the grinding wheel, and truing/dressing processes are important.

## 2.4 LAPPING

Lapping is a loose abrasive machining process that primarily relies on an abrasive mixture known as a slurry. The slurry consists of abrasive particles suspended in a water-based or oil-based solution. The fundamental principles of conventional mechanical lapping are shown in Figure 2.11. In this process, a platen rotates at an angular velocity $\omega_1$, while the workpiece rotates at an angular velocity $\omega_2$ in the same direction. Slurry is continuously fed between the workpiece and the lapping plate, usually made of metals such as cast iron or copper, to facilitate material removal. To enhance material removal, lapping pressure is required to be applied to the workpiece holder.

In batch processing, multiple parts are secured in place using retaining rings, while the abrasive mixture is applied to the flat lapping plate. The parts are then moved across the lapping plate in circular, reciprocating, or oscillating motions to ensure uniform material removal. Figure 2.12 presents a double-sided lapping machine that can process wafers in batches.

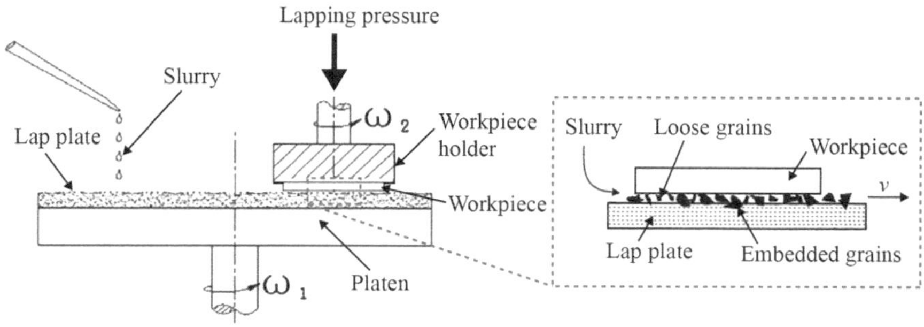


**FIGURE 2.11** Schematic of a lapping process.

**FIGURE 2.12** The experimental setup photo of double-sided planetary lapping [10].

The mechanism of lapping is distinctly different from grinding. As the lapping plate is hard, very few abrasive grains can be embedded into it. Instead, the loose abrasives are sandwiched between the hard-lapping plate and the workpiece surface thus the relative movement between the two flat surfaces drives the abrasive grains to roll against the workpiece surface. During rolling, the sharp corners of the abrasive grains act as indenters, making indents on the workpiece surface. For brittle materials, indentation causes median cracks during loading and lateral cracks during unloading, the propagation of the latter contributes to the material removal in lapping. Based on this material removal mechanism, the lapping process is usually used for flattening the wafers after grinding to remove the grinding-induced waviness and unevenness on the surface.

## 2.5 POLISHING

As another loose abrasive machining process, polishing is often carried out after a grinding or lapping operation to achieve an optimal surface finish. The difference between lapping and polishing lies in the pad hardness, and in turn, the material removal mechanism. Lapping is a method of efficiently shaping a material to the desired form and accuracy by using hard tools (plates, pads) and coarse abrasive grains. Polishing refers to the use of fine abrasive grains and soft tools (pads, media) made of felt, plastic, or fluids, to refine the surface texture following grinding. The final mirror finishing process is typically referred to as polishing.

According to the material removal mechanism, polishing can be further divided into a few subcategories, for example, mechanical polishing, chemomechanical polishing (CMP), float polishing, and elastic emission machining (EEM), etc. The brief introductions will be given in the following subsections.

### 2.5.1 Mechanical Polishing

Mechanical polishing, a loose abrasive process, operates on a principle similar to lapping, where abrasive grits suspended in a carrier fluid are introduced between a pad or tool and the workpiece. These abrasives are entrained by the tooling, and the applied load, combined with relative motion, removes material, resulting in a low roughness surface. The earliest forms of mechanical polishing date back to ancient civilizations, where artisans used natural abrasives like sand, pumice, and fine stones to smooth and shine materials such as metals, ceramics, and gemstones. However, controlling form accuracy relied heavily on human experience, making it difficult to achieve precision.

In the 1970s, computer-controlled optical surface (CCOS) technology was proposed. Aspden et al. [11] studied the material removal mechanism of CCOS and developed a mathematical model. This technology utilizes a CNC-controlled small grinding head, typically less than one-quarter of the workpiece diameter, to polish optical components. The grinding head makes contact with the workpiece surface under controlled pressure and moves relative to it. The computer precisely controls the machining path, motion speed, machining pressure, and dwell time of the grinding head to selectively remove material in different areas. This advancement significantly accelerated the development of modern optical technology.

As the polishing pad is made of soft materials, the abrasive grains can be partially embedded into the pad surface, thus the grain rolling, which dominates the mechanism of lapping, does not occur in polishing. Instead, sliding, ploughing and cutting take place like that in grinding. However, different from grinding, in polishing, the depth of micro cutting is extremely small and uniform among grains because the elasticity of the pad can flexibly adjust the protrusion height of each grain to fit the workpiece surface. This produces a mirror finish without significant scratches or polishing marks.

### 2.5.2 Chemomechanical Polishing

Chemomechanical polishing (CMP), a polishing technique derived from mechanical polishing, removes material from the workpiece surface through combined chemical and mechanical actions. It is widely regarded as a process capable of achieving global planarization, making it essential for obtaining ultrasmooth flat surfaces [12]. CMP is the most effective technique for wafer planarization in the semiconductor industry.

Figure 2.13(a) illustrates a typical rotary-type CMP system, in which the platen and wafer carrier rotate in the same direction, while the wafer carrier reciprocates synchronously along the platen's radial direction. The wafer is secured within a rotating carrier. Figure 2.13(b) shows a detailed schematic diagram of wafer–pad interactions. The carrier contains a membrane that applies downforce to the wafer's back, along with a retaining ring around its perimeter to keep it in place. A polishing pad is affixed to the rotating platen. The wafer surface is pressed against the polishing pad. The movements of the carrier and platen create the necessary relative motion for polishing. A slurry composed of abrasive particles and chemical solutions is applied to the pad to facilitate material removal. The particle-film interactions, and the SEM image of the pad top surface, are shown in Figure 2.13(c) and (d), respectively. The chemical reaction softens the wafer surface (in the case of Figure 2.13, the deposited film surface) to change it into a more easily removable layer. Through the combined chemical effects of the reagents and the mechanical actions of the abrasive particles, micro-scale material removal occurs via alternating chemical and mechanical actions, facilitating fine surface finishing.

The chemical reaction in CMP may take place either between the abrasive grains and the workpiece or between the chemical solution, or both. To activate the chemical reaction between the abrasive grains and the workpiece surface, normally soft abrasive grains with chemical affinity with workpiece material should be used.

For the CMP of silicon, diamond abrasives, which have been used popularly in mechanical polishing, are rarely used. Instead, silicon dioxide ($SiO_2$) abrasives,

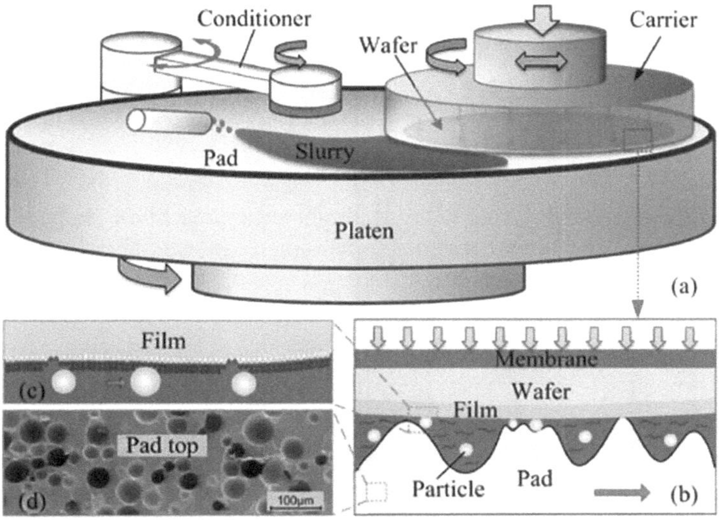


**FIGURE 2.13** Principle structure of CMP process: (a) CMP equipment, (b) wafer–pad interactions, (c) details of particle–film interactions, and (d) SEM image of pad top surface [46].

alumina ($Al_2O_3$) abrasives, and cerium oxide ($CeO_2$) abrasives which can provide the necessary mechanical actions with silicon, are adopted in semiconductor industry. The size of these particles typically ranges from 20 to 100 nanometers, depending on the required surface finish and material removal rate.

### 2.5.3 Float Polishing

Float polishing is one of the most promising techniques for producing super-smooth, damage-free surfaces of optical components. The main structure of a float polishing machine is illustrated in Figure 2.14 [13]. The polishing fluid consists of pure water mixed with extremely fine abrasive powder, typically at a concentration of less than 5 wt%. During polishing, the samples remain in a stable floating or suspended state due to the fluid's hydrodynamic behavior. This stability is achieved because the samples are sufficiently flattened through diamond lapping before polishing, and the tin lap features 2 mm pitch grooves along with finer grooves on their ridges. The samples and sample holders only rely on their weight to maintain stable movement. The combination of abrasives, fluid suspension, pressure, and sometimes chemical agents ensures that the workpiece surface is gradually polished to a fine finish. Due to the smoothing effect of the thin fluid film between the sample and the lap, the sample surface achieves better flatness than the lap itself. Additionally, larger particles in the fluid are pushed toward the vessel wall by centrifugal force during polishing [13].

Previous studies have successfully float-polished fused quartz and optical glass samples. In some cases, this method has achieved a good optical figure and an ultra-smooth surface finish with roughness below 0.5 nm rms. Additionally, float polishing has proven effective for polishing sapphire [14]. The limitation is that only flat surface can be polished by float polishing while it is difficult to polish curved or structured surfaces.

### 2.5.4 Elastic Emission Machining

Elastic emission machining (EEM) is an ultraprecision machining technology that utilizes chemical reactions between the workpiece and the abrasive to achieve

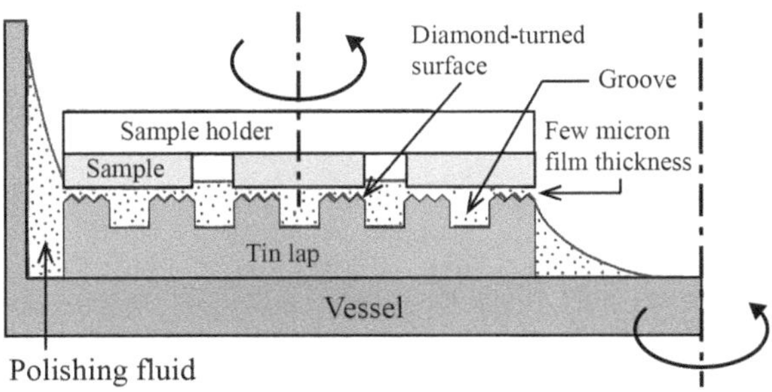


**FIGURE 2.14** Schematic of the structure of a float polishing machine.

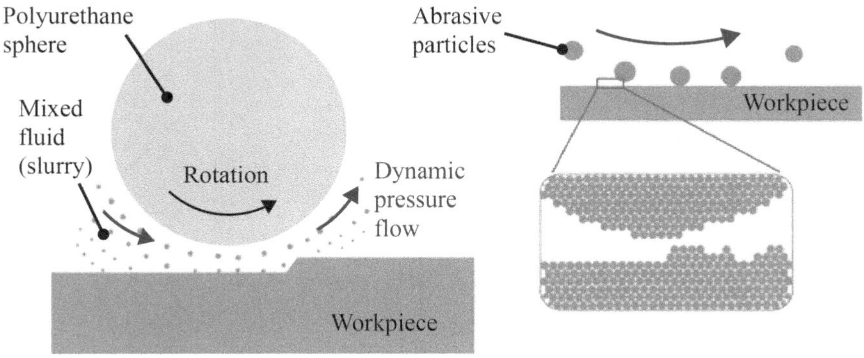


**FIGURE 2.15** Schematic of the principles of elastic emission machining [17].

atomic-scale polishing [15]. As shown in Figure 2.15, its working principle relies on a polyurethane polishing tool that maintains a fine gap of approximately 1 μm between the tool and the workpiece, preventing surface scratches during processing. In the EEM process, the rotating polishing tool generates a fluid dynamic pressure flow in the slurry. This fluid acts as a carrier, facilitating weak collisions between abrasive particles, which then form chemical bonds. As the particles flow over the workpiece surface, surface atoms migrate and transfer from one location to another. These chemically bonded atoms are subsequently removed, driven by chemical action and the principles of elastic hydrodynamic lubrication, enabling atomic-scale material removal [16].

EEM technology offers significant advantages: it produces surfaces free of damage and subsurface defects while reducing mid- and high-frequency roughness without compromising low-frequency surface form accuracy. Consequently, EEM is particularly suitable for atomic-scale machining of curved optical elements made from single crystalline materials, providing an advanced approach for achieving ultra-smooth and high-precision optical surfaces such as the reflecting mirrors for X-rays [18].

## 2.6 ELECTRICAL MACHINING

### 2.6.1 Electrical Discharge Machining

Electrical discharge machining (EDM) is a thermo-electrical process that removes material by thermal erosion due to repeated high-energy electrical sparks. The working mechanism of EDM is illustrated in Figure 2.16. The workpiece is immersed in an insulating dielectric fluid, typically ionized water or kerosene. A conductive tool electrode, such as a copper rod, is mechanically positioned near the workpiece surface. An electrode-driven mechanism precisely controls the gap between the tool electrode and the workpiece. When a voltage, usually around 100 V [19], is applied across this gap, the dielectric fluid becomes ionized as the

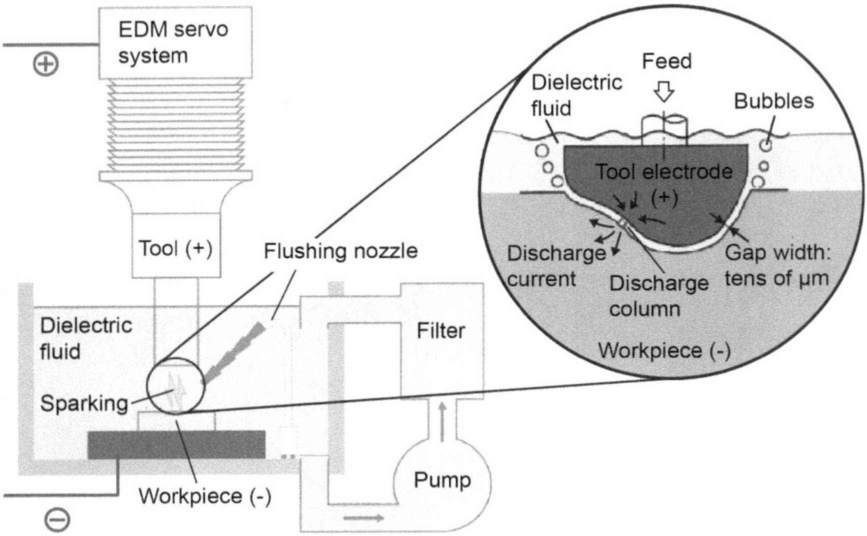


**FIGURE 2.16** Schematic illustration of the fundamental principles of the EDM process [21].

gap narrows, resulting in the formation of a plasma channel. The establishment of this plasma channel generates an immense amount of heat, raising the workpiece surface temperature to 4000–8000 K [20]. This extreme temperature melts the surface material and the subsequent sparks ablate it away as debris.

Since there is no physical contact between the workpiece and electrode in this process, the machinability depends only on the tool and the workpiece's electrical and thermal properties. Thus, this process can be used to machine any given conductive material regardless of its hardness. In addition, this method can also be used to directly transfer complex 3D shapes of the electrode with high accuracy onto the workpiece without force-induced deformation. Figure 2.17 shows the tool array of 9 rectangular electrodes with a size of 100 μm by 100 μm and a height of 200 μm. The electrodes were fabricated from a copper rod of 2 mm diameter by micro milling [22].

It should be pointed out that EDM process usually causes surface defects such as surface craters, resolidified layers, and microcracks; thus, generally speaking, the surface quality of EDM does not meet the requirement of ultraprecision machining, and a subsequent finishing process is needed to remove these surface and subsurface defects.

### 2.6.2 Electrical Chemical Machining

Electrochemical machining (ECM) is a promising surface finishing process for post-processing surfaces machined by EDM. The material removal mechanism of ECM

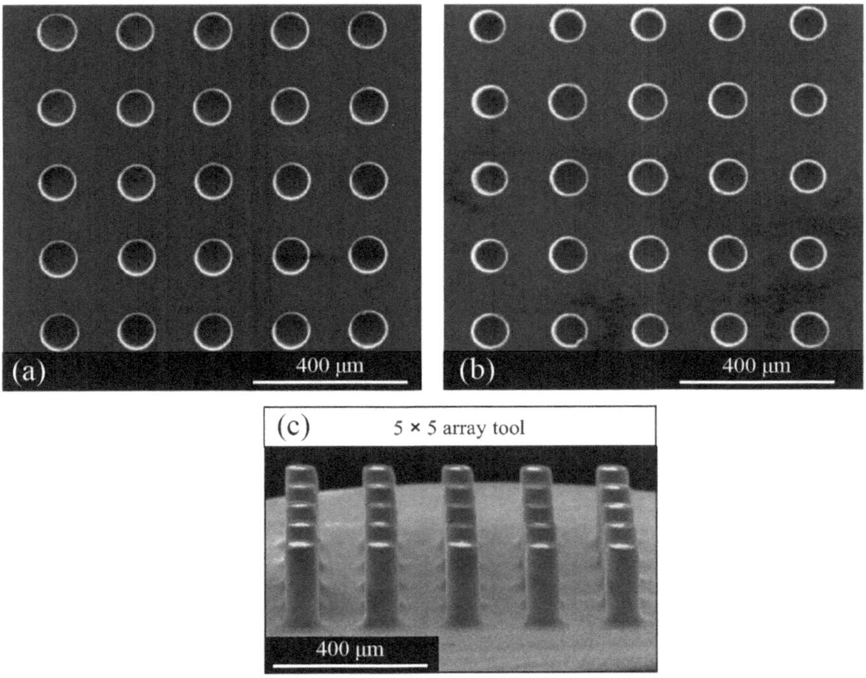


**FIGURE 2.17** SEM images of a 5 × 5 microhole array machined using vibration-assisted EDM: (a) at 400 Hz with a 5 μm amplitude, (b) at 1000 Hz with a 5 μm amplitude. (c) The 5 × 5 pillar array tool is used for machining the structures shown in (a) and (b) [22].

is shown in Figure 2.18. Initially, ECM induces the expansion of microcracks in the EDM-induced resolidified layer due to tensile residual stress and bubble action (Figure 2.18(a)). Bubbles generated during ECM further aid in detaching the resolidified layer fragments from the unaffected bulk workpiece and tend to remove these fragments (Figure 2.18(b)). Subsequently, ECM proceeds on the undulating surface of the bulk workpiece, generating electrolytic products such as $Mn^+$ and $O_2$ (Figure 2.18(c)). Eventually, the undulating surface of the bulk workpiece is smoothed (Figure 2.18(d)). The SEM images comparing surfaces machined by EDM and ECM are shown in Figure 2.19. The EDMed surface is covered by a resolidified layer, with craters and microcracks formed on the surface. After ECM, the resolidified layer is entirely removed, and the surface flatness is significantly improved.

While conventionally ECM has been performed by using a different machine from the EDM machine, recent research shows that ECM can also be performed after EDM without changing the machine and tool setup, which saves time and eliminates alignment errors caused by repositioning the workpiece and tools [23]. Figure 2.20 illustrates a hybrid EDM/ECM process for fabricating grooves on a workpiece. At first, EDM was performed by filling the tank with EDM oil. After EDM, the oil was drained, and the tank was refilled with an ECM electrolyte.

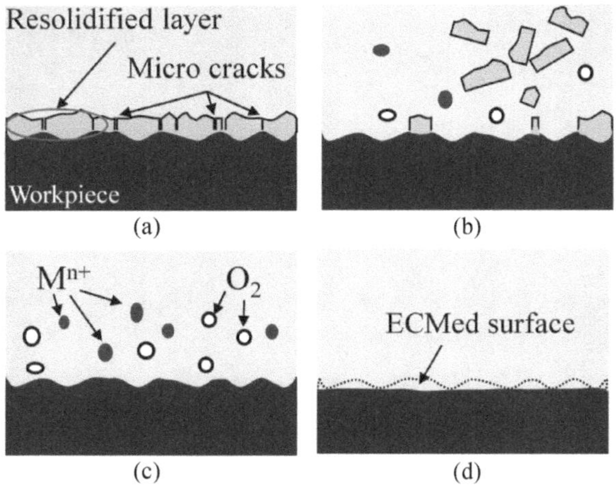


**FIGURE 2.18** Removal mechanism of resolidified layer: (a) after EDM, (b) early stage of ECM, (c) middle stage of ECM, and (d) end stage of ECM [23].

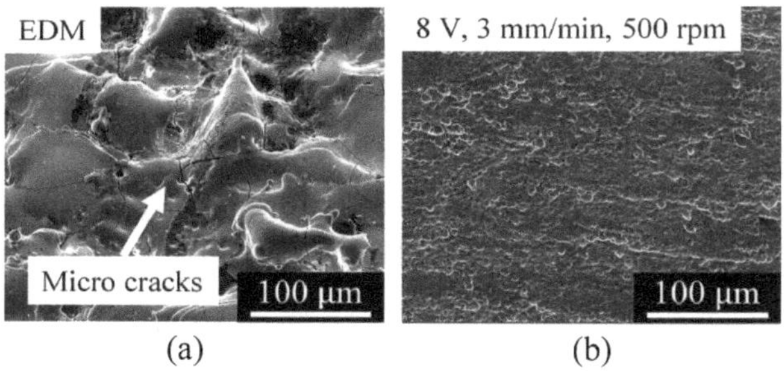


**FIGURE 2.19** SEM images of the surfaces machined by (a) EDM and (b) ECM [23].

Then, ECM was performed as the finishing process by using the same electrode and machine used for EDM.

ECM offers advantages such as good surface quality, no mechanical stress on the material, and the ability to machine hard-to-machine materials, although it has some limitations and challenges. ECM works by using an electrochemical reaction, which means it is only effective on conductive materials. It cannot be used to machine non-metallic materials or non-conductive materials like plastics, ceramics, or certain composites. Managing the electrolytes (including filtration, cleaning, and disposal) can be complex and costly, as the electrolytes may contain harmful chemicals depending on the material being machined. Although ECM provides a good surface finish, achieving the exact desired surface roughness can be challenging. Factors like electrolyte concentration, temperature, and machining parameters can influence the final finish the removed surface layer thickness, making it harder to control with consistency.

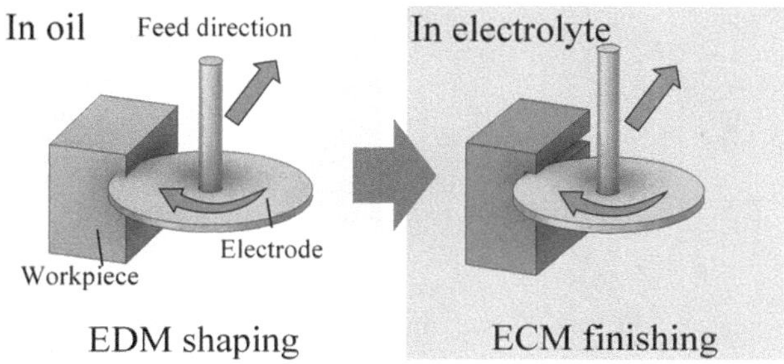


**FIGURE 2.20** EDM/ECM hybrid machining process [23].

## 2.7 LASER PROCESSING

### 2.7.1 Femtosecond Pulsed Laser Machining

Femtosecond lasers with ultrashort pulse durations provide a promising solution for achieving precise, high-quality machining of various materials. This is because femtosecond lasers can generate exceptionally high peak power over extremely short time intervals, facilitating multiphoton absorption and making them even suitable for machining transparent materials. This addresses the limitations of conventional continuous-wave or long-pulse lasers, which can cut materials into certain shapes but often cause pronounced thermal effects, resulting in thermal damage and hindering the formation of well-defined geometries and shapes. In comparison to continuous-wave or long-pulse lasers, the thermal effects induced by femtosecond lasers remain distinctly small, which helps to avoid unnecessary thermal damage and is beneficial for achieving high-precision machining of micro- and nanostructures. Meanwhile, femtosecond laser machining effectively mitigates plume formation and plasma shielding phenomena, which often occur during long-pulse laser processing, resulting in relatively high processing efficiency [24].

Owing to the advantages of non-contact operation, exceptional precision in controlling geometric patterns, inherent three-dimensional processing capability, and high flexibility, femtosecond pulsed laser machining technology has been extensively used to fabricate micro- and nanostructures on hard-brittle materials such as diamonds [25], zirconia [26], and silicon carbide [27]. Figure 2.21 presents a schematic of a femtosecond pulsed laser machining system and the corresponding pyramidal microstructures fabricated on zirconia ceramics using this system.

### 2.7.2 Laser Polishing

Laser polishing is an emerging surface finishing technology with advantages such as being environmentally friendly, applicable to a wide range of materials, providing stable polishing quality, and enabling easy automation. It has garnered

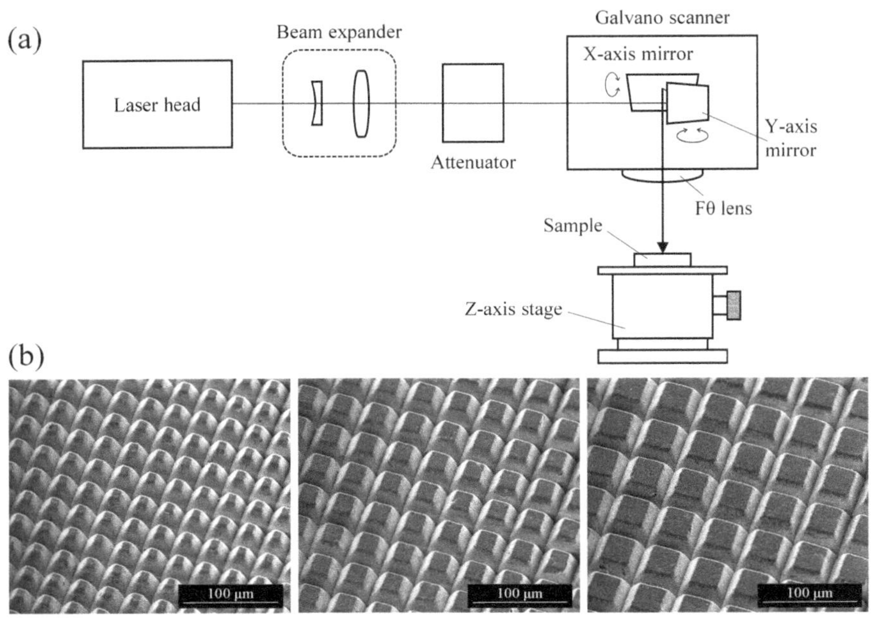


**FIGURE 2.21** (a) Schematic of a femtosecond pulsed laser machining system, (b) SEM images of the pyramidal microstructures machined on zirconia ceramics [26].

significant attention from researchers worldwide [28]. The underlying principles of laser polishing can be divided into two categories, as schematically illustrated in Figure 2.22(a). One is to use ultrashort (picosecond or femtosecond) pulsed lasers to selectively remove the surface peaks to smoothen the surface by the ablation effect. The other is to use a high-power (~300 W) CW laser to melt the material's surface, causing the molten metal to flow under the influence of surface tension, allowing surface peaks to fill the valleys and thereby producing a smooth surface. Laser polishing can process various materials, including metals, glass, and ceramics. In comparison to traditional polishing techniques, laser polishing offers inherent advantages for industrial applications and can be seamlessly integrated with technologies such as additive manufacturing, laser welding, and laser cleaning to achieve efficient and intelligent production processes. Figure 2.22(b) shows an example of laser polishing applied to ventricular assist devices.

### 2.7.3 Laser Recovery

Laser recovery is a surface-finishing method for single-crystal materials that eliminates subsurface defects generated in pre-processing steps such as cutting, grinding, lapping, and polishing [29]. The damaged areas are repaired using a laser beam, finally revealing a uniform surface with a perfect crystalline structure. A typical example is laser recovery of silicon single crystals. A nanosecond pulsed

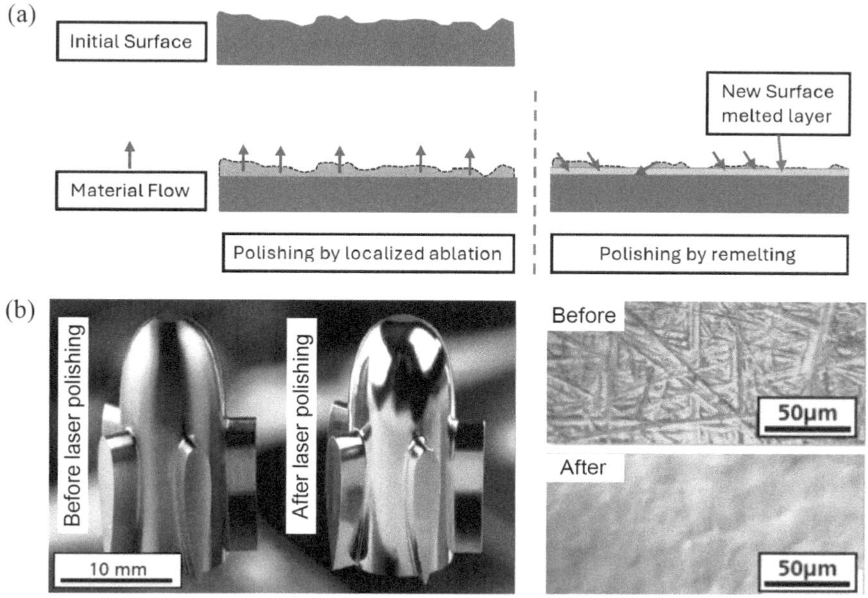


**FIGURE 2.22** (a) Schematic illustration of the fundamental principles of the laser polishing process, and (b) an example of laser polishing applied to ventricular assist devices [28].

laser was used to recover machined silicon wafers. At certain energy intensities, the laser caused recrystallization, turning the amorphized, dislocated, and cracked subsurface layers into the same single-crystal structure as the base material. An application of this method is to recover grinding-induced damage in boron-doped silicon wafers. By fine-tuning the pulse width and peak irradiance, a remarkably smooth surface was achieved with a low roughness of 1 nm Sa on curved surfaces of silicon edges and notches, as shown in Figure 2.23. This method can be also used for finishing micro lens arrays, grooves, and other micro-structured surfaces of single crystal materials. By controlling the shape, power, incidence angle, and scanning dynamics of the laser beam, laser recovery can be applied to various complex shapes including microscale surface structures.

## 2.8 ELECTRON BEAM POLISHING

The electron beam (EB) polishing method utilizes the high energy density of EB which has a diameter of a few tens of mm to instantly melt the metal surface. It is a highly efficient surface finishing technique for mold steels that require a precise, high-quality surface following electric discharge machining or milling.

Figure 2.24(a) schematically shows large-area EB irradiation equipment and its working principle. At first, a magnetic field is generated by the solenoid coil positioned on the outer side of the chamber. When the magnetic field

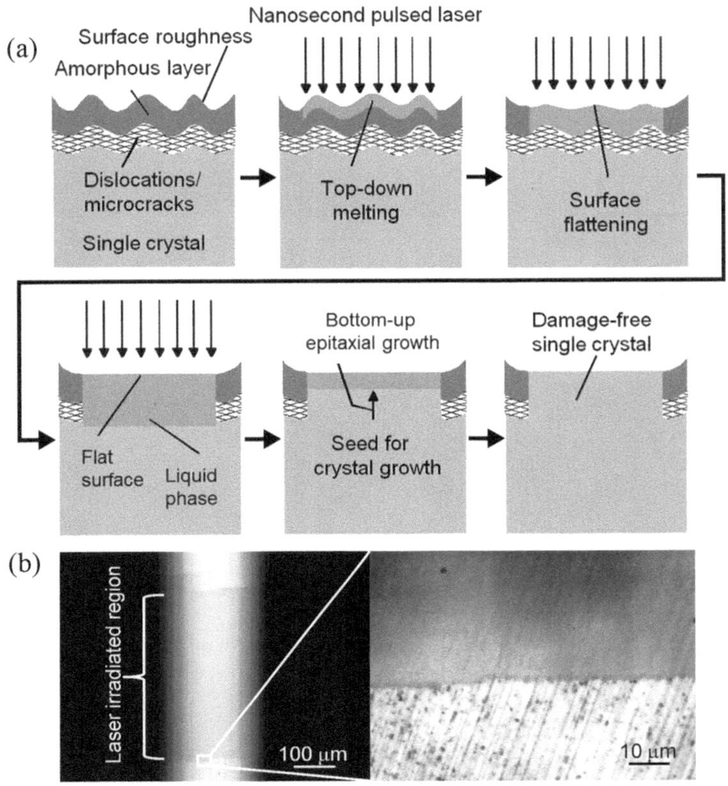


**FIGURE 2.23** Schematic illustration of principles of laser recovery and the boundary of original surface and laser-recovered region of single-crystal silicon surface [29].

reaches its maximum intensity, a pulse voltage is applied to the anode. Inside the chamber, electrons are generated via the Penning effect and begin moving toward the anode. Simultaneously, due to the Lorentz force, the electrons follow a spiral trajectory. Next, argon gas atoms are ionized through repeated collisions with these electrons, leading to the formation of plasma near the anode. When the plasma intensity reaches its peak, a pulse voltage is applied to the cathode, accelerating the electrons via the high electric field induced by the electric double layer formed near the cathode. Finally, this high-energy-density electron beam irradiates the workpiece surface with enough energy to induce melting [31].

Uno et al. [30] used 60 mm-diameter electron beams to irradiate mold steel (NAK80) and reduced surface roughness from 6 μm Rz to less than 1 μm Rz in just a few minutes. Figure 2.24(b) shows the effect of EB polishing on an electrical discharge machined surface of mold steel. The surface became distinctly smoother when an optimal energy density was used.

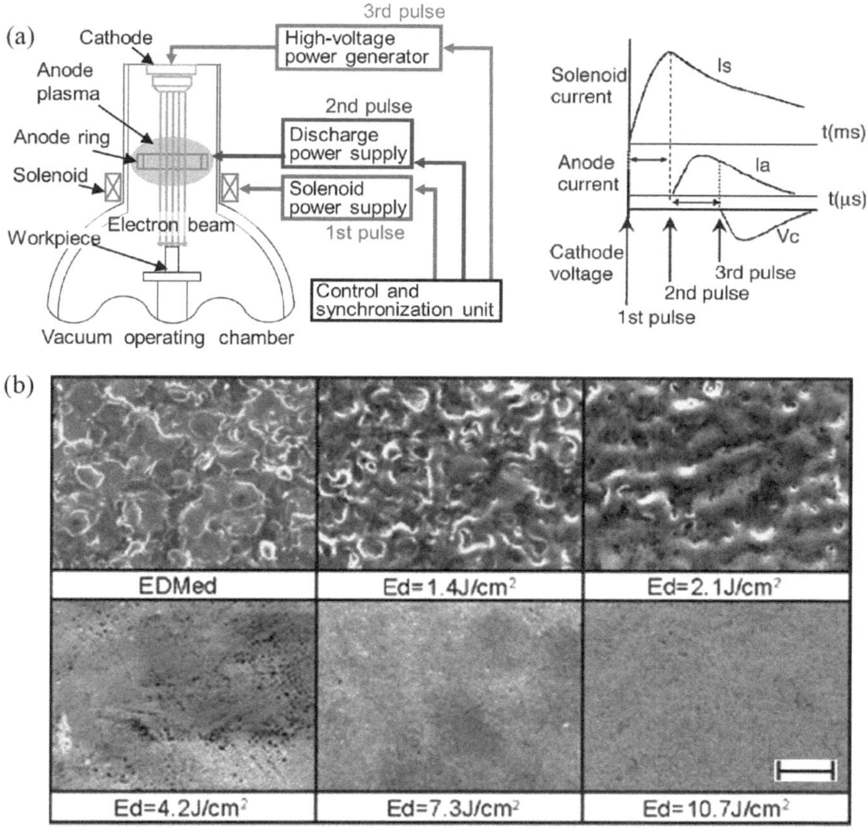


**FIGURE 2.24** (a) Schematic illustration of large-area EB irradiation equipment and timing chart of pulse inputs to solenoid coil, anode, and cathode, (b) EB polished mold steel surfaces [30]. (The scale bar is 50 μm).

## 2.9 ION BEAM FIGURING

Ion beam figuring (IBF) is an etching process under high vacuum conditions, which directs high-energy ions (typically argon) onto the workpiece surface, removing material through physical sputtering at the atomic level. The working principle of IBF is illustrated in Figure 2.25. A small beam of positively charged ions (e.g., argon) is directed onto the substrate surface. The collision of the ions with the substrate atoms initiates a "collision cascade". With an optimized process flow (adjusted energy and direction), the particles in this cascade are directed back to the substrate surface. In this way, the atoms at the end of the cascade can overcome the surface binding energy and detach from the substrate. This process is effective on all materials. The ion beam source is moved by an XY-axis system in a meander-shaped pattern across the substrate surface. An additional Z-axis is

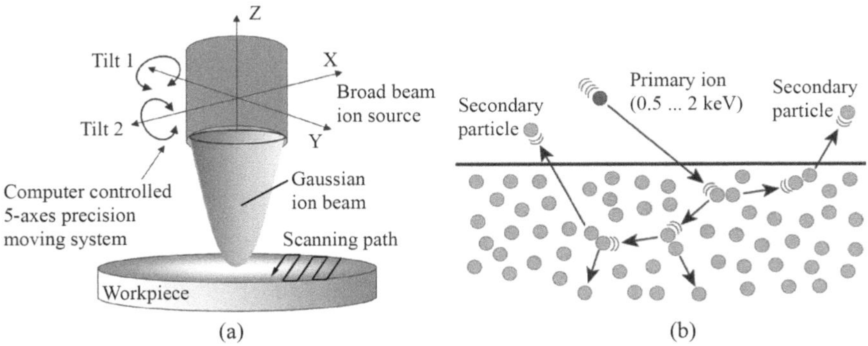


**FIGURE 2.25** Schematics of (a) ion beam figuring equipment and (b) collision cascade.

used to control the focus of the ion beam. For smaller substrates, it is also possible to use a static ion beam source and move the substrate in front of the ion beam. The amount of material removed is determined by the dwell time of the ion beam. When the ion beam moves quickly over a certain point, little material is removed, whereas slower scanning results in greater material removal. Other factors influencing the removal rate include ion beam energy, ion beam current, substrate material, and angle of incidence. Therefore, processing curved substrates requires a specialized approach that accounts for the different impact angles of the ion beam on the substrate surface [32].

One of the key advantages of IBF is its non-contact nature as a polishing tool, which eliminates edge roll-off effects commonly associated with mechanical tools. Moreover, the precise control of ion beam energy distribution enables highly stable atomic-level material removal [33], making IBF a deterministic polishing process for correcting surface shape errors. However, this method has significantly lower material removal efficiency than mechanical methods due to its inherently slow atomic-scale removal rate. Consequently, ion beam figuring is typically employed as the final step in a sequence of processing stages for finishing extremely high-accuracy optical components.

## 2.10 PLASTIC INJECTION MOLDING/IMPRINTING

Optical components such as Fresnel lenses, freeform lenses, aspheric lenses, microlenses, microlens arrays, and diffractive optical elements have been adopted across various fields, including automotive, lighting, photovoltaics, electronics, ophthalmology, and medical applications [34]. As it is expensive to manufacture these components from glass materials, they are popularly produced using plastic materials. Injection molding is the most efficient mass-production technology for manufacturing plastic components with complex geometries and high precision. Most plastics can be easily deformed and replicated at molding temperatures of 100–200°C using molds made of metal materials such as copper, nickel plating, or stainless steel.

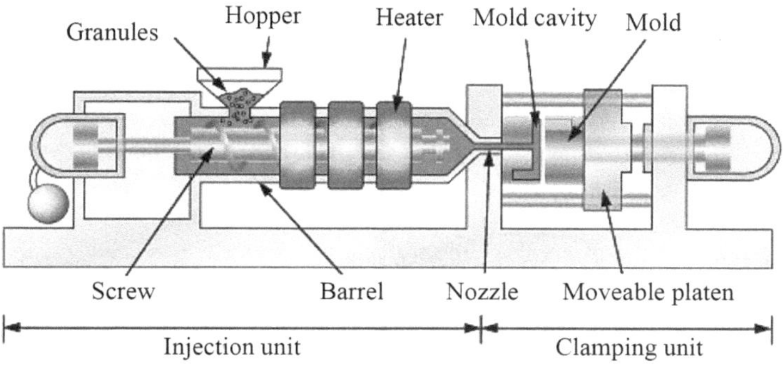


**FIGURE 2.26** A schematic drawing of a typical injection molding machine [35].

A schematic of an injection molding machine is shown in Figure 2.26. Thermoplastic material, typically in granule or pellet form, is fed into the machine through a hopper. A rotating screw transports the granules forward through its channels, pressing them against the heated cylinder wall. The material melts due to conduction from external heating units and frictional heat generated by the screw's rotation. The molten plastic accumulates at the screw tip until the desired volume is reached, at which point the screw stops rotating. The process then enters the injection and packing stages. The clamp unit holds the mold closed while the screw moves forward as a ram, forcing the molten plastic into the mold cavity. As the cavity fills, the screw maintains a holding pressure to compensate for volumetric shrinkage as the material cools. Once the cooling phase begins, the gate solidifies, causing a rapid drop in cavity pressure. The plastic continues to cool and solidify until it becomes rigid enough for the mold to open and the finished product to be ejected.

However, during the injection molding process, residual stresses may be generated in the plastic components due to thermal and mechanical reasons, leading to various issues in optical applications. The most significant of these is anisotropy, which affects optical properties such as birefringence and reduces mechanical strength. Additionally, residual stresses can cause warpage, non-uniform refractive index distribution, variations in curvature radius, and form deviations. These factors directly impact optical performance by altering the focal length, principal plane location, and wavefront aberration. Therefore, achieving consistent quality in optical components requires a highly stable process with precise control over its parameters.

## 2.11 GLASS MOLDING

The industrial needs for glass components are increasing. Glass has predominant advantages over plastic on aspects of hardness, refractive index, light permeability, stability to environmental changes in terms of temperature and humidity, and so on. However, there are technological challenges in glass injection molding and nanoimprinting. The molding temperature of the glass is usually higher than 400°C, which causes difficulties in mold making, environmental control, and mold release, etc. As an alternative approach, glass molding has been developed to produce precision optical elements such as aspherical lenses, Fresnel lenses, diffractive optical elements (DOEs), micro lens arrays as mentioned in Chapter 1.

The process flow of glass molding is shown in Figure 2.27. According to its thermal cycle, a typical glass molding process can be divided into four stages: heating, pressing, annealing, and cooling. Firstly, a glass preform is placed on the lower mold, and inert gas, such as nitrogen ($N_2$), is flowed to purge the air in the machine chamber; then the molds and glass preform are heated to the molding temperature by a heat source, such as infrared lamps. Secondly, the glass preform is pressed by closing the two mold halves. Thirdly, while a small pressing load is maintained, the molded lens is slowly cooled down to release the internal stress, namely, annealing. Finally, the glass lens is cooled rapidly to ambient temperature and released from the molds. Through these four stages, the shapes of the mold cores are precisely replicated to the glass lens. In recent years, glass molding under vacuum conditions has been popular, as it can solve the problem of residual gas bubbles between the mold and the glass lens when using an inert gas to purge the air in the machine chamber.

In glass molding, the fabrication of the molding die is an important step. Hard materials such as silicon carbide (SiC) and tungsten carbide (WC) are preferable mold materials for continuous surfaces and surface structures with large feature sizes. However, they have not been commonly used for molding micro/nanoscale surface structures because it is very difficult to generate such small features on these hard materials with satisfactory surface quality. To solve this problem, nickel-phosphorous (NiP) electroless plating, which has been popularly used for molding plastic optical parts, has been attempted for molding glass microstructures such as microgrooves and micro pyramids by Yan et al [6]. It was also

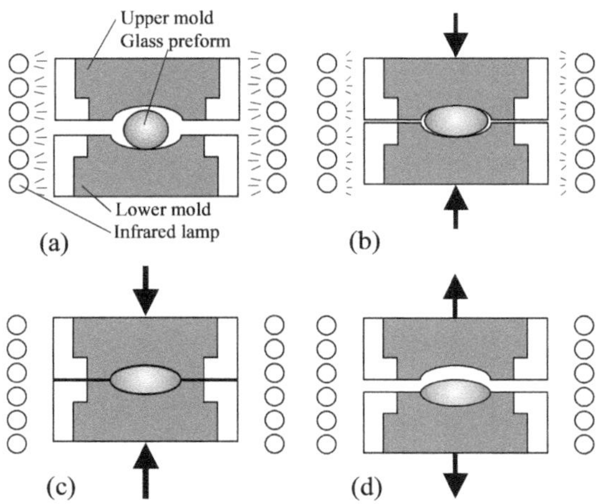


**FIGURE 2.27** Schematics of four stages of a glass lens molding process: (a) heating, (b) pressing, (c) annealing, and (d) cooling [36].

demonstrated that after heat treatment NiP plating was useable for molding aspherical and diffractive glass lenses with a practically long mold life [37].

## 2.12 HYBRID PROCESSES

A hybrid method combines different machining processes and mechanisms to take advantage of their best features, leading to better performance, accuracy, efficiency, and flexibility. Hybrid processes offer significant merits in processing complex geometries, challenging materials, composite structures, and achieving superior surface finishes at a low cost. The following subsections present two examples of hybrid processes.

### 2.12.1 EDM-Grinding Hybrid Process

Single-crystal SiC is difficult to machine owing to its extreme hardness and brittleness. It is typically finished through chemomechanical polishing (CMP), which exhibits a significantly low material removal rate (MRR). Since SiC is electrically conductive, it can also be machined via electrical discharge machining (EDM), which offers a higher MRR but compromises surface quality. To enable high-efficiency, high-precision, and low-damage micromachining of SiC, Yan et al. [38] proposed EDM of SiC utilizing a sintered diamond (polycrystalline diamond, PCD) tool electrode, and following EDM shaping, the same PCD tool was subsequently employed for grinding to eliminate the EDM-induced subsurface damage and recast layers. Figure 2.28 illustrates the principle of the hybrid EDM-grinding process.

Figure 2.29(a) presents a micrograph of an EDMed micro-cavity on 4H-SiC. The surface exhibits significant roughness with dense discharge craters. Figure 2.29(b) shows a micro-cavity after hybrid process of EDM and grinding. The surface appears as smooth as the surrounding area, which was finished by

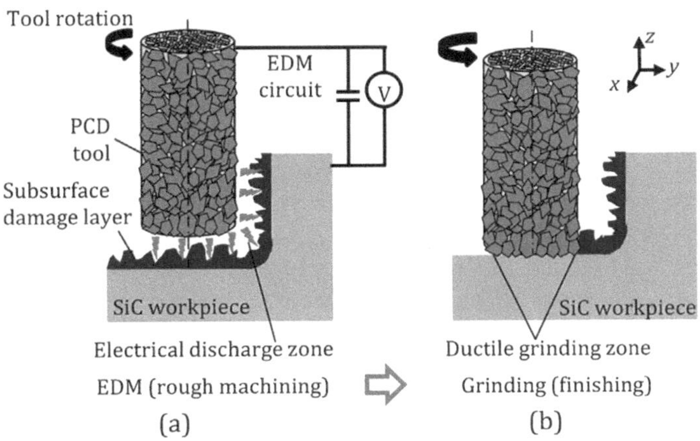


**FIGURE 2.28** Schematic of micromachining of SiC using a PCD hybrid tool [38].

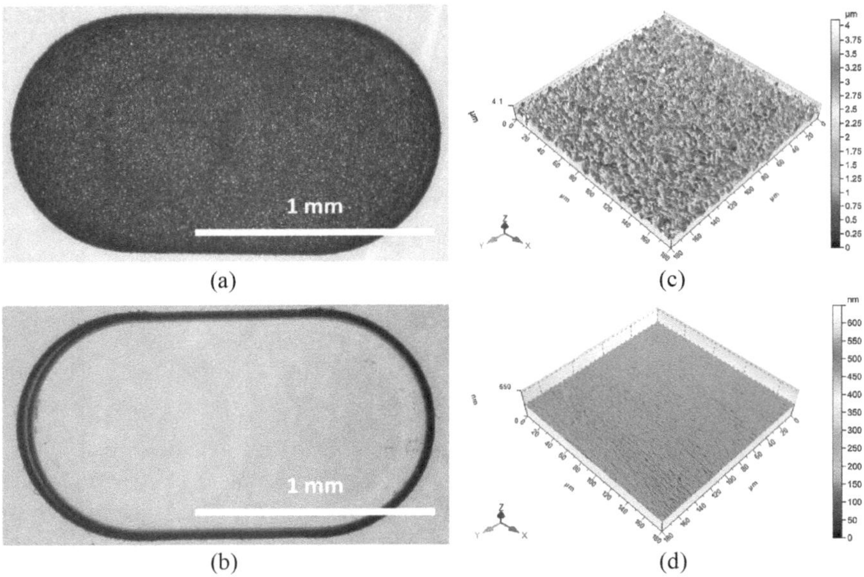


**FIGURE 2.29** (a) and (b) present micrographs of surfaces machined by EDM and the hybrid EDM-grinding process, respectively. (c) and (d) depict the three-dimensional topographies of the bottom surfaces shown in (a) and (b) [38].

CMP. Figure 2.29(c) and (d) show the three-dimensional topographies of the bottom surfaces of the corresponding cavities, measured using a white light interferometer. The surface roughness Ra after EDM was 417 nm (Sa = 439 nm). After grinding, however, Ra decreased to 1.85 nm (Sa = 2.60 nm), a reduction of a factor of 200 compared to the EDMed surface. The surface quality resulting from the proposed process approaches that of CMP, while the processing time is significantly shorter than that of CMP.

### 2.12.2 Cutting and Photolithography Hybrid Process

Machining microstructures on thin films presents a challenge as thin film substrates are highly susceptible to deflection during clamping and machining due to external forces. To address the workpiece deflection issue, Yan et al. [39] proposed fabricating thin-film optics by combining diamond turning with photolithographic processes. The targeted component shape and the key steps of the proposed process are illustrated in Figure 2.30. First, thin films of Cr (50 nm thick) and Au (100 nm thick) are deposited on a silicon wafer as interlayers, and electrolytic plating of Cu (30 µm thick) is performed. Second, diamond turning of the Cu layer is performed to generate the Fresnel lens structure. Third, a photoresist layer is spin-coated on the back of the silicon wafer, followed by photolithography to remove the portion of the photoresist corresponding to the Fresnel lens. Fourth, the exposed silicon substrate is removed using reactive ion etching (RIE). Fifth, the residual photoresist layer is removed via chemical etching to obtain a thin-film Fresnel lens on Cu.

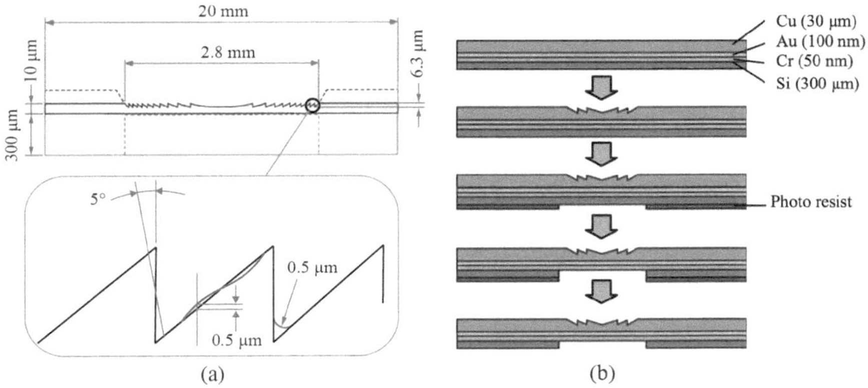


**FIGURE 2.30** Schematic diagrams of fabricating a thin-film Fresnel lens by combining diamond turning and photolithographic processes: (a) design of the thin-film Fresnel lens, (b) flow of the hybrid fabrication process [39].

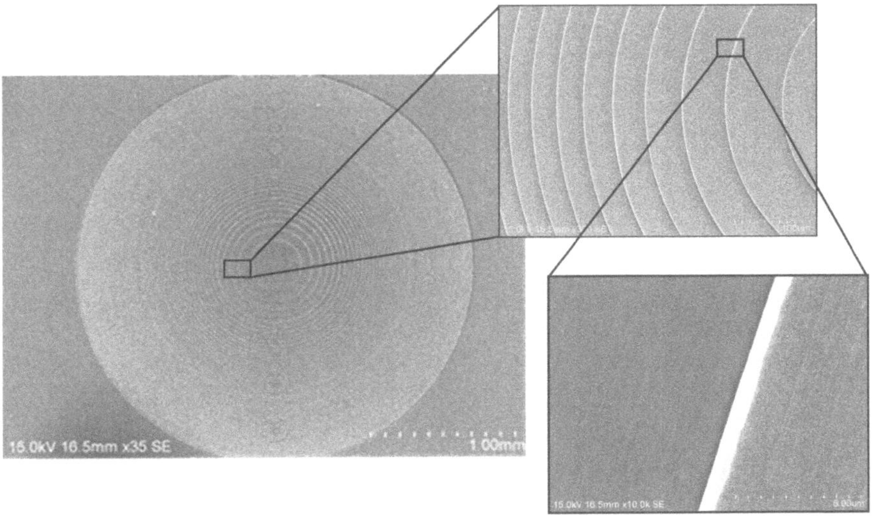


**FIGURE 2.31** SEM images of a micro Fresnel lens fabricated on thin copper film by a hybrid process of diamond turning and photolithography [39].

Figure 2.31 illustrates the SEM images of the center and edge regions of a Fresnel lens fabricated by the hybrid process. From these results, it is clearly confirmed that the lens was precisely fabricated with a smooth surface and sharp zone steps. Attributed to the support of the surrounding silicon substrate, the lens shape keeps flat despite that it is machined on a thin film of Cu. Neither protrusion due to tool decentering at the lens center nor burr formation due to workpiece material side flow at the edge of the zone step is present.

## REFERENCES

[1] Taniguchi, N. (1983) Current status in, and future trends of, ultraprecision machining and ultrafine materials processing. *CIRP Annals* 32 (2): 573–582.
[2] Ohta, T., Yan, J., Yajima, S., Takahashi, Y., Horikawa, N., Kuriyagawa, T. (2007) High-efficiency machining of single-crystal germanium using large-radius diamond tools. *International Journal of Surface Science and Engineering* 1 (4): 374–392.
[3] Yan, J., Syoji, K., Kuriyagawa, T. (2002) Fabrication of Large-Diameter Single-Crystal Silicon Aspheric Lenses by Straight-Line Enveloping Diamond-Turning Method. *Journal of the Japan Society for Precision Engineering* 68 (4): 561–565.
[4] Yan, J., Maekawa, K., Tamaki, J., Kuriyagawa, T. (2005) Micro grooving on single-crystal germanium for infrared Fresnel lenses. *Journal of Micromechanics and Microengineering* 15 (10): 1925–1931.
[5] Yan, J., Zhang, Z., Kuriyagawa, T., Gonda, H. (2010) Fabricating micro-structured surface by using single-crystalline diamond endmill. *The International Journal of Advanced Manufacturing Technology* 51 (9): 957–964.
[6] Yan, J., Oowada, T., Zhou, T., Kuriyagawa, T. (2009) Precision machining of micro-structures on electroless-plated NiP surface for molding glass components. *Journal of Materials Processing Technology* 209 (10): 4802–4808.
[7] Brinksmeier, E., Mutlugünes, Y., Klocke, F., Aurich, J.C., Shore, P., Ohmori, H. (2010) Ultra-precision grinding. *CIRP Annals* 59 (2): 652–671.
[8] Yamamoto, Y., Suzuki, H., Okino, T., Moriwaki, T., Fukuta, M., Nishioka, M., Kojima, Y. (2006) Study on Precision Grinding of Micro Aspherical Surface (4th Report). *Journal of the Japan Society for Precision Engineering, Contributed Papers* 72 (1): 84–88.
[9] Kuriyagawa, T. (2011) Introduction to precision engineering: Fundamental of aspherical grinding. *Journal of the Japan Society for Precision Engineering* 77 (10): 927–932.
[10] Yang, L., Guo, X., Kang, R., Zhu, X., Jia, Y. (2022) Effect of kinematic parameters considering workpiece rotation on surface quality in YAG double-sided planetary lapping with the trajectory method. *The International Journal of Advanced Manufacturing Technology* 123 (7): 2679–2690.
[11] Aspden, R., McDonough, R., Nitchie, F.R. (1972) Computer Assisted Optical Surfacing. *Applied Optics* 11 (12): 2739–2747.
[12] Lee, H., Lee, D., Jeong, H. (2016) Mechanical aspects of the chemical mechanical polishing process: A review. *International Journal of Precision Engineering and Manufacturing* 17 (4): 525–536.
[13] Namba, Y., 1984, *Mechanism of Float Polishing, Topical Meeting on the Science of Polishing*, Monterey, California, ppTuBA2.
[14] Bennett, J.M., Shaffer, J.J., Shibano, Y., Namba, Y. (1987) Float polishing of optical materials. *Applied Optics* 26 (4): 696–703.
[15] Mori, Y. (1980) Elastic Emission Machining and its surface. *Journal of the Japan Society of Precision Engineering* 46 (6): 659–666.
[16] Mori, Y., Yamauchi, K., Endo, K. (1987) Elastic emission machining. *Precision Engineering* 9 (3): 123–128.
[17] Xie, M., Pan, Y., An, Z., Huang, S., Dong, M. (2022) Review on surface polishing methods of optical parts. *Advances in Materials Science and Engineering* 2022 (1): 8723269.
[18] Yamauchi, K., Mimura, H., Inagaki, K., Mori, Y. (2002) Figuring with subnanometer-level accuracy by numerically controlled elastic emission machining. *Review of Scientific Instruments* 73 (11): 4028–4033.

[19] Gattu, S.D., Yan, J. (2022) Micro electrical discharge machining of ultrafine particle type tungsten carbide using dielectrics mixed with various powders. *Micromachines* 13 (7):998.
[20] Natsu, W., Ojima, S., Kobayashi, T., Kunieda, M. (2004) Temperature distribution measurement in EDM Arc plasma using spectroscopy. *JSME International Journal Series C Mechanical Systems, Machine Elements and Manufacturing* 47 (1): 384–390.
[21] Zhu, X., Li, G., Mo, J., Ding, S. (2023) Electrical discharge machining of semiconductor materials: A review. *Journal of Materials Research and Technology* 25: 4354–4379.
[22] Esser, G., Yan, J. (2022) Direct observation of discharge phenomena in vibration-assisted micro EDM of array structures. *Micromachines* 13 (8):1286.
[23] Kishi, R., Yan, J. (2020) Electrical discharge/electrochemical hybrid machining based on the same machine and tool electrode. *Journal of Micro and Nano-Manufacturing* 8 (1):010906.
[24] Brinkmann, M., Qian, Y., Huang, H., Yan, J. (2024) Femtosecond pulsed laser machining of fused silica for micro-cavities with sharp corners. *Journal of Manufacturing Science and Engineering* 147 (1):011007.
[25] Takayama, N., Ishizuka, J., Yan, J. (2018) Microgrooving of a single-crystal diamond tool using a picosecond pulsed laser and some cutting tests. *Precision Engineering* 53: 252–262.
[26] Yamamuro, Y., Shimoyama, T., Yan, J. (2022) Microscale surface patterning of zirconia by femtosecond pulsed laser irradiation. *International Journal of Precision Engineering and Manufacturing-Green Technology* 9: 619–632.
[27] Meshram, T., Yan, J. (2023) Formation of laser-induced periodic surface structures on reaction-bonded silicon carbide by femtosecond pulsed laser irradiation. *Nanomanufacturing and Metrology* 6 (1): 4.
[28] Gisario, A., Barletta, M., Veniali, F. (2022) Laser polishing: A review of a constantly growing technology in the surface finishing of components made by additive manufacturing. *The International Journal of Advanced Manufacturing Technology* 120 (3): 1433–1472.
[29] Yan, J., Kobayashi, F. (2013) Laser recovery of machining damage under curved silicon surface. *CIRP Annals - Manufacturing Technology* 62 (1): 199–202.
[30] Uno, Y., Okada, A. (2005) Surface modification of EDMed surface by electron beam irradiation. *Journal of the Japan Society for Precision Engineering* 71 (5): 553–556.
[31] Uno, Y., Okada, A., Uemura, K., Raharjo, P., Furukawa, T., Karato, K. (2005) High-efficiency finishing process for metal mold by large-area electron beam irradiation. *Precision Engineering* 29 (4): 449–455.
[32] Scia Systems Gmbh, The Technology of Ion Beam Figuring. https://www.scia-systems.com/learning/ionbeamfiguring (accessed Feb. 2025).
[33] Zhang, Z., Yan, J., Kuriyagawa, T. (2019) Manufacturing technologies toward extreme precision. *International Journal of Extreme Manufacturing* 1 (2): 022001.
[34] Peixoto, C., Valentim, P.T., Sousa, P.C., Dias, D., Araújo, C., Pereira, D., Machado, C.F., Pontes, A.J., Santos, H., Cruz, S. (2022) Injection molding of high-precision optical lenses: A review. *Precision Engineering* 76: 29–51.
[35] Svečko, R., Kusić, D., Kek, T., Sarjaš, A., Hančič, A., Grum, J. (2013) Acoustic emission detection of macro-cracks on engraving tool steel inserts during the injection molding cycle using PZT sensors. *Sensors* 13, 6365–6379.
[36] Yan, J., Zhou, T., Masuda, J., Kuriyagawa, T. (2009) Modeling high-temperature glass molding process by coupling heat transfer and viscous deformation analysis. *Precision Engineering* 33 (2): 150–159.

[37] Zhou, T., Yan, J., Masuda, J., Oowada, T., Kuriyagawa, T. (2011) Investigation on shape transferability in ultraprecision glass molding press for microgrooves. *Precision Engineering* 35 (2): 214–220.
[38] Yan, J., Tan, T. (2015) Sintered diamond as a hybrid EDM and grinding tool for the micromachining of single-crystal SiC. *CIRP Annals* 64 (1): 221–224.
[39] Yan, J., Watanabe, K., Nakagawa, Y. (2013) Fabrication of thin-film fresnel optics by combining diamond turning and photolithographic processes. *International Journal of Automation Technology* 7 (4): 385–390.

# 3 Ultraprecision Machine Tool System

## 3.1 INTRODUCTION

Machine tools are sometimes called "mother machines". According to the maternal principle, the accuracy of the machined parts directly reflects the accuracy of the machine being used. In other words, the precision of the child parts cannot exceed the precision of the mother machine. Although the basic structure and appearance of machine tools have not changed so much over these decades, the precision of machine tools has shown a great improvement. The advance in machine precision directly leads to an increase in the added value of the products.

Precision and ultraprecision machine tools are typical high-tech products that involves the latest scientific and technological advances in mechanical design, material development, sensors and actuators, signal processing and control, and so on. A precision machine tool must be accurate in positioning, stable in movement, robust to vibration and environmental changes, and rigid to static/dynamic loads. Compared to conventional machine tool design, special attention must be paid to the design of precision and ultraprecision machine tools in terms of the machine structure, mechanical elements, and the sensing and controlling systems. In this chapter, some fundamental aspects of ultraprecision machine tools will be introduced.

Precision machine tools can be huge or compact, according to the size range of the machined parts. In recent years, the machine tools for precision and ultraprecision machining have undergone miniaturization. These machine tools are very compact and sometimes are desktop type. For example, the ultraprecision lathe Nanoform X (AMETEK Precitech Inc.) occupies an area of less than 2 square meters, while can be configured from 2 to 4 axis to produce spherical, aspherical, and freeform surfaces of up to 440 mm diameter [1]. This compact footprint provides users, especially laboratories in universities and research institutions, with a powerful machine that minimizes the use of valuable floor space.

The machine tools used for diamond cutting usually consist of ultraprecision air-bearing spindles, pneumatic/hydrostatic slides, positioning and control systems, vibration isolation systems, temperature control systems, etc. In recent decades, significant advances have been made in the ultraprecision machine tool manufacture industry, and the movement accuracy of the machines has been improved from ten-nanometer to nanometer level and the stepping resolution has been developed from nanometer to sub-nanometer (angstrom) level.

DOI: 10.1201/9781003292388-3

## 3.2 MACHINE STRUCTURES

Typical ultraprecision machine tool structures comprise mechanical units, driving units, control units, and attachment units. Additionally, a primary component known as the machine bed provides support and connects the aforementioned units, as illustrated in the chart in Figure 3.1. This chart also delineates detailed subcategories within each unit. Further explanations of these units will be provided in the following sections.

The machine bed, alternatively called machine base, is important in terms of stiffness, thermal stability, and damping characteristics. Typical materials for precision machine bases include cast iron, granite, and resin concrete [2]. Cast iron is commonly used for machine bases due to its ease of machining into various shapes and relatively high damping characteristics. However, cast iron used for precision machines must undergo annealing or seasoning to relieve residual stress in the material.

Granite stands out as an ideal material for ultraprecision machine bases due to its superior damping characteristics compared to cast iron. Particularly noteworthy is its exceptional long-term dimensional stability, crucial for maintaining repeatable high machining accuracy. However, granite presents challenges in machining. It proves highly resistant to drilling small or screw holes essential for mechanical assembly.

Resin concrete has emerged as a machine base material in recent years. It is a composite material comprising hard powders like sand, gravel, ceramics (alumina, silicon carbide, zirconia, etc.) as filler, and resin (organic polymer) as a binder. It boasts extremely high damping characteristics, approximately 10 times that of cast iron and 3 times that of granite, respectively [3]. Moreover, it offers high thermal stability and short production lead time. However, its Young's modulus is low (about one-seventh that of cast iron), resulting in low static stiffness. Consequently, resin concrete is not suitable for heavy machining tasks involving significant forces.

Recently, a few alloys including nickel-ferrous alloy, nickel-ferrous-cobalt alloy, and low thermal expansion cast iron, have emerged as materials for

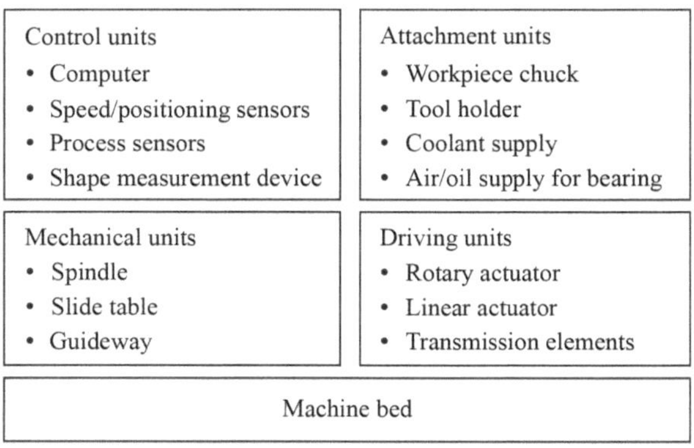


**FIGURE 3.1** Main units of an ultraprecision machine tool.

precision machine bases [4]. These materials exhibit significantly lower thermal expansion rates, approximately one-hundredth that of standard steels. Despite their relatively high production costs, they have found applications in precision machines, particularly desktop models used for precision micro-machining. It is worth noting that machine bases are typically supported by a set of air mounts or cushions to isolate vibrations from the ground.

## 3.3 SLIDES, GUIDEWAYS, AND SPINDLES

The movement of a machine tool can be divided into two kinds: linear movement and rotational movement. Linear movement is provided by slides and guideways, and rotational movement is provided by spindles, as shown in Figure 3.2. The motion accuracy of these movements is essential for the machine tool's performance and machining accuracy. Thus, various advanced technologies are required in the design and manufacturing of these key parts.

As the slides and spindles move and rotate, they inevitably encounter friction with immovable parts. Hence, bearings play a crucial role in supporting rotating shafts or moving slides, guiding rotation or movement, and withstanding the load transmitted by the shaft or slide, which contributes to reducing the friction-induced vibration, heat generation, and wear. According to the mechanisms of friction and lubrication, bearing can be divided into several categories, including sliding bearings, rolling bearings, hydrostatic bearings, and aerostatic bearing, as shown in Figure 3.3.

Each type of bearing has its own advantages. The criteria used for selecting bearings and guideways for machine building and factors for evaluating their performance of service include:

- Friction coefficient
- Load capacity and stiffness
- Vibration damping characteristics

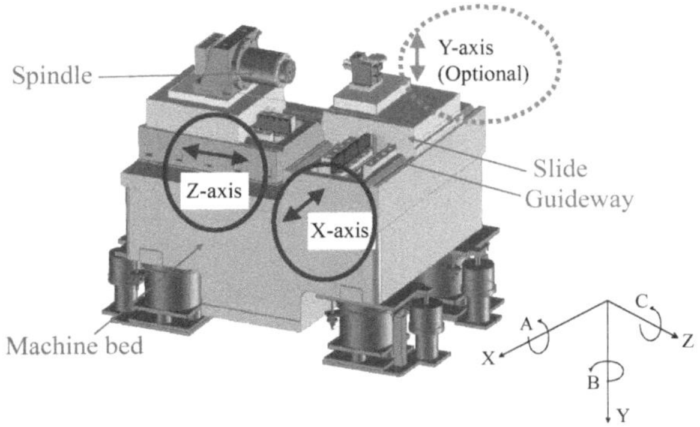


**FIGURE 3.2** Schematic of the structure and movement of a machine tool. (Courtesy of Shibaura Machine.)

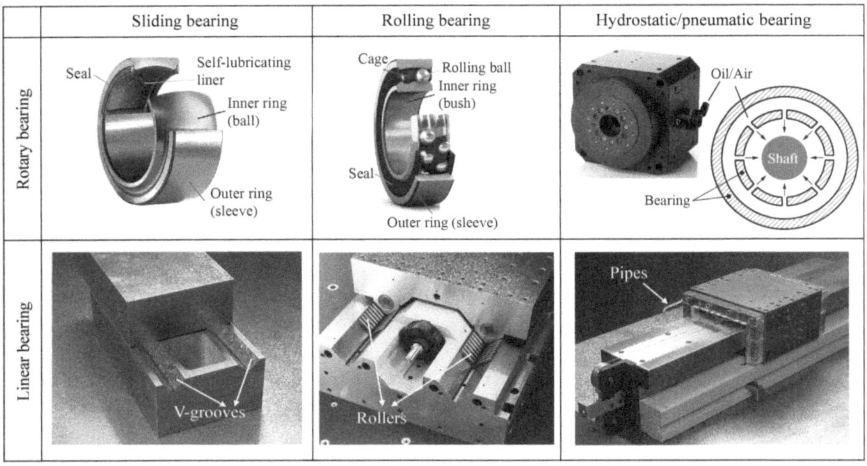


**FIGURE 3.3** Typical types of bearings [5].

- Speed capacity
- Thermal stability
- Service life
- Maintenance
- Manufacturing cost

Table 3.1 lists a comparison of the main features of typical types of bearings. For ultraprecision machine tools, the stiffness usually ranges from $10^8$ to $10^9$ N/m, and the friction coefficient ($\mu$) should be as low as possible. As the speed for the linear movements is far lower than that of the rotation movement, generally speaking, thermal expansion of straight bearing elements is smaller, and less friction wear takes place than the spindle bearings. Instead, attention should be paid to load capacity, stiffness, and vibration damping characteristics. The straightness of a slide in an ultraprecision machine tool is below 0.1 μm in stroke of 100 mm [6]. This requires very high-level precision machining and assembling technologies for the slide table elements, which is sometimes strongly dependent on the personal skills and experience of engineers.

### 3.3.1 Sliding Bearing

Sliding bearings offer several advantages: they feature a straightforward bearing and housing design, occupy less radial space due to their compactness, and are cost-effective. Additionally, they facilitate a simple shaft design, operate quietly, and exhibit resilience against shocks. These bearings excel in operations at medium and high speeds. However, they come with their downsides. The disadvantages include higher frictional power loss, which necessitates careful lubrication. Furthermore, sliding bearings are typically designed to carry either radial or axial loads exclusively.

**TABLE 3.1**
**The Main Features for Various Types of Bearings**

| Bearing Type | Friction Coefficient | Load Capacity | Damping Property | Speed Capacity | Service Life | Maintenance Cost |
|---|---|---|---|---|---|---|
| Sliding | + | ++ | + | + | + | +++ |
| Rolling | ++ | +++ | + | ++ | + | ++ |
| Hydrostatic | +++ | +++ | +++ | +++ | +++ | + |
| Aerostatic | +++ | + | +++ | +++ | +++ | ++ |

(+: poor, ++: average, +++: excellent).

### 3.3.2 Rolling Bearing

Rolling bearings offer many advantages in engineering applications. Their design simplicity and compactness make them highly desirable, occupying minimal radial space within machinery. Additionally, they boast cost-effectiveness, contributing to overall efficiency in various mechanical systems. Furthermore, one advantage of rolling element bearings lies in their ability to accommodate both radial and axial loads efficiently. This versatility simplifies the design process, allowing for more streamlined shaft configurations and enhancing overall system performance. Moreover, these bearings operate with reduced frictional power loss, particularly advantageous in applications demanding high speeds. While they excel in many aspects, they are not without their limitations. One consideration is the necessity for meticulous lubrication to ensure optimal performance and longevity. Furthermore, compared to sliding bearings, rolling bearings may exhibit increased vibration and noise levels during operation.

### 3.3.3 Hydrostatic Bearing

The structure of a hydrostatic bearing is shown in Figure 3.3. The high-pressure lubricating oil is provided by an oil pump to the recess between the moving part (shaft) and the static part (bearing). There is a small gap (a few tens of microns) between the moving part and the static part, through which the oil flows out of the bearing and is recycled and reused.

The high-pressure oil is usually provided to the bearing at a constant pressure. The oil passes through a throttle valve before entering the recess. The throttle valve is composed of capillary material/structure or an orifice. This enables the automatic adjustment of the oil pressure inside the recess if there is a change in load. That is, if the load increases, the gap between the moving part and the static part will tend to decrease, causing an increase in the resistance of oil flow. As a result, the oil pressure within the recess will increase to sustain the increasing load, keeping the gap between the moving part and the static part almost constant.

The performance of the throttle valve is very important for maintaining the stiffness of the hydrostatic table. The aperture of the throttle valve can also be automatically controlled according to the load acting on the moving part. This kind of hydrostatic bearing is called "self-adjusting hydrostatic bearing", which enables a hydrostatic bearing to have infinitely high stiffness [7].

There are many advantages of the hydrostatic bearing. As the rotator and the stator are separated by an oil film, the micro unevenness or runout error of the bearing elements can be absorbed by the oil film, enabling a very steady rotation. The stiffness and the load capacity of the hydrostatic spindle are constant and independent of the rotation speed. However, for hydrostatic bearing, it is essential that the temperature of the oil be strictly constant to prevent machining errors due to the thermal expansion of the bearing elements. As the thermal exchanging coefficient between oil and bearing elements (usually steel or cast ion) is high, the temperature change in the oil might result from the high rotation speed of the spindle and environmental temperature changes. Usually, for ultraprecision machine tools, a special oil handling system is used to control the temperature of the hydrostatic oil within ±0.01°C.

For example, if cast iron (thermal expansion coefficient: $12 \times 10^{-6}$/°C) [8] is used as the material of the bearing elements the length of which is 1 m, a temperature rise of 0.01°C may cause a thermal expansion of 0.12 μm. This error is considerably big and cannot be neglected in ultraprecision machining. In some cases, the bearing elements are made of Super Invar, which has a coefficient of thermal expansion of $\sim 0.1 \times 10^{-6}$/°C, about 1/100 of that of cast iron [9]. Also, there are hydrostatic bearings made of ceramics or glass ceramics, such as Zerodur (Schott, Germany), which has an extremely low coefficient of thermal expansion of $\sim 0.01 \times 10^{-6}$/°C, approaching zero.

### 3.3.4 Pneumatic Bearing

Air bearings utilize a thin film of pressurized air to provide an exceedingly low friction interface between the moving part and the static part. The two surfaces don't touch. Being non-contact, air bearings avoid problems of friction, wear, thermal expansion, particulates, and lubricant handling, and offer distinct advantages in precision positioning, such as lacking backlash and stiction, as well as in high-speed applications.

The design of air bearing is to make the pressure between the surfaces of the bearing enough to support the working loads while the air constantly escapes from the bearing gap. The gap between the moving part and the static part is a few microns. Due to that the viscosity of air is low, about 1/1000 that of a lubricant oil, air bearing has low stiffness, low load capacity, and low damping capacity. Air bearing is especially suitable for diamond turning machines where the cutting force is small (a few Ns).

The structure of air bearing is very similar to that of hydrostatic bearing. The only difference is that unlike the hydrostatic bearing, there is no recess on the surface of an air bearing, and the gap between the rotator and the stator (a few microns) is smaller than that of the hydrostatic bearings. This is because the pressured air accumulated in a recess, or a big gap, may cause self-excited vibration, alternatively termed "pneumatic hammering", and instability of the bearing, due to the compressibility of air [10]. Air is easier to handle than oil when considering the aspect of environmental protection. As the thermal capacity of air is low, the temperature control tolerance of air (±0.1°C) is wider than oil. Therefore, the initial equipment cost of air bearings is lower than that of hydrostatic bearings.

### 3.3.5 Magnetic Bearing

Magnetic bearings are a type of bearing system that uses magnetic fields to levitate and support a rotating shaft, eliminating the need for physical contact between the shaft and the bearing components, as shown in Figure 3.4. Instead of traditional mechanical components like balls, rollers, or fluid (oil, air) films, magnetic bearings rely on electromagnetic forces to suspend the shaft in place.

In a magnetic bearing, the rotation error of the rotator is measured, and the magnetic force of the bearing is dynamically controlled to correct the error based on the feedback of the rotating error signal. For this reason, the stiffness of the magnetic bearing is very high, and the rotation error can be real-time compensated even for a varying load. However, the magnetic bearing cannot yet be used in high-speed spindles up to date due to that the response frequency of the feedback control system is limited [11]. It is expected that in the near future this problem can be solved, and the magnetic bearings can be widely used in precision machine tools.

## 3.4 AXIS DRIVES

### 3.4.1 Servo Motor

A spindle rotation accuracy of 0.1 μm, or much higher (~0.01 μm), is required in ultraprecision machining. The rotation accuracy of a spindle not only depends on

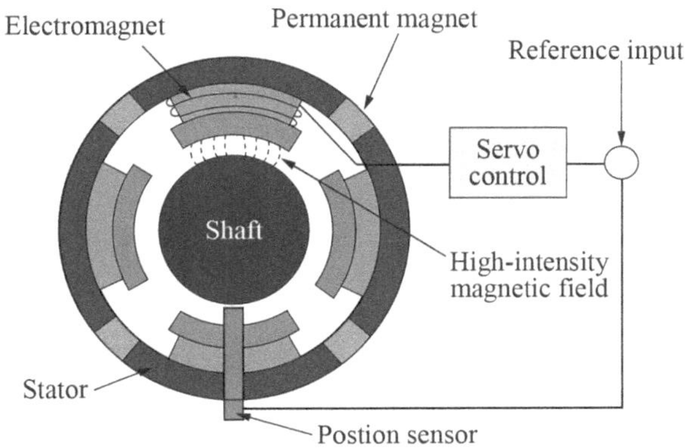


**FIGURE 3.4** Schematic structure of magnetic bearings.

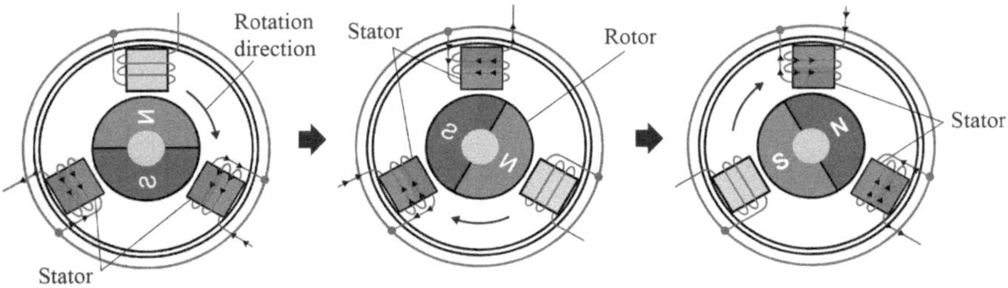


**FIGURE 3.5** Schematic of the structure of a brushless DC motor.

the accuracy of the bearing itself, but also depends on the accuracy of the spindle driving units. High-performance brushless direct current (BLDC) motors are usually used as spindle driving motors. As their name suggests, BLDC motors operate without the use of brushes. The rotor consists of a permanent magnet, while the coils remain stationary, fixed in place on the stator, as illustrated in Figure 3.5. Since the coils remain stationary, there is no requirement for brushes and a commutator. Rotation is achieved by altering the direction of the magnetic fields produced by the surrounding stationary coils. To regulate rotation, it is necessary to adjust both the magnitude and direction of the current flowing into these coils.

The first significant advantage of BLDC motors lies in efficiency, as they can continuously control at maximum rotational force (torque). In contrast, brushed motors only reach maximum torque at specific points during rotation. To match the torque output of a brushless motor, a brushed motor requires larger magnets. This explains why even small BLDC motors can deliver substantial power. The second major advantage is controllability. BLDC motors can be precisely controlled using feedback mechanisms to deliver the desired torque and rotation speed. Such precision control, in turn, reduces energy consumption and heat generation.

### 3.4.2 Linear Motor

For linear movement, AC servo motors are usually used as drives, and the rotation movement of which is transmitted to linear movement by the use of screws. Ball screws are often used for the movement transmission for their high transmission efficiency. By applying a preload to the ball screw, the backlash movement can be significantly eliminated. However, the assembly of the ball screw unit to the motor and the table is a difficult procedure, which determines the resulting accuracy of the table movement.

In recently developed ultraprecision machine tools, linear movement driven by linear motors are usually adopted. Linear motors are derived from rotary motors, sharing similar working principles. Conceptually, a linear motor can be envisioned as a rotary motor flattened along its radius direction, transitioning its motion mode from rotational to linear, as shown in Figure 3.6. Its key components typically include a workbench, a slide rail, a mover, a mover seat, and a driver.

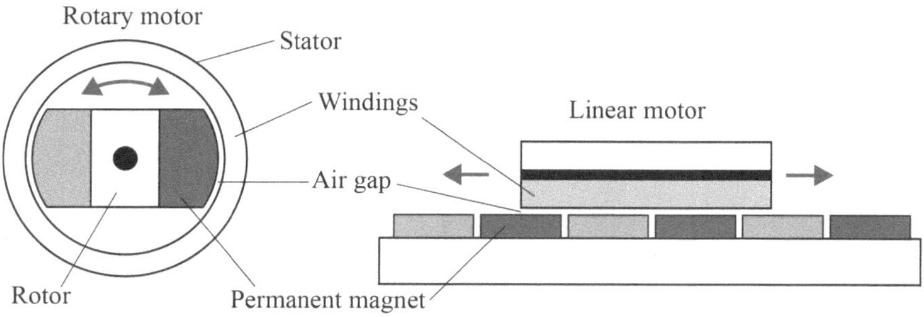


**FIGURE 3.6** Comparison between structures of rotary motors and linear motors.

Linear motors employing permanent magnet technology offer higher precision and shorter setup times compared to rotary servo motors paired with screw rods.

Generally, linear motors can be classified into two types based on whether iron cores are used in the movers. Iron core linear motors wrap coils around silicon steel plates to maximize force generation through a single-sided magnetic circuit. This design employs special electromagnetic configurations to deliver maximum rated thrust with minimal cogging force, making them an ideal choice for moving heavy objects while maintaining rigidity. The advantages of iron core linear motors include low-cost core iron design, high thrust density, and good heat dissipation. However, they suffer from cogging effects that limit smooth motion, and the attractive force can reach 5–13 times the motor's rated thrust, posing challenges during installation [12].

In contrast, ironless linear motors use movers made of windings wrapped around laminated steel and embedded in epoxy resins [13]. As a result, cogging effects are eliminated, and they exhibit lightweight construction with no attraction between the coil assembly and the magnetic circuit. These characteristics make them particularly suitable for applications requiring minimal bearing friction, high acceleration with relatively light loads, and maintaining constant velocity even at ultra-low speeds. The modular magnetic circuit design consists of two rows of magnets to maximize thrust generation and provide a magnetic flux path for the circuit. The advantages of ironless linear motors include absence of attraction, fixed air gaps, easy alignment, and installation, lack of cogging effects, stable operation, low mass, and high acceleration of the mover. However, they tend to have lower full-load power compared to iron core motors, which require twice the number of magnets, leading to higher costs and higher thermal resistance. The I-shaped structure design helps mitigate this issue. Ironless linear motors are better suited for high-speed, light-load applications requiring high repeatability in positioning.

## 3.5 POSITIONING AND FEEDBACK CONTROL

Even if the accuracy of motors, transmission units, and tables is guaranteed, the final positioning accuracy of the system may not be satisfactory. Therefore, usually closed-loop feedback control is necessary for an ultraprecision machine. That is, the motion error of the machine must be real-time measured and compensated. Laser interferometers and laser scales are two major kinds of sensors for precision positioning.

### 3.5.1 Laser Interferometer

Laser interferometer has been the most popular positioning sensor in early ultraprecision machine tools. The effective measurable distance is long (~1 m), and the resolution is high (~1 nm). One of the great benefits of laser interferometry for position measurement is that multiple axes can be measured using the same interferometer system by splitting the laser beam and directing the multiple beams to different targets and back to different receivers [14]. For example, Figure 3.7 shows the configuration of a two-axis plane mirror interferometer, where a single beam is split into two by a polarizing beam splitter. Each of these beams is reflected off of a planar mirror on the side of a moving stage, resulting in ultraprecision position measurements for the x and y axes simultaneously.

However, because the laser wavelength is significantly affected by temperature, pressure, humidity, and composition of the media (usually air), the environmental control of laser interferometers is very strictly required. Usually, the variations in temperature, pressure, and humidity are real-time measured and controlled to guarantee the stability of the laser interferometer.

### 3.5.2 Grating Scale

A grating scale unit is composed of two parts: a scale with gratings and a reading head. Figure 3.8(a) shows the structure of a typical grating scale [15]. The key part of the grating detection device is the grating reading head, which consists of a light source, a converging lens, an indicating grating, a photoelectric

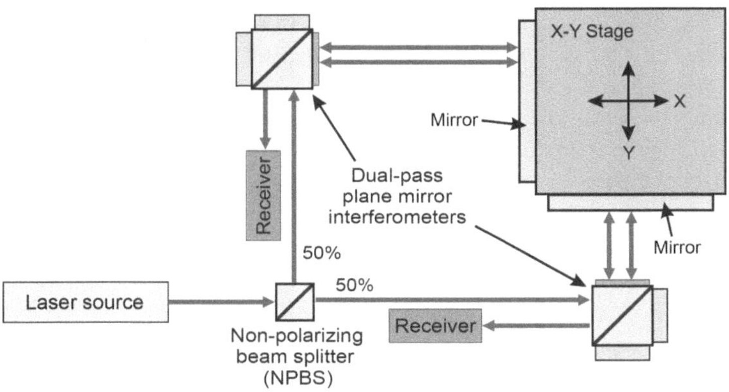


**FIGURE 3.7** Two-axis plane mirror interferometer configuration.

component, and an adjusting device. If the light source is laser, the grating scale is usually called laser scale. Laser scales possess excellent capabilities such as high precision, high resolution, and rapid response without strict requirements for environmental temperature, pressure, and humidity. Ultraprecision machine tools commercially available in recent years are mainly equipped with laser scales. It should be pointed out that a laser scale is used for the positioning of a single axis. Thus, a multi-axis machine tool requires multiple laser scales, which may increase the production cost of the machine.

The working principle of a grating scale is primarily based on optical and electronic technologies. It determines position and displacement by measuring the diffraction of light passing through the grating. The grating scale consists of a series of alternating transparent and opaque stripes, known as grating lines, as illustrated in Figure 3.8(b). When light illuminates the grating, diffraction occurs

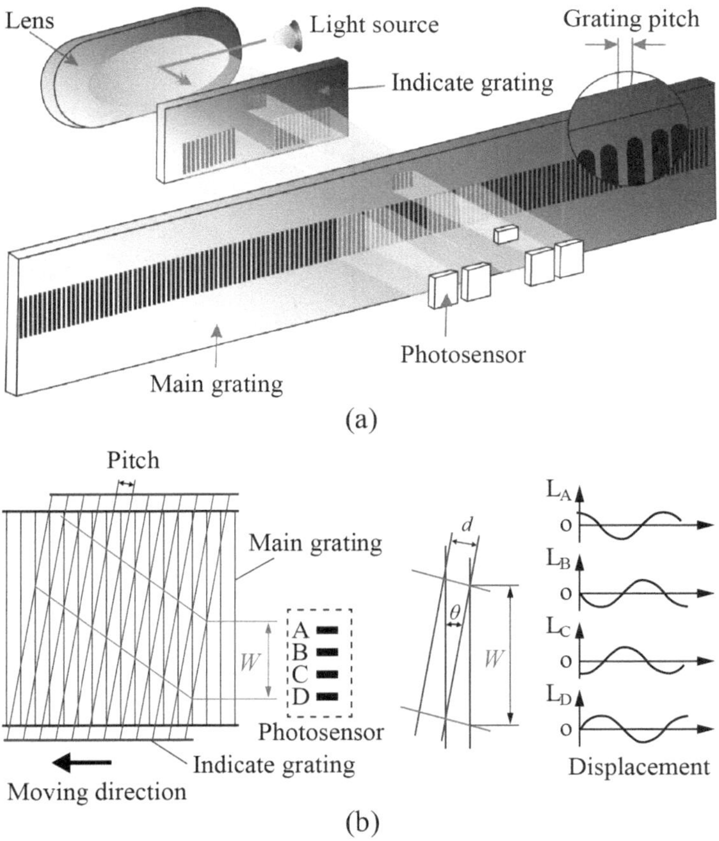


**FIGURE 3.8** Schematics of a grating scale: (a) typical structural configuration and (b) working principle.

on the grating lines, creating a pattern of alternating bright and dark stripes, which is known as Moire fringe. The reading head on the grating scale utilizes photoelectric conversion technology to capture these brightness changes and convert them into changes of electrical signals, thereby obtaining precise position information.

The Moiré fringe exhibits the following properties:

1. When illuminated by parallel beams of light, the intensity distribution of the light passing through the Moire fringe approximates a cosine function.
2. If $W$ represents the width of the Moire fringe, $d$ represents the grating pitch, and $\theta$ represents the angle between the lines of two gratings, then their geometric relationship is $W = d/2*\sin(\theta/2)$. When the angle is small, the equation can be approximated as $W = d/\theta$.

If we take $d = 0.01$ mm and $\theta = 0.01$ rad, then from the above equation, we get $W = 1$ mm. This indicates that, without the need for complex optical and electronic systems, utilizing the interference phenomenon of light, the grating pitch can be converted into the width of the Moiré fringe amplified by a factor of 100. This amplification effect is an important characteristic of the grating.

Recently, ultraprecision machines equipped with rotary tables (for example, B-axis tables) are increasing. The positioning of the rotary table is also important. In this case, to prevent possible backlash movements due to transmission elements and to minimize the unit size, direct drive motors are used, where the rotator of the motor is directly used as the rotary table and the stator of the motor is used as the housing of the table unit. The positioning sensors used for the rotary tables are rotary encoders, which have a positioning resolution of ~1 second.

Rotary encoders can convert rotary motion into electrical signals for detecting rotational displacement, velocity, and direction. They mainly consist of an encoder disk, a photoelectric sensor, and a signal decoder. The encoder disk typically has a central axis surrounded by concentrically arranged opaque and transparent lines, which can be either patterns blocked by light or magnetic pole patterns, as illustrated in Figure 3.9.

In photoelectric rotary encoders, the photoelectric sensor reads these lines, converting rotary displacement into periodic electrical signals, which are then

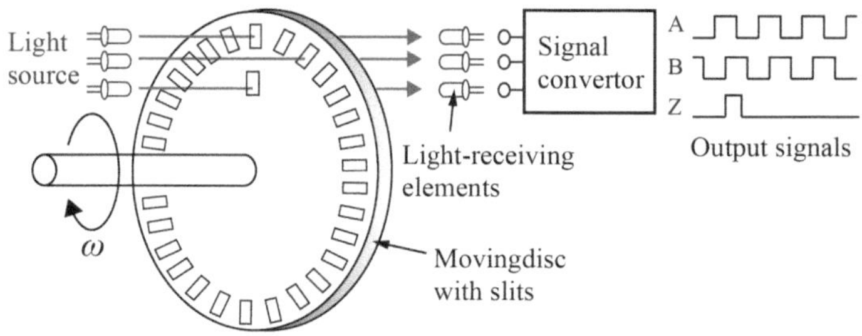


**FIGURE 3.9** Working principle of the optical incremental encoder.

transformed into counting pulses representing the magnitude of displacement. Another type of rotary encoder is the absolute encoder, which records the absolute position of rotation, with each position corresponding to a specific digital code. Therefore, its output is only related to the measured starting and ending positions, not the intermediate process. The output of a rotary encoder typically includes several sets of signals, such as A and B phase signals and Z-phase pulses. The A and B phase signals have a 90-degree phase difference and can be used to determine the direction of rotation. When A is ahead, it indicates forward rotation; when B is ahead, it indicates reverse rotation. The Z-phase pulse represents the zero-reference position, used to obtain zero reference of the encoder.

## 3.6 CHUCKING AND CLAMPING

In ultraprecision machining, particularly in the turning process, the workpiece must be securely held in place and precisely aligned with the lathe spindle axis, because even minor misalignments can introduce inaccuracies into the machining process, ultimately compromising the final dimensions and tolerances of the part. Vacuum chucking and mechanical clamping are two common methods used to secure workpieces in machining processes, each with its own advantages and applications.

### 3.6.1 Vacuum Chucking System

Vacuum chucking relies on creating a vacuum between the workpiece and the chucking surface to hold the workpiece securely. The workpiece is placed on a flat or contoured surface with channels through which a vacuum is applied, pulling the workpiece onto the surface. Figure 3.10(a) depicts an example of a vacuum chucking system equipped with a machine spindle, which is used to chuck an aluminum fixture for holding workpiece. Adjustment screws are typically provided to adjust the diameter of the effective vacuum channels, allowing the contact area of the workpiece to be properly fitted. To secure the workpiece to the spindle, a vacuum pump is essential for creating and sustaining the vacuum pressure within the chucking system, as schematically shown in Figure 3.10(b). A rotary adapter facilitates the transmission of vacuum pressure while accommodating rotational movement. This adapter guarantees consistent vacuum pressure, enabling uninterrupted machining operations without compromising the holding force.

For vacuum chucking, the workpiece and chuck surfaces that come into contact with each other must be flat and smooth to prevent leakage of air. Moreover, vacuum chucking is typically suitable for thin and lightweight workpieces with large surface areas. Complex or irregularly shaped workpieces are difficult to be securely fixed by vacuum chucking alone. Vacuum chucking is quick to set up by simply placing the workpiece on the chucking surface and activating the vacuum system. Additionally, it is less likely to cause workpiece distortion, as the holding force is evenly distributed across the entire chuck surface. However, adjusting the position of the workpiece rotation center to align with the spindle axis may require time and skill.

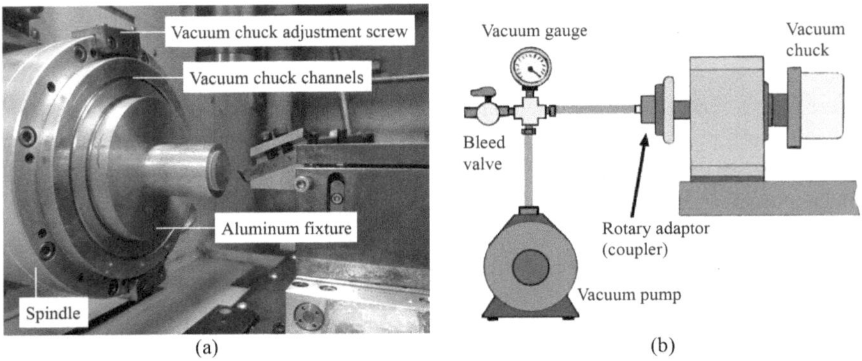


**FIGURE 3.10** (a) Photo of a vacuum chuck and (b) schematic of a vacuum system.

### 3.6.2 Mechanical Clamping System

While vacuum chucking is used for fixing a workpiece by holding its end face, a mechanical clamping system holds the workpiece through mechanical forces exerted on cylindrical surface of the workpiece by the jaws. In some cases, particularly when securing a lengthy cylinder, a tailstock may also be employed to support the opposite end of the workpiece, guaranteeing stability and proper centering throughout the machining process, as shown in Figure 3.11(a). The clamping system adapts to a diverse array of workpiece geometries, sizes, and materials through the utilization of different types of clamps and fixtures, accommodating a wide range of shapes and sizes. Figure 3.11(b) shows a sealed 6-jaw power lathe chuck, which offers improved precision, stability, and versatility compared to conventional 3-jaw chucks [16]. A clamping system involves small-area contacts between the workpiece and the clamping mechanism, which may leave marks or

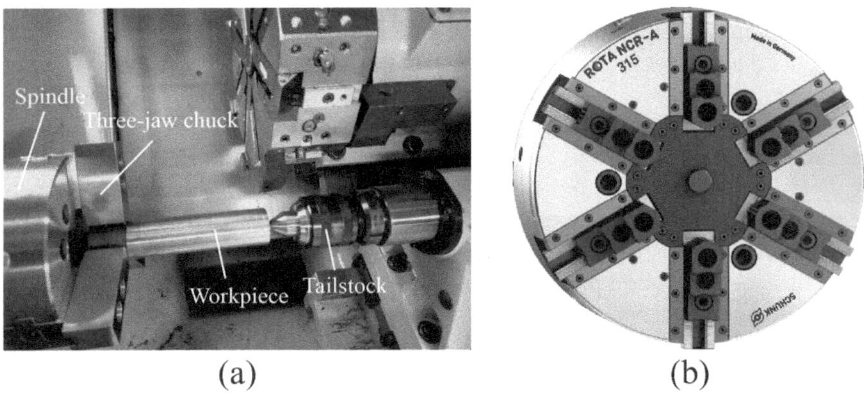


**FIGURE 3.11** Photos of (a) a clamp chuck fixing a cylinder workpiece and (b) a sealed 6-jaw power lathe chuck [16]. (Courtesy of SCHUNK.)

indents on the workpiece surface and cause deformation and internal stress in the workpiece. Nevertheless, a clamping system can provide a higher clamping force compared to vacuum chucks. This makes them suitable for heavy-duty machining operations or when working with tough materials that require a strong grip. Moreover, clamping systems do not require an additional pump to generate the chucking force. Therefore, compared to vacuum chucking systems, their setup is less complex and does not suffer from fluctuations in vacuum pressure or potential vacuum system failures.

## 3.7 COOLING AND LUBRICATING

In ultraprecision machining, it is extremely important to keep the geometry and sharpness of the tool edge stable by preventing wear. Any defects like chipping or edge recession due to gradual wear will cause machining errors. Thus, cooling and lubricating of the cutting tools are important issues in durable machining processes.

In conventional metal cutting, where material removal occurs at a size scale of several microns or more, a lot of cutting heat is generated, which can accelerate both the tool wear and the workpiece deformation. Therefore, cryogenic coolant (such as liquid nitrogen) [17] and flood lubrication (such as lubricating oil) [18] are commonly used to provide both cooling and lubricating during the machining process. However, in ultraprecision machining, the low temperature caused by a cryogenic coolant may cause significant thermal deformation in the workpiece material, leading to form errors in the final machined surface. Additionally, in flooding lubrication, excessive coolant is often used compared to the amount necessary to cool the tool tip, which may cause environmental problems. Therefore, as an alternative, minimum quantity lubrication (MQL), a micro-lubrication technique facilitating near-dry machining, is widely used in ultraprecision machining [19]. It replaces large quantities of water and mineral oil-based cutting fluids with fine mists of environmentally friendly lubricant, primarily vegetable oils, mixed with air. The lubricant flow rate in MQL application is typically less than 50 mL/h, which is a reduction in fluid flow of over 20,000 times compared with conventional flood application. A number of advantages are possible with MQL compared to conventional wet machining, including an improved environment, reduced infrastructure demands in the plants, and lower overall costs [20].

The MQL systems typically consist of an air compressor, tubing (usually metal or flexible tubing designed for high-pressure applications), a fluid reservoir, precision nozzles, and a control unit. The configuration of such pneumatic-driven MQL systems is shown in Figure 3.12. The control unit regulates the flow of both oil and compressed gas, atomizing the lubricant into a fine mist at a controlled pressure. Recently, ultrasonic [21] and electrostatic-driven [22] MQL systems have also been developed.

Cooling and lubricating settings are flexible for users to modify by changing the type of cutting fluid, adjusting flow rate parameters, and so on. Due to their environmentally friendly characteristics, vegetable oils, such as soyabean,

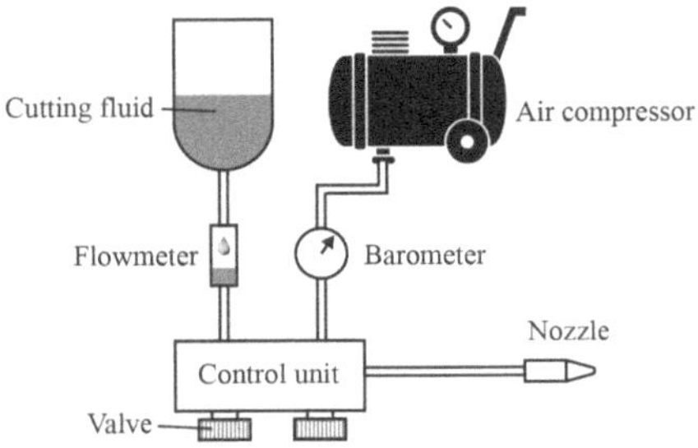


**FIGURE 3.12** Schematic of an MQL supply system.

**FIGURE 3.13** SEM photographs of nanoparticles: (a) $MoS_2$, (b) graphite fiber, (c) copper, and (d) copper oxide [26].

sunflower, and rapeseed oils, emerged as promising alternatives to mineral-based oils [23]. They have been utilized in the development of biodegradable lubricants for various industrial applications, extending the trend toward formulating environmentally friendly cutting fluids.

In addition, adding nanoparticles into lubricants and coolants appears to be a preferred option for improving machining performance [24]. Figure 3.13 shows scanning electron microscope (SEM) photographs of four types of typical nanoparticles used for this purpose, including molybdenum disulfide ($MoS_2$), graphite fiber, copper, and copper oxide. It has been reported by lots of research that the addition of nanoparticles to conventional cutting fluids not only significantly improves their thermal conductivity, leading to better cooling capabilities, but also results in a noticeable reduction in friction and tool wear and an increase in the load-bearing capacity of friction parts [25].

Peng et al. [27] specified four mechanisms of anti-wear and friction reduction by mixing nanoparticles into ordinary cutting fluid. These are as follows: (1) The nano-sized spherical particles are more likely to roll between two surfaces and flip the sliding friction into a combination of rolling and sliding frictions; (2) The nano-sized particles have a tendency to interact with friction pairs to develop a surface protective film; (3) The nano-sized particles are deposited on the contact surface to form a physical tribo-film that compensates for the loss of mass, which is known as "mending effect"; and (4) Compressive stress concentrations due to high contact pressure can be decreased by a number of nanoparticles which uniformly bear the compressive force.

## 3.8 ENVIRONMENTAL CONTROLLING TECHNOLOGY

### 3.8.1 Thermal Control

Thermal expansion of machine elements, tools, and workpieces is a critical problem for ultraprecision machining. For example, the thermal expansion of a 1 m-long machine element made of a metal, the coefficient of thermal expansion of which is $10 \times 10^{-6}$/°C, will be 10 μm when a temperature change of 1°C takes place. The machining error due to this thermal expansion will fatally destroy the form accuracy of the produced components. For this reason, the temperature change of environment should be controlled with ±0.1°C for precision machining, and within ±0.01°C for ultraprecision machining, respectively, to suppress the thermal expansion of elements during machining. Temperature stability during a long time, for example, a few hours or even a few days, is essential for ultraprecision machining of large workpieces.

Major heat sources in ultraprecision machining include: (1) moving machine elements such as bearings, and their driving units, such as motors, (2) mankind around the machine tool, and illuminating device, (3) heat flux in the environment due to the non-uniformity of temperature distribution and convection, and (4) the cutting heat generated during machining. During a cutting process, a significant portion, approximately 60%–70%, of the electrical energy input into the machine tool is consumed as cutting energy. This energy contributes to heating the tools and workpieces directly, leading to machining errors. The remaining 30%–40% is converted into heat within the machine tool, resulting in thermal deformation and further contributing to machining inaccuracies [28]. To minimize machining errors caused by temperature changes, the workpiece, the machine tool, and its components should be at thermal equilibrium before the machining process begins.

In the early 1980s, Lawrence Livermore National Laboratory (LLNL) developed an ultraprecision machine tool called Large Optics Diamond Turning Machine (LODTM) which was capable of cutting metal materials with tolerances

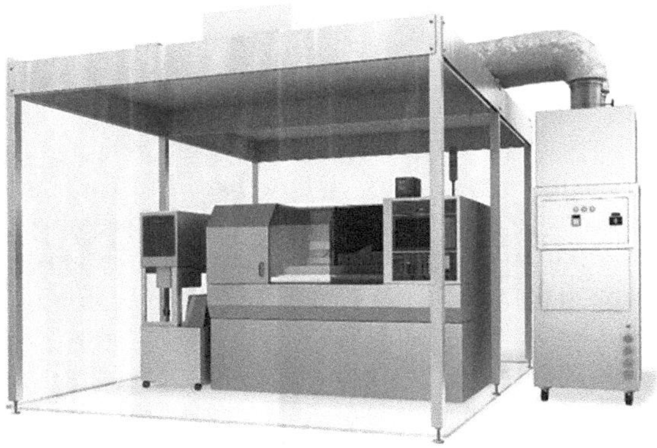

**FIGURE 3.14** Precision air controller for an ultraprecision machine tool [30]. (Courtesy of Nihon Spindle Manufacturing.)

as fine as 28 nanometers [29]. To mitigate temperature fluctuations, the LODTM enclosure maintains an exact air temperature of 20°C. Prior to commencing machining, the machine is allowed to stabilize for at least 12 hours to dissipate any residual heat from the machinist's body. Throughout the machining process, all personnel remain outside the LODTM enclosure to avoid possible temperature changes. Any heat generated by the cutting tool was efficiently dissipated by cutting oil, also maintained at 20°C. In recent industrial applications, ultraprecision machine tools are usually placed within an enclosed space equipped with a precision air control system to precisely regulate the surrounding temperature, as depicted in Figure 3.14.

### 3.8.2 Vibration Control

Environmental vibrations, typically at the micron level or higher, pose significant disturbances to ultraprecision machining processes. Therefore, it is imperative to isolate machine tools from these vibrations. In general, there are two types of vibration isolators categorized by their operating principles: passive vibration suppression systems and active vibration suppression systems.

Passive vibration suppression systems employ damping materials or devices to dissipate vibrational energy without requiring external power sources. These systems commonly utilize techniques such as viscous damping, elastic damping, and tuned mass dampers [31]. Viscous damping involves using fluids or materials that convert kinetic energy into thermal energy as vibrations occur, while elastic damping utilizes elastic materials or structures to absorb vibrations through deformation. Tuned mass dampers add secondary mass-spring-damper systems tuned to resonate at the frequency of the primary vibrating system, reducing its amplitude. Passive systems are generally simple, cost-effective, and

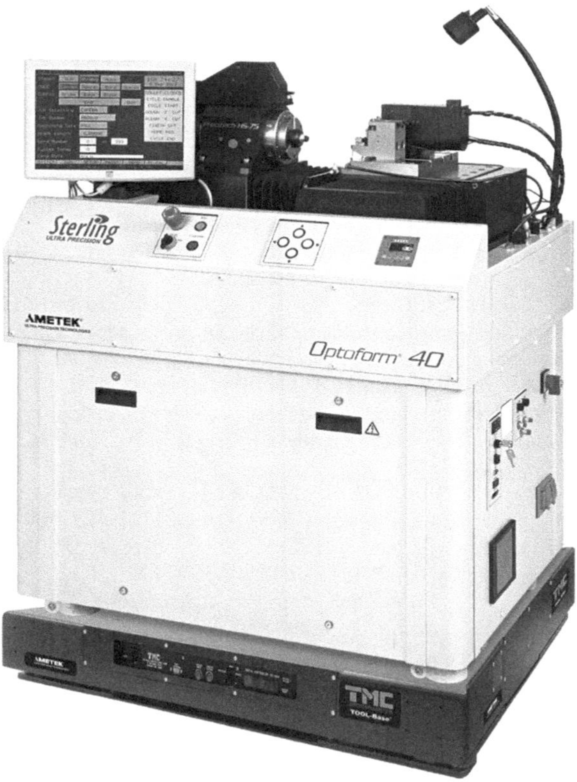


**FIGURE 3.15** Photo of an active piezoelectric floor vibration cancellation platform [33]. (Courtesy of AMETEK.)

highly reliable but lack adjustability and adaptability to changing vibration characteristics.

In contrast, active vibration suppression systems employ external energy sources and feedback control mechanisms to actively counteract vibrations in real time. These systems typically consist of sensors to detect vibrations, a controller to process sensor data, and actuators to apply forces to suppress vibrations. Actuators in active systems can be electromagnetic, piezoelectric, hydraulic, or pneumatic devices capable of generating forces to counteract vibrations [32]. Active systems offer the advantage of adaptability and adjustability, allowing for real-time control over vibration suppression characteristics. However, they are often more complex, expensive to implement, and require additional maintenance and power sources compared to passive systems. Figure 3.15 shows an example of an active vibration suppression system, on which a machine tool is placed.

## REFERENCES

[1] AMETEK Precitech Inc. (2024) Nanoform X. https://www.precitech.com/product/smallframelathesoverview/nanoform-x (accessed 28th April 2024).
[2] Rahman, M., Mansur, A., Karim, B. (2001) Non-conventional Materials for Machine Tool Structures. *JSME International Journal Series C Mechanical Systems, Machine Elements and Manufacturing* 44 (1): 1–11.
[3] McKeown, P. A., Morgan, G.H. (1979) Epoxy granite: a structural material for precision machines. *Precision Engineering* 1 (4): 227–229.
[4] Möhring, H., Brecher, C., Abele, E., Fleischer, J., Bleicher, F. (2015) Materials in machine tool structures. *CIRP Annals* 64 (2): 725–748.
[5] Ding, L. (2021) Surface texturing on dry-lubricated transmissions and power systems to deliver better performance and simplified systems in aerospace engineering [PhD thesis, University of Nottingham]. Nottingham eTheses.
[6] Gao, W., Lee, J. C., Arai, Y., Noh, Y. J., Hwang, J. H., Park, C. H. (2010) Measurement of slide error of an ultra-precision diamond turning machine by using a rotating cylinder workpiece. *International Journal of Machine Tools and Manufacture* 50 (4): 404–410.
[7] Kane, N.R., Sihler, J., Slocum, A.H. (2003) A hydrostatic rotary bearing with angled surface self-compensation. *Precision Engineering* 27 (2): 125–139.
[8] Möhring, H., Brecher, C., Abele, E., Fleischer, J., Bleicher, F. (2015) Materials in machine tool structures. *CIRP Annals* 64 (2): 725–748.
[9] TDC CO., LTD. (2025) Invar/Super Invar: Properties, Uses and Processing Examples. https://mirror-polish.com/en/material_knowledge/invar (accessed 22nd May 2025).
[10] Wang, P., Zhang, Y., Feng, L., Hou, W., Wang, J., Li, W., Feng, K. (2023) Study on the pneumatic hammer phenomenon of aerostatic bearings based on the empirical mode method: Numerical and experimental analysis. *Tribology International* 181: 108305.
[11] Xu, C., Zhao, F., Xu, G., Lian, G. (2023) General-purpose controller for multiple types of magnetic bearing rotor systems. *Energy Reports* 9: 345–354.
[12] Parker Hannifin Corp. (2025) I-Force and Ironcore Linear Motors. https://www.parker.com/content/dam/Parker-com/Literature/South-Korea/Parker-Linear-Motors_MSDK_EN.pdf (accessed 22nd May 2025).
[13] Collins, D. (2025) What are ironless linear motors? https://www.linearmotiontips.com/ironless-linear-motors (accessed 22nd May 2025).
[14] Loughridge, R., Abramovitch, D.Y., (2013) *A tutorial on laser interferometry for precision measurements, Proceeding of 2013 American Control Conference*, IEEE, pp. 3686–3703.
[15] Sinowon Ltd. (2025) Precision Linear scales. https://www.sinowon.com/precision-linear-scales.html. (accessed 22nd May 2025).
[16] SCHUNK SE & Co. KG Sealed 6-jaw power lathe chuck. https://schunk.com/ca/en/workpiece-clamping-technology/lathe-chucks/power-lathe-chucks-without-through-hole/rota-ncr-a/c/PGR_4122 (accessed 29th April 2024).
[17] Yildiz, Y., Nalbant, M. (2008) A review of cryogenic cooling in machining processes. *International Journal of Machine Tools and Manufacture* 48 (9): 947–964.
[18] Singh, G., Aggarwal, V., Singh, S. (2020) Critical review on ecological, economical and technological aspects of minimum quantity lubrication towards sustainable machining. *Journal of Cleaner Production* 271: 122185.
[19] Weinert, K., Inasaki, I., Sutherland, J.W., Wakabayashi, T. (2004) Dry machining and minimum quantity lubrication. *CIRP Annals* 53 (2): 511–537.

[20] Tai, B.L., Dasch, J.M., Shih, A.J. (2011) Evaluation and comparison of lubricant properties in minimum quantity lubrication machining. *Machining Science and Technology* 15 (4): 376–391.
[21] Katna, R., Agrawal, N., Suhaib, M. (2020) A short review on machining with ultrasonic MQL method. *IOP Conference Series: Materials Science and Engineering* 802 (1): 012011.
[22] Xu, W., Li, C., Zhang, Y., Ali, H.M., Sharma, S., Li, R., Yang, M., Gao, T., Liu, M., Wang, X., Said, Z., Liu, X., Zhou, Z. (2022) Electrostatic atomization minimum quantity lubrication machining: from mechanism to application. *International Journal of Extreme Manufacturing* 4 (4): 042003.
[23] Shashidhara, Y.M., Jayaram, S.R. (2010) Vegetable oils as a potential cutting fluid—An evolution. *Tribology International* 43 (5): 1073–1081.
[24] Said, Z., Gupta, M., Hegab, H., Arora, N., Khan, A.M., Jamil, M., Bellos, E. (2019) A comprehensive review on minimum quantity lubrication (MQL) in machining processes using nano-cutting fluids. *The International Journal of Advanced Manufacturing Technology* 105 (5): 2057–2086.
[25] Sharma, A.K., Tiwari, A.K., Dixit, A.R. (2016) Effects of Minimum Quantity Lubrication (MQL) in machining processes using conventional and nanofluid based cutting fluids: A comprehensive review. *Journal of Cleaner Production* 127: 1–18.
[26] Yan, J., Zhang, Z., Kuriyagawa, T. (2011) Effect of nanoparticle lubrication in diamond turning of reaction-bonded SiC. *International Journal of Automation Technology* 5 (3): 307–312.
[27] Peng, D.X., Kang, Y., Hwang, R.M., Shyr, S.S., Chang, Y.P. (2009) Tribological properties of diamond and SiO2 nanoparticles added in paraffin. *Tribology International* 42 (6): 911–917.
[28] Nishiwaki, N. (1990) Heat transfer in high precision machining. *Journal of the Society of Mechanical Engineers* 93 (864): 928–929.
[29] Lawrence Livermore National Laboratory (2001) Research Highlights - The world's most accurate lathe.
[30] Nihon Spindle Manufacturing Co., Ltd Precision Air Controller. https://www.spindle.co.jp/product/kuutyou/thermalcube.html (accessed 3rd May 2024).
[31] Ibrahim, R.A. (2008) Recent advances in nonlinear passive vibration isolators. *Journal of Sound and Vibration* 314 (3): 371–452.
[32] Karnopp, D. (1995) Active and Semi-Active Vibration Isolation. *Journal of Vibration and Acoustics* 117 (B): 177–185.
[33] TMC AMETEK, Inc Active piezoelectric floor vibration cancellation platform. https://www.techmfg.com/products/stacis/tool-base (accessed 3rd May 2024).

# 4 Tooling Technologies

## 4.1 INTRODUCTION

Cutting is a machining process for removing materials from workpieces in the form of chips using single-point or multipoint tools with clearly defined geometries. Cutting tools are securely mounted onto tool spindles or tool posts of machine tools, making direct contact with the workpieces through machine motions. Consequently, they are often regarded as the "teeth" of machine tools. To a significant extent, the performance of a cutting tool plays a pivotal role in determining the machining quality, while also exerting influence on efficiency, productivity, and cost-effectiveness of the machining process. Over the past decades, tooling technology has undergone substantial development. Consequently, cutting tools are now widely recognized as standard products for supporting various precision machining applications. Nonetheless, with so many options, including various tool materials and tool geometries, in the market, selecting the most suitable cutting tool for a particular cutting process is not easy. In addition, to meet the demands of emerging manufacturing technologies, tool materials, and their corresponding fabrication technologies are continually advancing.

## 4.2 TOOL MATERIALS

Cutting is a high-speed material removal process that involves high temperature, high pressure, and high shear stress. It is known that the cutting temperature during machining of a few hard materials can be higher than 1000°C and the pressure acted on the tool face can reach a few GPa. For this reason, the ideal cutting tool should exhibit superior performance in the following five aspects:

1. Hardness: crucial for maintaining a sharp and consistent cutting edge at elevated temperatures encountered during machining.
2. Toughness: essential for the cutting edge to withstand the cutting forces and shocks during machining, particularly in intermittent cutting of hard brittle materials.
3. Resistance to thermal shock: essential for overcoming the thermal impacts from the continuous cycle of heating and cooling which is typically encountered in milling operations or during intermittent cutting operations, such as eccentric turning on a lathe or fly cutting.
4. Affinity between the tool and the workpiece: low affinity is desired to prevent the formation of a built-up edge (BUE) and reduce tool wear.
5. Resistance to oxidation: essential for preventing chemical wear.


DOI: 10.1201/9781003292388-4

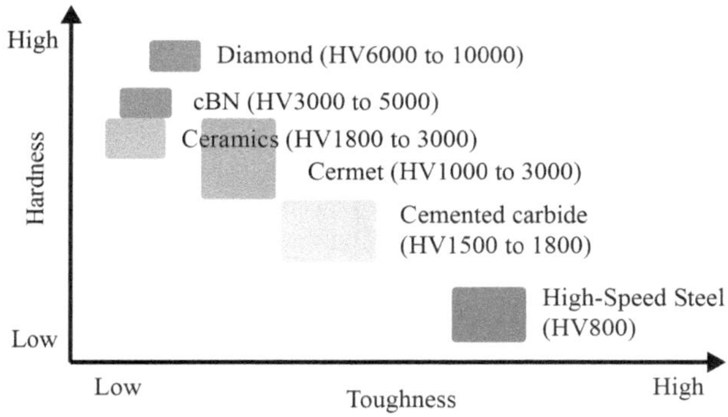


**FIGURE 4.1** Comparison of hardness and toughness of various tool materials.

It should be pointed out that not all these aspects can be satisfied by a single tool material. In other words, it is extremely difficult to find an ideal cutting tool material, for which all these five factors are superior at the same time. In industrial applications, tool selection for machining a specific workpiece material is based on deciding the major factor with synthetic consideration of other factors.

The first factor for selecting tool material is hardness. For practical machining, the hardness of the tool should be at least 3 times higher than that of the workpiece material [1]. Moreover, higher hardness enables faster cutting speed at a low tool wear rate. The other factor is fracture toughness. Higher toughness allows for discontinuous cutting with no edge chipping and breakage. The simultaneous attainment of high hardness and high toughness when selecting cutting tool materials is desirable, yet they often exhibit an inverse relationship. In Figure 4.1, a comparison of tool material is highlighted where fracture toughness is plotted against hardness to indicate the range of applicability for each tool material. Diamond (single/polycrystal) exhibits the highest hardness and heat conductivity among all tool materials and has excellent chemical stability and wear resistance. Because of these superior characteristics, diamond is usually used as tool material for precision and ultraprecision machining. A diamond tool is applicable to turning, milling, and grinding, producing a good surface finish. In addition to diamond, cubic boron carbide (cBN) and tungsten carbide (WC) are also popular tool materials for precision machining although they are rarely used for ultraprecision machining. The following sections will briefly introduce these typical tool materials.

### 4.2.1 Single-Crystal Diamond

Single-crystal diamond (SCD) possesses the highest level of hardness and thermal conductivity among all known materials, along with a very low coefficient of friction. These properties make it an ideal tool material for achieving the highest precision and accuracy in machining. The covalent structure of diamond with exceptional hardness allows for obtaining an extremely sharp tool edge, particularly at the nanoscale. This unique feature positions diamond as the exclusive tool material suitable for ultraprecision machining. The estimated edge sharpness, characterized by edge radius, is in the level of a few nanometers to a few tens of nanometers. Based on the source of production, SCD can be categorized into two types: natural diamond and artificially synthesized diamond. In the past, only natural diamond was used for ultraprecision machining. In recent years, however, artificially synthesized diamond has been commercialized as a popular cutting tool material for ultraprecision cutting processes, offering a cost-effective alternative to natural diamond. In addition, for artificially synthesized diamond, it is possible to control the purity and dopant type as well as unify the crystal quality and its stableness during mass production. Chemical vapor deposition (CVD) and high-pressure high-temperature (HPHT) processing are two major methods for fabricating single-crystal diamond. Figures 4.2(a) and (b) show a natural single-crystal diamond tool and a synthetic single-crystal diamond tool, respectively. The former exhibits a transparent color, indicating a very high purity of diamond, while the latter exhibits a yellow color due to the dopant type used. Figure 4.2(c) shows SEM images of the cutting edge of a single-crystal diamond tool. An extremely smooth and sharp edge is achieved around the central part of the tool, while some submicron-level defects can be observed at the corner part, which might have been caused by mechanical lapping during tool fabrication.

SCD tools can be used to machine various nonferrous metals such as copper, aluminum, and electroless plated nickel etc., which are used as the laser mirrors,

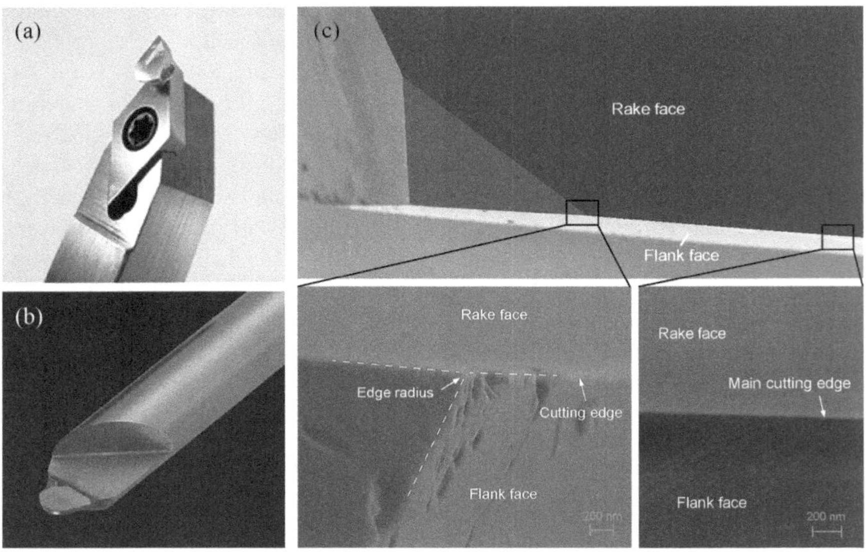


**FIGURE 4.2** (a) Photo of natural diamond tool, (b) Photo of CVD diamond tool, and (c) SEM images of the cutting edge of a single-crystal diamond tool.

magneto-optical discs, optical lenses and their molds. Under specific conditions, SCD tools can also be used to machine brittle crystalline materials such as germanium (Ge), silicon (Si), calcium fluoride ($CaF_2$), and a few kinds of optical glasses in a ductile mode. This technology is termed ductile regime machining (DRM) or simply ductile machining (DM) [2, 3]. High-quality surfaces obtained on these materials provide key parts for the advanced optoelectronic and optical technologies [4, 5]. However, SCD is a highly brittle material with low fracture toughness; as a result, it is prone to fracture under shocks and impacts. Figure 4.3(a) shows micro chippings along the cutting edge of an SCD tool when cutting a typical hard-brittle material, single-crystal silicon. For this reason, cutting operations must be performed to minimize such impacts in order to avoid edge chipping and prolong tool life. The impacts may be caused by either the tool vibration or the runout and unevenness of the workpiece.

Another important point is that an SCD tool is not suitable for machining ferrous materials, such as iron and steel. This is attributed to the fact that the carbon (C) element in diamond reacts with elements like Fe, Cr, Ti, Ni, etc., present in ferrous materials, leading to accelerated tool wear [6]. Figure 4.3(b) shows a diamond tool after cutting the high-entropy alloy FeCrCoMnNi for a cutting distance of 88 m, observed from the rake face and flank face sides, respectively. Although the hardness of the high-entropy alloy is much lower than that of silicon, the tool

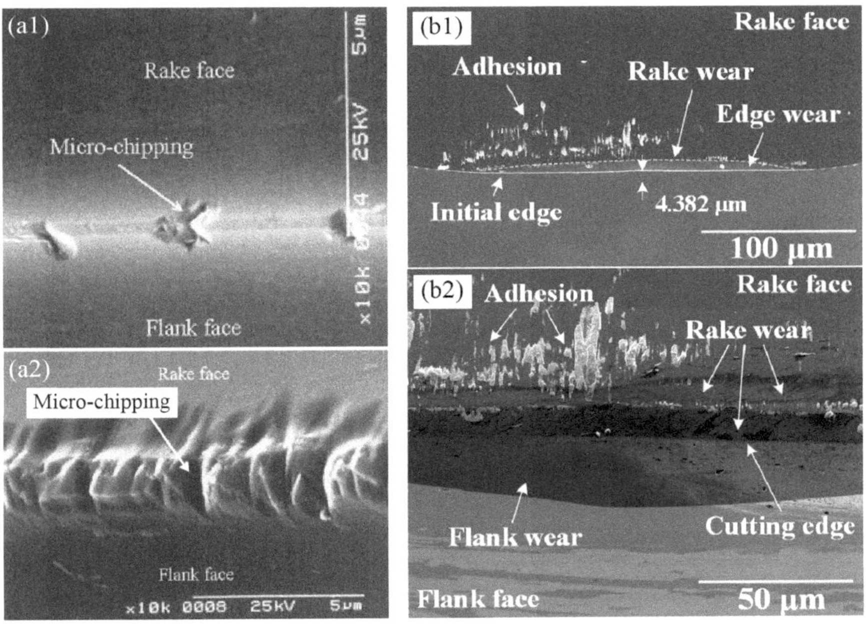


**FIGURE 4.3** SEM images of single-crystal diamond tools: (a1-2) chippings occurred along the edge after cutting silicon for 1.27 km and 7.62 km [7]; (b1-2) tool wear observed on the rake face and flank face sides after cutting high-entropy alloy for 88 m [8].

wear is much more severe, even though the cutting distance is far shorter than that of silicon cutting (7.62 km).

### 4.2.2 Poly-Crystal Diamond

Poly-crystal diamond (PCD) is sintered by diamond powders with binders, such as cobalt powders, under high pressure (~5000 atmospheric pressure) and high temperature (~several thousand °C). This type of sintered diamond shows excellent abrasion resistance when cutting nonferrous metals and other hard materials and is taking an active part in precision machining. The fracture toughness of PCD tools is also higher than those of SCD due to the presence of metal binder in the material.

It should be pointed out that the surface roughness achievable by PCD tools is in the submicron level (~0.1 μm Ry), higher than that by an SCD tool ( ~0.01 μm Ry). The surface roughness is generated by both the tool feed marks and the edge unevenness caused by micro chippings or diamond grain dislodgement on the cutting edge. Figure 4.4 shows the cutting edges of an SCD cutting tool and a PCD cutting tool, viewed from the tool rake face. After tool polishing, the SCD tool exhibits a uniform and smooth cutting edge. However, the cutting edge of the PCD tool is relatively rough due to it being susceptible to the detachment of diamond particles and minute irregularities at the edge. If the cutting edge has sub-micrometer-scale irregularities or detached parts, the unevenness will be transferred to the workpiece surface, making it challenging to achieve a nanometric surface finish.

The hardness of PCD tools is only slightly less than that of SCD tools, but the hardness is isotropic, i.e., consistent in all directions, which is different from SCD where the hardness is anisotropic, i.e., crystal orientation dependent. It is widely applied to cutting tools for non-ferrous workpieces in mass production. Recently, attempts have been made to use a single-crystalline diamond end mill for generating microstructures. However, when machining super hard materials, such as silicon carbide, cleavage-induced edge chipping and flank wear may occur [10]. Considering that PCD tools have higher toughness, they might be more suitable than SCD tools for milling hard material molds.

### 4.2.3 Nano-Polycrystal Diamond

Recently, a new kind of diamond called nano-polycrystal diamond (NPD) has been developed by Sumitomo Electric Industries, Ltd., which has a binderless polycrystal structure with the diamond grain size down to ~10 nm level. Synthetically

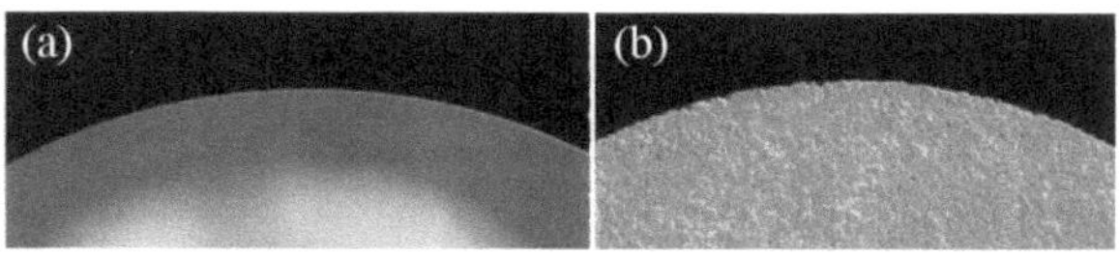


**FIGURE 4.4** Optical micrographs of cutting edges viewed from the rake face side: (a) SCD tool and (b) PCD tool [9]. (Courtesy of Sumitomo Electric Industries.)

produced nano-sized diamond crystalline grains were formed into a bulk using the high-pressure high-temperature (HPHT) method [11]. At such ultra-high pressure and high temperature, nano-sized diamonds could be synthesized through the graphite-diamond transformation from graphite powders. NPD tools have also been made, as shown in Figure 4.5. Due to the randomly distributed nano-sized diamond grains, NPD tools feature superior mechanical properties, namely, higher hardness and higher toughness, which could overcome the disadvantages of SCD and PCD tools. The grain structure imparts isotropic material behavior, in contrast to the anisotropic behavior of the SCD, as the microstructure with trans-granular boundaries prevents crack propagation and improves fracture toughness [12].

Figure 4.6 shows a comparison of a NPD tool, a PCD tool (average grain size of 5 μm), and an SCD tool in the turning of high-strength Al–Si alloy (A390;

**FIGURE 4.5** Photo of nano-polycrystal diamond tools. (Courtesy of Sumitomo Electric Industries.)

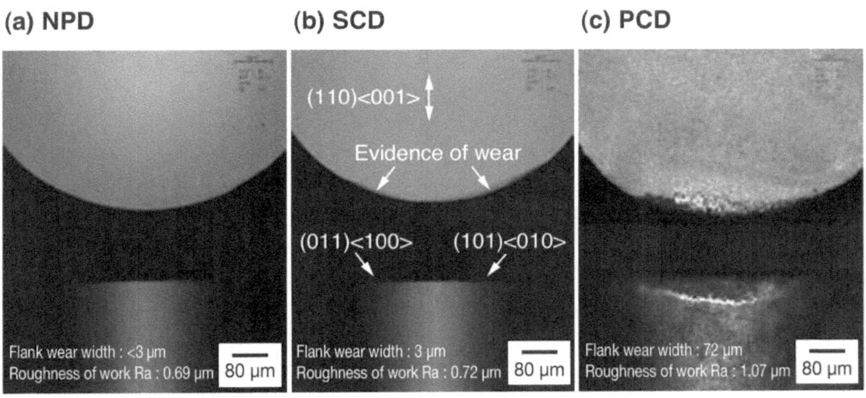


**FIGURE 4.6** Cutting edges of diamond tools after cutting tests on Al–Si alloy: (a) SCD (type Ib), (b) PCD, and (c) NPD [13]. (Courtesy of Sumitomo Electric Industries.)

Al-17Si-5Cu) after 30 km of cutting. The wear resistance of the NPD tool was approximately 10–20 times higher than that of the conventional fine-grain PCD tool and equivalent to that of the SCD tool. The SCD tool was disfigured by uneven wearing due to its anisotropic wear resistance, while the NPD tool retained its shape. The tool wear of PCD progresses mainly due to thermal deterioration influenced by metal binders such as Co. This gives rise to micro-cracking and grain-falling because of the difference of thermal expansion between diamond and metal binder, and graphitization by catalytic reaction of metal binder depending on the rise in temperature at the cutting point. On the contrary, NPD showed excellent wear resistance (because it has no metal binders), excellent high thermal resistance, and high bonding strength between diamond grains.

For example, the surface quality of ZnS ceramics, machined by NPD and SCD tools, is shown in Figure 4.7(a). In the case of the natural SCD tool, the cutting surface roughness exceeded 50 nm, indicating the end of tool life, after machining only five ZnS lenses. On the other hand, with the NPD tool, the surface roughness remained less than 20 nm even after machining 20 lenses. Figure 4.7(b) is another example showing the flank wear widths of NPD, SCD, and PCD tools in cutting WC-alloy, a difficult-to-cut material. The SCD tool suffered serious breakage and chipping damage due to the cleavages in the early stage after just 20 m cutting, whereas the NPD tool showed no remarkable chipping even after 280 m cutting. The wear resistance of the NPD tool was found to exceed that of PCD-B tool (average grain size: 5 μm) by more than five times and surpass that of PCD-C tool (average grain size: 30–50 μm) by three times. These results reveal that NPD tools have outstanding potential for various cutting applications.

### 4.2.4 Cubic Boron Nitride

Cubic boron nitride (cBN), which is synthesized artificially under the conditions of high pressure and high temperature similar to the production process of diamond, possesses the second-highest hardness and thermal conductivity after diamond. It exhibits superior chemical and thermal stability compared to diamond, particularly at temperatures exceeding 700°C. The cBN tools play an important role in various cutting applications. Specifically, owing to its lower reactivity with the iron-system metals, cBN has become the primary tool material for machining ferrous materials that prove challenging for diamond tools.

The mechanical and thermal properties of cBN are influenced significantly by the kinds and quantities of the binder materials. In general, cBN can be divided into three categories: single crystals, polycrystals (PcBN), and binderless polycrystals (BcBN). Low binder content PcBN tool has an excellent property for cutting cast iron materials and heat-resisting alloys; while high binder content PcBN tool is suitable for machining quenched steels. Compared to PcBN tools, BcBN tools can create a sharper tool edge and achieve an exceptionally low degree of roughness on the machined surface [14]. By making full use of the characteristics of cBN tools, high-efficiency cutting of hard ferrous components such as automobile parts, engine blocks, gears, shafts, and bearings is possible.

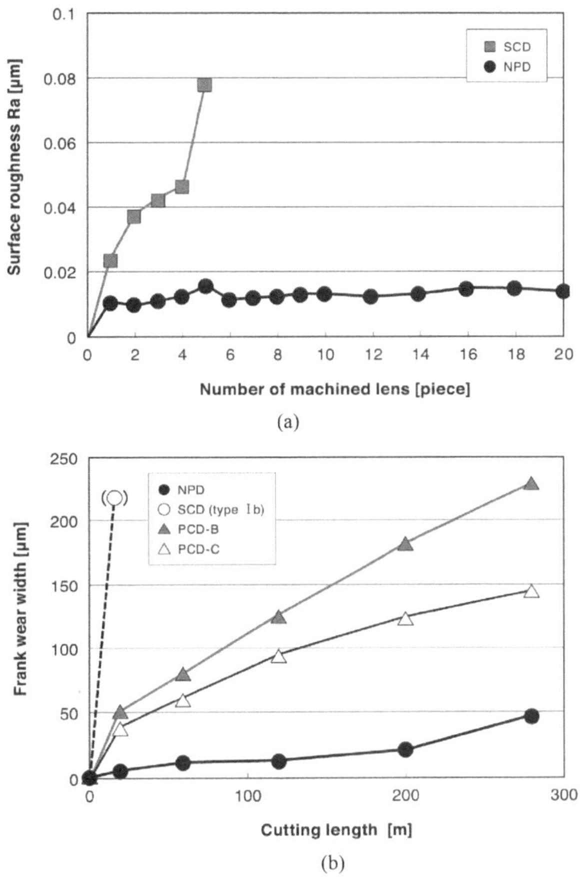


**FIGURE 4.7** Cutting performances of (a) NPD and SCD tools in high precision cutting of ZnS ceramics and (b) NPD, SCD, and PCD tools in cutting of WC-alloy. (Courtesy of Sumitomo Electric Industries.)

Recently, single-crystal cBN has also been attempted for making tools [15]. For cutting hardened stainless steel (HRC 52), the surface roughness of the machined workpiece was less than 50 nm $R_{max}$. Despite the superior cutting performance exhibited by single-crystal cBN compared to PcBN and BcBN, its application to cutting tools is still confined to the laboratory level. The development of commercially available single-crystal cBN is still ongoing, as the growth of large single-crystal bulk remains challenging.

Table 4.1 compares the material properties of BcBN and PcBN. It is evident that BcBN has better mechanical and thermal properties than PcBN. After cutting stainless steel for a cutting distance of 2 km, the wear of the BcBN and PcBN tools was observed, as shown in Figure 4.8. For both tools, flank wear was the

**TABLE 4.1**
**Properties of Binderless cBN and Conventional cBN [17]**

| | Binderless cBN | Conventional cBN |
|---|---|---|
| cBN contents (vol.%) | >99.9 | 85–90 |
| cBN grain size (μm) | <0.5 | 3–5 |
| Other constituents | Hexagonal boron nitride | Binder (Co etc.) |
| Process | Direct conversion | cBN + binder |
| Hardness (GPa) at room temperature | 50–55 | 35–40 |
| Hardness (GPa) at 1273 K | 20 | 12 |
| Thermal conductivity (W/m K) | 360–400 | 100–130 |
| Thermal stability (K in air) | 1620 | 1270 |

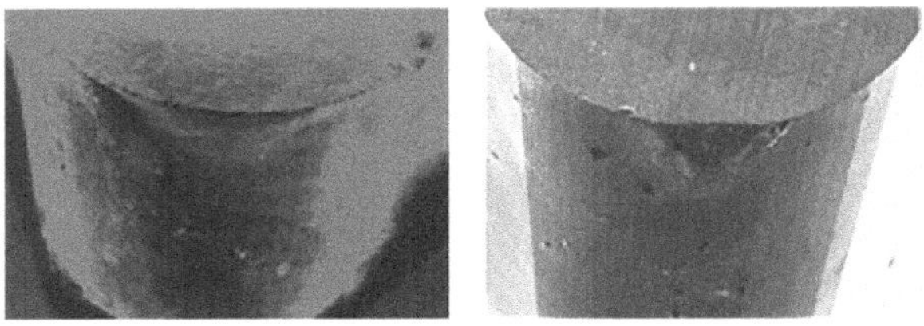

(a) (b)

**FIGURE 4.8** Flank wear of (a) BcBN and (b) PcBN tools after cutting stainless steel [16].

main form of tool wear, while the wear on the rake faces was minimal. However, the BcBN tool exhibited greater wear resistance compared to the PcBN tool under the same cutting conditions.

### 4.2.5 Tungsten Carbide

Tungsten carbide (WC) cutting tools, also known as cemented carbide cutting tools are crafted from composite materials comprising a hard phase tungsten carbide, and a ductile phase cobalt-based binder. Tungsten carbide tools are widely utilized in industry for cutting iron materials, Ni-based superalloys [17], and titanium alloys [18], because of the characteristics of high hardness, high strength, high wear resistance, and high modulus of elasticity, as well as good impact toughness and corrosion resistance.

Table 4.2 compares the properties of WC, PcBN, and PCD cutting tools. Although WC cutting tools have lower hardness than PcBN and PCD, they exhibit excellent performance in the machining of titanium alloys compared to other material cutting tools, due to better characteristics in chemical inertness to depress the tendency to react with titanium; toughness and fatigue resistance to withstand the chip segmentation process; and high compressive, tensile and shear strength [19].

**TABLE 4.2**
**Basic Properties of WC, PcBN, and PCD Cutting Tools**

| | WC (K10) [19] | cBN (Amborite) [20] | PCD (CTB010) [21] |
|---|---|---|---|
| Knoop hardness (GPa) | 13 | 28 | 50 |
| Modulus of elasticity (GPa) | 620 | 680 | 776 |
| Modulus of rigidity (GPa) | 258 | 279 | 363 |
| Compressive strength (GPa) | 4.5 | 3.8 | 7.6 |
| Poisson ratio | 0.22 | 0.22 | 0.07 |
| Thermal conductivity (W/m K) | 100 | 100 | 560 |

To further improve the abrasion resistance, hardness, and chemical inertness of WC tools to prevent chemical interaction between the tool and the work material during machining, coating technologies, such as physical vapor deposition (PVD) and chemical vapor deposition (CVD), have been used for coating the WC tools, which has marked a significant advancement in cutting tool technology [22]. The WC cutting tools are coated with one or more thin layers of wear-resistant material, such as titanium carbide (TiC), titanium nitride (TiN), and/or aluminum oxide ($Al_2O_3$). Various coatings are employed for different purposes. For example, the $Al_2O_3$ coating has proven chemical stability at high temperatures and resists abrasive wear while the TiN coating gives the insert a lower coefficient of friction and therefore reduces adhering tendency on the rake face [23]. The benefits of thin, hard coatings in reducing tool wear and enhancing tool life and productivity are widely recognized. Consequently, the majority of WC tools utilized in the metal cutting industry are coated, despite the additional cost associated with coating.

Compared with PcBN and PCD, WC makes it easier to shape the edges of a tool by grinding and polishing, thus the production cost is relatively lower. It is especially popularly used for end milling tools. However, similarly to PcBN and PCD, the edge sharpness and smoothness of WC tools are not enough to be used for ultraprecision cutting where a nanometer-level surface roughness is required.

## 4.3 TOOL DESIGN

Tool design is another important factor that determines the machining quality and efficiency as the tool geometry directly impacts the following machining characteristics. The following aspects should be considered in tool design:

1. Uncut chip thickness: In machining processes, a larger uncut chip thickness is desired to maximize material removal rate and prevent burnishing and rubbing instead of cutting.
2. Direction of chip flow: A proper direction is essential for controlling chip breakage and evacuation, thereby preventing burr formation, avoiding chips scratching the finished surface, and reducing cutting forces.

3. Cutting force: A lower cutting force is desired to prevent excessive bending, buckling, and deformation of long and non-rigid tools and workpieces, ensuring the stability and accuracy of machining.
4. Surface integrity: The tool geometry defines surface topography (surface quality) and, to a great extent, the state of stress in the materials around the tool (subsurface quality).
5. Tool life: Tool geometry determines the direction of chip flow and influences the cutting force, thus impacting friction coefficient, cooling performance, and temperature distribution in the cutting zone.

As the parameters of tool geometry can be determined by referring to handbooks and databases or based on the experience of the designer, understanding how these parameters affect the above-mentioned cutting characteristics is necessary.

### 4.3.1 General Shape Design

Since single-crystal diamond tools have sharper cutting edges and can be machined with higher precision than tools made from other materials, they are designed in some simple and universal shapes to achieve ultraprecision turning of surfaces with different shapes. Figure 4.9 presents three typical shapes for commercially available diamond tools, that is, round-nosed, straight-nosed, and V-shaped tools. For creating a specific surface shape on a workpiece, the selection of tool shape is important. For example, a round-nosed tool is used for generating a curved surface, such as aspherical or free-form surfaces [24], while a straight-nosed tool is used for cutting flat surfaces or convex aspherical surfaces [25, 26]. When fabricating a Fresnel structure, groove zones with sharp corners and straight walls are required. Thus, a round-nosed tool is inappropriate to cut zones, even if the nose radius is small, because it generates large shadow areas [27, 28], as shown schematically in Figure 4.10(a). In such a case, a V-shaped tool that has a near-zero nose radius is used for cutting the Fresnel structures without shadows [29], as shown schematically in Figure 4.10(b).

Tool shape not only affects the geometric accuracy of the workpiece but also alters the machining mechanisms and characteristics. For example, by using a round-nosed tool, the effective chip thickness $t_{\text{eff}}$ (also known as uncut chip thickness, undeformed chip thickness $h$) varies from zero at the tool center to a maximum value ($t_{\text{m}}$) at the top of the uncut shoulder as shown in Figure 4.11(a). $t_{\text{eff}}$ can be approximately estimated by the following equation:

$$t_{\text{eff}} = f\left(z+f\right)/R \tag{4.1}$$

where $f$ is the feed rate, $R$ is the tool nose radius. When using a round-nosed tool for machining brittle materials, a ductile response will be achieved around the tool center, where $t_{\text{eff}}$ is sufficiently small, provided that a small feed rate or a large tool nose radius is used.

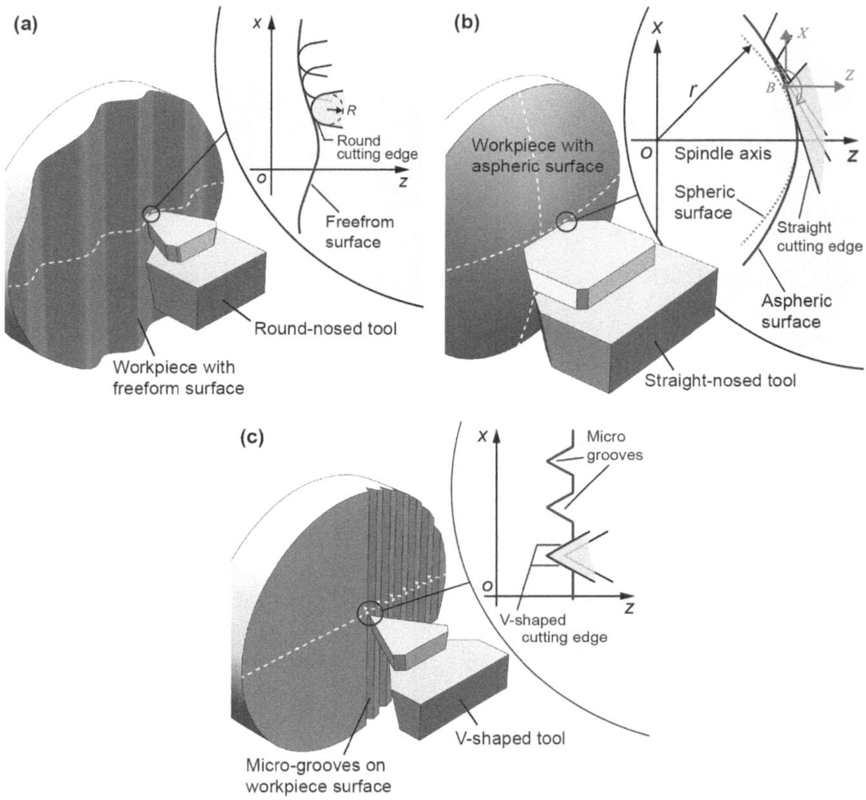


**FIGURE 4.9** Schematics of typical shapes for diamond tools: (a) round-nosed, (b) straight-nosed, and (c) V-shaped tools.

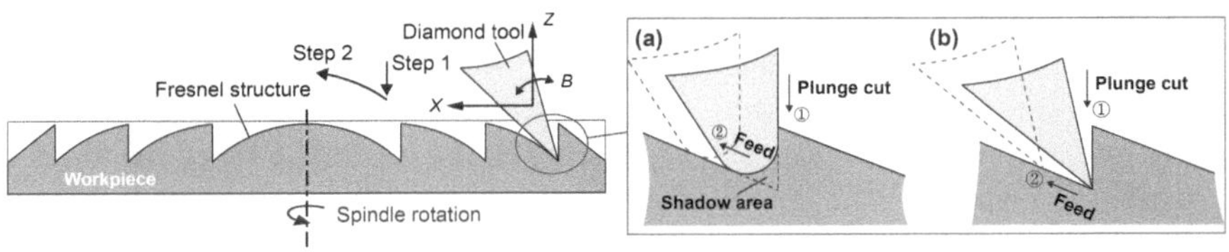


**FIGURE 4.10** Machining models for the Fresnel structures using (a) around-nosed tool and (b) a V-shaped tool.

When using a straight-nosed tool, however, as shown in Figure 4.11(b), the undeformed chip thickness ($h$) remains almost constant along the entire main cutting edge. The $h$ can be estimated by the following equation [30]:

$$h = f \cdot \sin\kappa \tag{4.2}$$

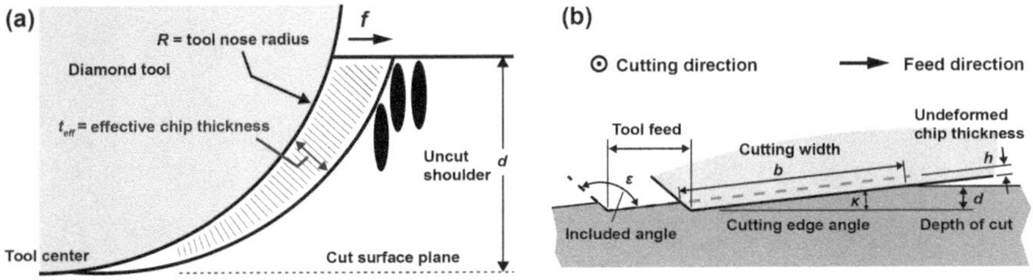


**FIGURE 4.11** Schematic model of uncut chip thickness for (a) round-nosed tool and (b) straight-nosed tool.

where $f$ is the feed rate, and $\kappa$ is the cutting-edge angle. Therefore, when using the straight-nosed tool for cutting a brittle material, an extremely small $h$ can be achieved by setting a sufficiently small cutting edge angle even a large feed rate is used. This tool shape ensures higher efficiency of ductile machining compared to the round-nosed tool.

The machining model of a V-shaped tool is similar to the model of the straight-nosed tool in which the undeformed chip thickness ($h$) is unvarying along the main cutting edge, except that the V-shaped tool has two symmetrical main cutting edges. The chips formed along two main cutting edges will interfere with the tool tip, resulting in a material stagnation region on the tool tip (see Figure 5.17).

Apart from turning, end milling is another effective method for fabricating three-dimensional surfaces [10, 31]. Different from the turning process, during which the tool is non-rotational and continues to cut the material with a stable cutting force, in the milling process the tool intermittently cuts the material with multiple cutting edges due to the rotation of the tool. As an end milling cutter has an effective diameter of hundreds of microns to a few millimeters, it is suitable for fabricating components with micro-structured surfaces like dimple arrays, and micro three-dimensional curved shapes [32]. However, considering that the diameter of end milling cutter is small, the cutting forces must be controlled small to prevent the end milling cutter from being broken or deflected [33]. Moreover, the micro end milling process is usually time-consuming especially when machining a large-area surface.

For micro end milling, there are many types of cutter geometries developed and optimized by researchers and companies [32–34]. The tool materials used for milling are primarily sintered materials, such as PCD, PcBN, and WC. Figure 4.12(a)–(c) shows milling cutters have flat ends. Such flat-end cutters can generate flat surfaces and slots with sharp perpendicular edges in vertical micro milling. In contrast, the (d) and (e) are cutters having hemispherical cutting edges at the tool end. Such ball-end cutters are suitable for machining parts with curved micro features. Single-crystal diamond has also been used as a milling cutter material, but the tool shape is typically simple, such as a half-round nose tool, as shown in Figure 4.12(f).

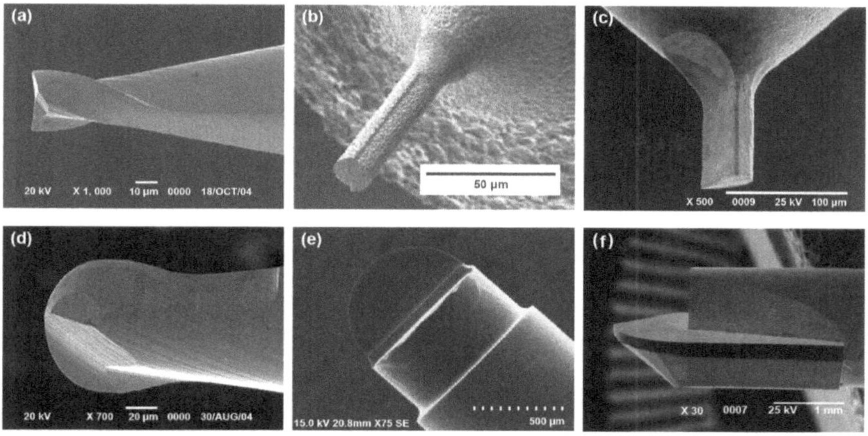


**FIGURE 4.12** Micro end milling cutters: (a) two-flute flat end cutter with a diameter of 30 μm [31], (b) cylindrical end cutter with a diameter of 20 μm [35], (c) D-shaped end cutter with a diameter of 50 μm [36], (d) two-flute ball-end cutter with a diameter of 100 μm [31], (e) single-flute spherical end cutter with a radius of 400 μm [37], and (f) single-crystal diamond ball endmill with a diameter of 1 mm [10].

### 4.3.2 Edge Angle Design

The rake angle and relief angle are two important parameters for a tool. The rake angle is the angle between the rake surface and perpendicular to the workpiece. This angle is provided for ease of chip flow and overall machining. For machining of ductile materials, a negative rake angle may increase the difficulty of chip formation, resulting in increased burr sizes. However, for machining of brittle materials, a properly designed negative rake angle (usually from −20° to −40°) can increase the hydrostatic pressure in the cutting zone, which can suppress brittle fractures, resulting in a crack-free surface [39].

The relief angle is the angle between the flank face of a cutting tool and the surface of the workpiece. The relief angle significantly influences the strength of the tool. A smaller relief angle corresponds to a higher strength. However, if the relief angle is too small, the flank face may collide with the machined surface when cutting micro-structures on workpieces. The discussion on the effects of rake and relief angles will be presented in detail in Chapter 6.

For micro end milling cutters, the helix angle is an additional important design parameter, affecting the chip evacuation. Positive helix angles would guide chips flowing upwards along the rake face, forming large flip burrs, while negative helix angles would suppress the flip burrs [32].

### 4.3.3 Crystal Orientation (of Diamond)

For SCD tools, the crystal orientation of diamond is an essential aspect of tool design. Similar to other crystals, single-crystal diamond also exhibits anisotropic

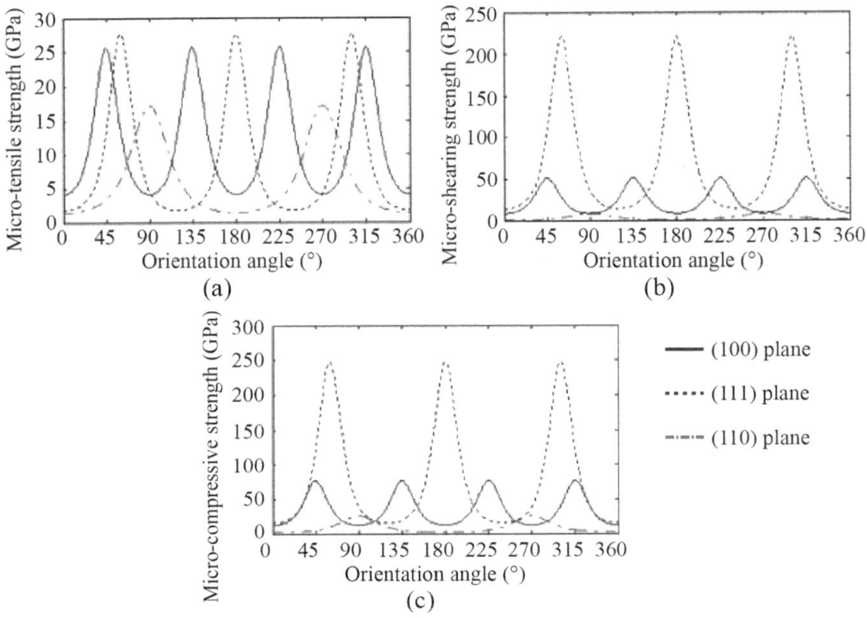


**FIGURE 4.13** Dynamic micro-mechanical strengths of single-crystal diamond in different orientations and on different planes. (a) micro-tensile strength, (b) micro-shearing strength, and (c) micro-compressive strength [38].

mechanical properties. The order of the Knoop hardness among the three typical planes is (111) > (110) > (100). The theoretical results of tensile strength, shear strength, and compressive strength on the three planes are presented in Figure 4.13. The (110) plane presents a twofold symmetry, and the (100) plane shows a fourfold symmetry, while the (111) plane exhibits a threefold symmetry. Therefore, for a specific crystal plane, there are some directions that exhibit the highest strength and some other directions that exhibits the lowest strength, as summarized in Table 4.3.

Zong et al. [40] designed four diamond cutting tools, which have different crystallographic orientations for the rake and flank faces. The orientation details of these tools are shown in Figure 4.14, where "R" and "F" denote rake face and flank face, (100) and (110) denote the (100) plane and (110) plane of diamond crystal. The surface roughness Ra (average roughness) and Rms (root mean square roughness) of the workpiece machined with these four tools are plotted in Figure 4.15. As the cutting distance increases, the surface roughness with all tools inevitably increases due to progressive tool wear. The fluctuating curves demonstrate the surface roughness achieved by these four tools has no visible differences while the cutting distance is less than 3.75 km, being less than 10 nm in Ra or 20 nm in Rms, but considerable anisotropy appears once the cutting distance goes beyond 3.75 km. The tool oriented with R(100) F(110) produces the worst surface

**TABLE 4.3**
**Micro-strengths in Various Directions of Three Typical Planes of Diamond [38] (Unit: GPa)**

| | "Soft" direction | | | "Hard" direction | | |
|---|---|---|---|---|---|---|
| | Tensile | Shearing | Compressive | Tensile | Shearing | Compressive |
| (110) | 1.40 | 0.70 | 2.10 | 17.2 | 8.59 | 25.8 |
| (100) | 4.11 | 8.21 | 12.3 | 25.7 | 51.3 | 77.0 |
| (111) | 1.73 | 13.8 | 15.5 | 27.7 | 221.0 | 248.7 |

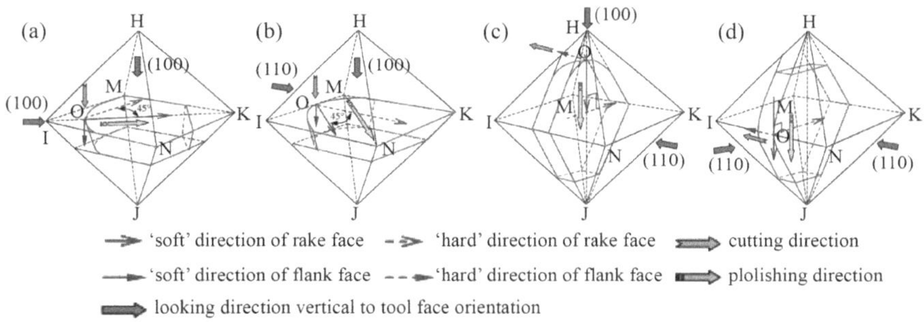


**FIGURE 4.14** Schematic diagram of diamond cutting tools with different orientations on rake and flank faces: (a) R(100) F(100), (b) R(100) F(110), (c) R(110) F(100), and (d) R(110) F(110) [40].

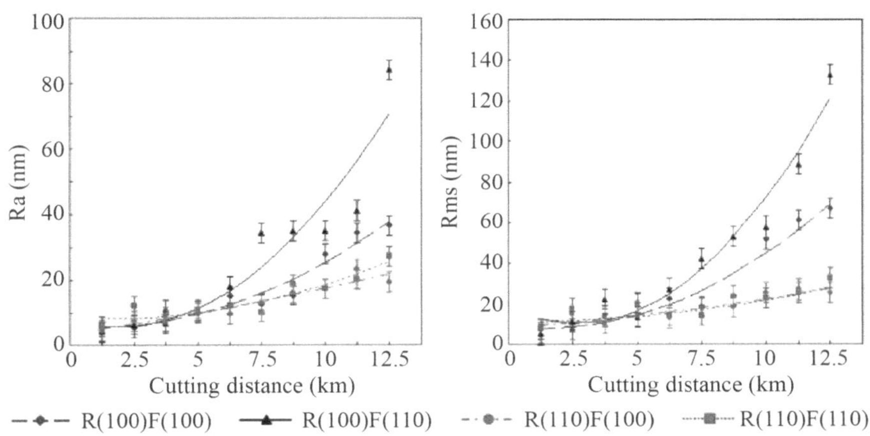


**FIGURE 4.15** Change of surface roughness versus cutting distance for various tools [40].

roughness, while the R(110) F(100) and R(110) F(110) oriented tools achieve a similar level of finish, both resulting in better surface roughness. As a result, the reliability of these four tools can be classified based on the machined surface roughness. Specifically, the R(110) F(100) and R(110) F(110) oriented tools exhibit the best surface roughness, followed by the R(100) F(100) oriented tool; and finally, the R(100) F(110) oriented tool is found to have the worst surface roughness. Considering the material properties listed in Table 4.3, it can be concluded that maximizing the tensile strength of the flank face along the cutting direction is crucial for ensuring the cutting tool's durability.

### 4.3.4 Edge Sharpness

Edge sharpness is one of the most important parameters for ultraprecision cutting where the edge radius effects become very significant and cannot be neglected as in traditional machining. Edge radius effects include many aspects. One of the effects is that the effective rake angle induced by the edge radius becomes negative, thus the material in front of the cutting edge is downward suppressed, and after the tool has passed, recovers due to elastic deformation. This suppressing/recovering deformation will cause machining errors in terms of the change in effective cutting depth. More discussion on cutting-edge radius will be presented in detail in Chapter 6.

A commercially available high-quality diamond tool usually has an edge radius of a few tens of nanometers. A deliberately polished diamond tool may achieve an edge radius in the order of several nanometers. However, the quantitative measurement of the edge radius of every tool in industry is technically difficult, thus only limited to research level. To characterize the edge radius of a cutting tool, an atomic force microscope (AFM) is typically used for direct measurement of the tool edge [41], or for measuring the replica of the tool edge after indenting it on an ultrasmooth copper [42]. Additionally, a scanning electron microscope with two electron detectors can also be employed to observe and measure the cross-sectional profile of a tool edge [43]. The methodologies of ultraprecision metrology will be introduced in detail in Chapter 7.

## 4.4 TOOL FABRICATION

To achieve high-precision machining, fabricating tools with high accuracy is an essential step. Among various tool materials, a diamond tool is the most difficult to fabricate due to its high hardness. Here, we mainly introduce the fabrication methods for diamond tools. Some of the methods are also useable for fabricating cBN and WC tools.

### 4.4.1 Mechanical Polishing

Mechanical polishing is the most popular method for achieving smooth diamond surfaces. The conventional mechanical polishing of diamond is conducted on a

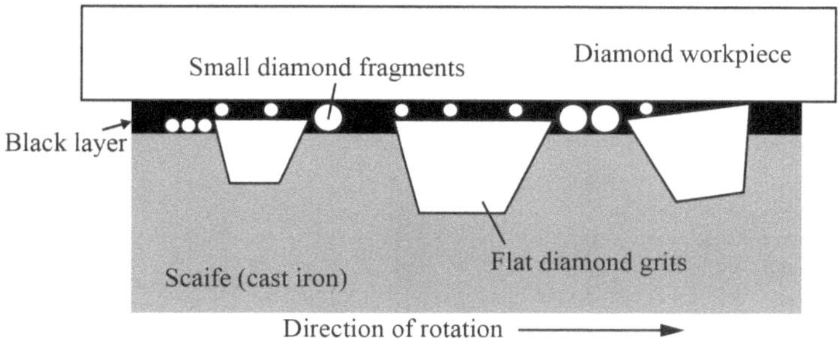


**FIGURE 4.16** Schematic of mechanical polishing process for diamond (not to scale).

rotating disk charged with diamond abrasive grains. The rotating disk is typically made of cast iron (scaife) and has a diameter of approximately 300 mm. The rotating speed of the disk is 2500 $min^{-1}$, and the applied contact pressure is several mega Pascals [44].

Figure 4.16 shows a schematic of the diamond polishing process. Prior to polishing, the scaife is cleaned with alcohol or methylated spirits and then dried. A thin film of olive oil is applied to the metal surface, with excess oil being wiped off. One to two carats (0.2–0.4 g) of diamond grits are then distributed on the scaife. The surface of the scaife should now appear uniformly grey. The scaife is then rotated at the polishing speed, and a piece of boart (a rough, or inferior quality diamond) is used to work the powder into the scaife, a procedure appropriately known as "boarting". At some points, the scaife surface transitions from its original matte grey color to a dark grey or black color. It is this change that a diamond polisher uses as an indicator that the scaife is now ready for polishing. The choice of grit size depends to some extent on the quality of polishing required. However, a typical grit size distribution would range from below 1 μm to around 20 μm. The flat diamond grit particles act as "anvils", allowing smaller diamond fragments present in the black layer to exert high pressure on the diamond workpiece. Additionally, the flat diamond grit particles can pivot to present sharp edges to the oncoming diamond workpiece [45].

Since single-crystal diamond is an anisotropic material, as mentioned in Section 4.3.3, the mechanical polishing direction needs to be carefully considered. In the case of polishing with a high material removal rate, the brittle-ductile transition in the <110> direction becomes the main factor affecting the surface quality. To achieve a balance between material removal rate and surface quality, it is advisable to control the polishing depth near the brittle-ductile transition in the <110> direction of the (100) single-crystal surface [46].

### 4.4.2 Thermo-Chemical Polishing

It is well known that thermo-chemical reactions between a diamond tool and a work material cause tool wear. The mechanisms of tool wear are classified into

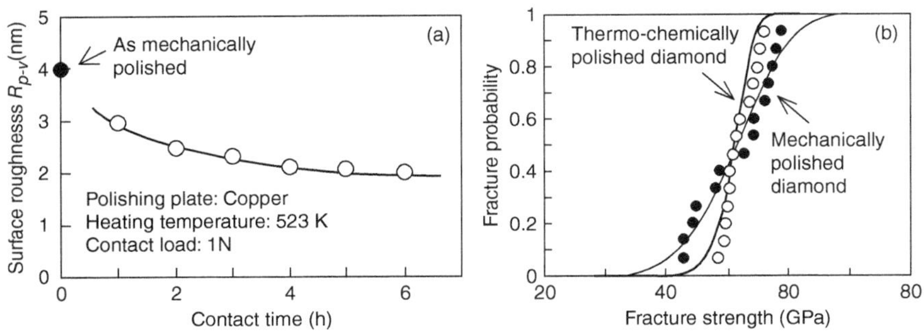


**FIGURE 4.17** (a) Variation in surface roughness of thermo-chemically polished diamond with contact time. (b) Cumulative distribution of Hertzian strength of diamonds polished thermo-chemically and mechanically [48].

three types, i.e., graphitization, oxidization–deoxidization reaction, and carbonization. This thermos-chemical reaction mechanism provides useful hints on how to polish diamond tools. The graphitization of diamond and carbide formation between the diamond and work material are not applicable to the polishing of diamonds, because the tool, polished by carbon diffusion, has low resistance to edge chipping [47]. However, it is possible to polish diamond tools based on the principle of oxidizing the work material and deoxidizing the oxidized work material with the diamond. Furushiro et al. [48] studied the oxide formation of copper and carbon based on thermodynamics and the equilibrium state of chemical reaction systems. When carbon atoms come into proximity with copper oxides, they deoxidize the copper oxides and themselves undergo oxidation. Therefore, by bringing a diamond into contact with copper oxides in air at temperatures higher than room temperature, it is possible to remove carbon atoms from the surface of the diamond.

As shown in Figure 4.17(a), thermo-chemical polishing decreases the surface roughness, because peaks on the diamond surface in contact with the copper plate are removed as carbon oxides, one after another. In addition to the better surface roughness, the thermo-chemically polished surface has improved ruggedness. Figure 4.17(b) shows the cumulative distribution of the Hertzian strength of diamonds polished thermo-chemically or mechanically. The scatter in the strength of the diamonds polished mechanically is attributed to microscopic cracks induced on the diamond surface. In contrast, the removal of the diamond surface layer by thermo-chemical polishing reduced the cracks, so the scatter in strength resulted in reduction of the fracture probability.

### 4.4.3 Ion Milling

Mechanically shaping of microtools is challenging due to the presence of forces. The slim body of the microtool will be easily broken by the force, especially if the tool material used is a brittle material. Errors will also arise because of elastic

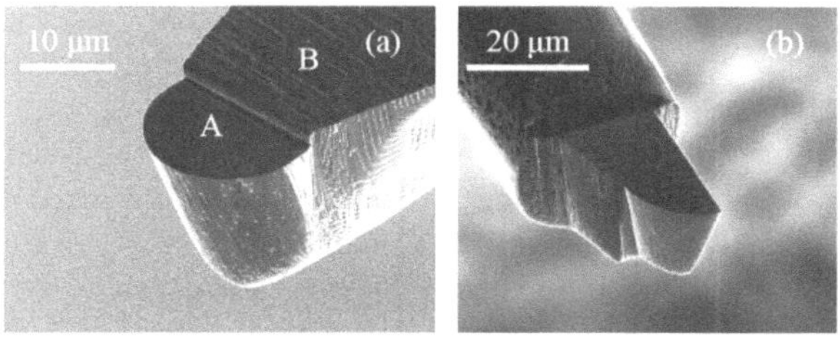


**FIGURE 4.18** Micro-tool fabricated by focused ion beam viewed at various directions and magnifications [51].

bending. As a result, currently the fabrication of cutting tools with diameters less than 50 micrometers is not commercially available [49].

Ion milling is a tool fabrication method based on the sputtering phenomenon of materials on a workpiece by the bombardment of energized ions projected at several hundred electron volts up to several kilo electron volts. Sputtering essentially involves knocking out surface atoms of a workpiece through kinetic momentum transfer from the projected ions to the surface atoms. Since there is no mechanical contact during tool fabrication, this technique can be applied to the machining of hard and brittle materials such as diamond. In 1980s, Miyamoto and Taniguchi [50] investigated the crystalline structure and surface topography of diamond processed by ion milling. They found the angular dependence of the sputtering yield and sputter-machining rate of diamond, as well as a method for predicting the desired profile. Such sputter-machining of diamonds has potential applications, for example, to the forming of sharp edges for diamond knives used in ultramicrotomy and ultra-precise diamond turning tools.

Figure 4.18 demonstrates a micro diamond tool fabricated by a focused ion beam. Tool facets have noticeable nanometer-size ripples when viewed at high magnification with scanning electron microscope, which is a characteristic of a sputtered surface. These ripples are aligned with the direction of the incident ion beam. The origin of this ripple formation can be traced to a surface instability caused by the competition between roughening (curvature-dependent sputtering) and smoothening (surface diffusion) processes [51]. Increasing the ion incident angle to greater than 40° leads to an increased sputtering rate. However, it also results in increased surface roughness due to the formation of more significant ripples.

### 4.4.4 Laser Machining

Laser is another effective method for fabricating tools, especially small tools or tools with small surface structures. Despite the hardness and brittleness of the tool material, a high-power laser beam can heat and vaporize the material selectively from the tool surface without forces. By using ultrashort pulses, such as picosecond and femtosecond pulses, the laser-induced thermal damage to the tool material can be minimized and a smooth and low-damage tool surface can be obtained. In recent years, laser has been used for tool shape creation, surface texturing of a tool, and edge profile modification, etc.

An example is given here on edge profile modification by microgrooving of the cutting edge of a diamond tool, as schematically shown in Figure 4.19(a). As an extremely hard and brittle material, single-crystal diamond cannot be grooved using conventional mechanical methods. Though a focused ion beam can be used to fabricate micro grooves, it is very expensive and time-consuming. Instead, laser machining is suitable for microgrooving of a diamond tool due to its high processing speed and relatively low cost. Figure 4.19(b) shows the cutting edge of a single-crystal diamond tool grooved using a picosecond pulsed laser. The depth of the grooves ranged from 3 to 5 μm. The dimensional variation of the groove depth may be due to self-focusing properties in the bulk material. As diamond is a transparent material at the laser wavelength used, laser machining mainly commences from impurities, crystal defects, and micro-cracks. Therefore, the size and depth of irradiation points depend on the distribution of such impurities and defects in the material [52].

Laser modification of cutting edges has also been validated on cBN tools, as shown in Figure 4.20. The depth of the grooves ($h$) is 11 μm, the width of the grooves ($w$) is 22 μm, and the pitch of the groove is 40 μm. Unlike diamond, cBN is a non-transparent material without self-focusing phenomenon. Therefore, uniform grooves were fabricated on the cBN tool edge using a femtosecond laser.

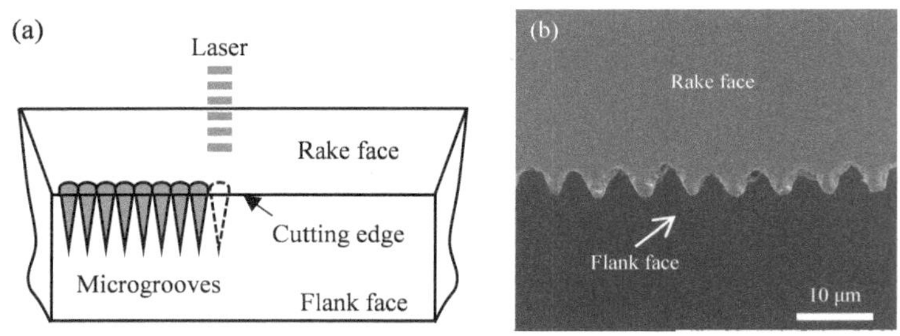


**FIGURE 4.19** (a) Schematic of laser grooving of a diamond tool and (b) SEM image of the edge of a grooved tool [53].

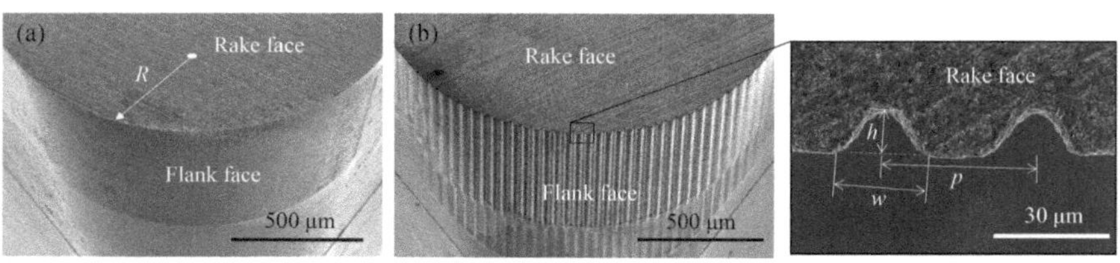


**FIGURE 4.20** Periodic V-grooves on the cutting edge machined by using a femtosecond pulsed laser: (a) before machining and (b) after machining [53].

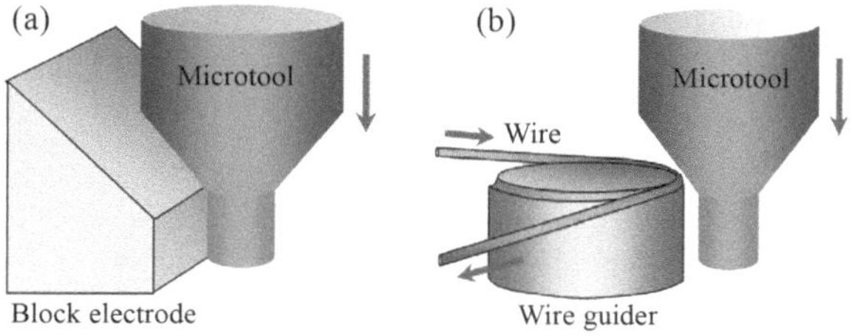


**FIGURE 4.21** Two major EDM methods for fabricating end mills using (a) block electrode and (b) wire electrode [54].

### 4.4.5 Electro-Discharge Machining

Electro-discharge machining (EDM) provides another effective way to micro tool fabrication. EDM is a noncontact process without problems of causing mechanical stress, tool bending/breaking, and chattering, but it requires the tool materials to be electrically conductive. EDM has been used for producing WC and PCD tools, as shown in Figure 4.12(c) and (e). The fabricated WC microtool has a diameter of 50 μm, with an edge radius of approximately 1 μm (Figure.4.12(c)). The spherical PCD microtool has a profile radius of 400 μm, with an edge radius of around 1.8 μm (Figure.4.12(e)).

Two major EDM methods can be used to fabricate a micro end mill. One is to use a shaped block electrode, as shown in Figure 4.21(a). When feeding the rotating tool against the block electrode, electro-discharges occur at the contact area and a reverse shape can then be transcribed to the tool. This method is suitable for fabricating cylindrical and conical tools having circular cross-sections. To make noncircular tools, the fabrication of the block electrode will be difficult. Even if the electrode has been precisely fabricated, during EDM the electrode will be unevenly worn, and its shape will change. As a result, the form accuracy of the tool is very difficult to guarantee. The other method is to use a fine wire as an electrode which is guided by a rotator, as shown in Figure 4.21(b). This method is termed wire EDM or wire electro-discharge grinding (WEDG). In contrast to block electrode EDM, the contact area between the wire electrode and the tool in wire EDM is extremely small, approaching a single point. By precisely controlling the relative movement between the wire guider and the tool, various complex shapes can be generated onto the tool. In wire EDM, since the wire is fed continuously, the region in contact with the tool is always new. Therefore, a highly precise tool can be fabricated, provided the EDM machine has a high repetitive accuracy and a high step resolution.

## REFERENCES

[1] Iwasaki, T., Aiso, T., Watari, K. (2019) Effect of Si in Workpiece Material on Tool Wear in Hard Turning. *ISIJ International* 59 (11): 2062–2072.

[2] Blake, P.N., Scattergood, R.O. (1990) Ductile-regime machining of germanium and silicon. *Journal of the American Ceramic Society* 73 (4): 949–957.
[3] Schinker, M.G. (1991) Subsurface damage mechanisms at high-speed ductile machining of optical glasses. *Precision Engineering* 13 (3): 208–218.
[4] Yan, J., Yoshino, M., Kuriagawa, T., Shirakashi, T., Syoji, K., Komanduri, R. (2001) On the ductile machining of silicon for micro electro-mechanical systems (MEMS), opto-electronic and optical applications. *Materials Science and Engineering: A* 297 (1): 230–234.
[5] Floyd, E.J., Theodore, T.S. (1977) Applications of Diamond Turning to Infrared Optical Systems. *Optical Engineering* 16 (4): 164387.
[6] Guo, J., Zhang, J., Pan, Y., Kang, R., Namba, Y., Shore, P., Yue, X., Wang, B., Guo, D. (2020) A critical review on the chemical wear and wear suppression of diamond tools in diamond cutting of ferrous metals. *International Journal of Extreme Manufacturing* 2 (1): 012001.
[7] Yan, J., Syoji, K., Tamaki, J. (2003) Some observations on the wear of diamond tools in ultra-precision cutting of single-crystal silicon. *Wear* 255 (7–12): 1380–1387.
[8] Zhang, L., Hashimoto, T., Yan, J. (2021) Machinability exploration for high-entropy alloy FeCrCoMnNi by ultrasonic vibration-assisted diamond turning. *CIRP Annals* 70 (1): 37–40.
[9] Obata, K. (2016) Single-crystal diamond cutting tool for ultra-precision processing. *SEI Tech. Rev* 82: 82–88.
[10] Yan, J., Zhang, Z., Kuriyagawa, T., Gonda, H. (2010) Fabricating micro-structured surface by using single-crystalline diamond endmill. *The International Journal of Advanced Manufacturing Technology* 51 (9): 957–964.
[11] Sumiya, H. (2012) Novel development of high-pressure synthetic diamonds "Ultra-hard Nano-polycrystalline Diamonds". *SEI Technical Review* 74: 15–23.
[12] Mizumoto, Y., Amano, H., Kangawa, H., Harano, K., Sumiya, H., Kakinuma, Y. (2018) On the improvement of subsurface quality of CaF2 single crystal machined by boron-doped nano-polycrystalline diamond tools. *Precision Engineering* 52: 73–83.
[13] Harano, K., Satoh, T., Sumiya, H. (2012) Cutting performance of nano-polycrystalline diamond. *Diamond and Related Materials* 24: 78–82.
[14] Kato, H., Shintani, K., Sumiya, H. (2002) Cutting performance of a binder-less sintered cubic boron nitride tool in the high-speed milling of gray cast iron. *Journal of Materials Processing Technology* 127 (2): 217–221.
[15] Nishiguchi, T., Masuda, M. (1988) Precision Cutting of Ferreous Metals with Single Crystal CBN Cutting Tools. *Journal of the Japan Society for Precision Engineering* 54 (2): 384–389.
[16] Neo, K.S., Rahman, M., Li, X.P., Khoo, H.H., Sawa, M., Maeda, Y. (2003) Performance evaluation of pure CBN tools for machining of steel. *Journal of Materials Processing Technology* 140 (1): 326–331.
[17] Fan, W., Ji, W., Wang, L., Zheng, L., Wang, Y. (2020) A review on cutting tool technology in machining of Ni-based superalloys. *The International Journal of Advanced Manufacturing Technology* 110 (11): 2863–2879.
[18] Ezugwu, E.O., Wang, Z.M. (1997) Titanium alloys and their machinability—a review. *Journal of Materials Processing Technology* 68 (3): 262–274.
[19] Wood, R.J.K., Herd, S., Thakare, M.R. (2018) A critical review of the tribocorrosion of cemented and thermal sprayed tungsten carbide. *Tribology International* 119: 491–509.
[20] Abrão, A.M., Aspinwall, D.K., Wise, M.L.H. (1993). A review of polycrystalline cubic boron nitride cutting tool developments and application. In: A. K. Kochhar (eds) *Proceedings of the Thirtieth International MATADOR Conference: held in Manchester 31st March – 1st April 1993*. Macmillan Education UK, London.

[21] Sadik, M.I., Coronel, E., Lattemann, M. (2019) Influence of characteristic properties of PCD grades on the wear development in turning of β-titanium alloy (Ti5Al5V5Mo3Cr). *Wear* 426–427: 1594–1602.
[22] Konyashin, I.Y. (1995) PVD/CVD technology for coating cemented carbides. *Surface and Coatings Technology* 71 (3): 277–283.
[23] Ciftci, I. (2006) Machining of austenitic stainless steels using CVD multi-layer coated cemented carbide tools. *Tribology International* 39 (6): 565–569.
[24] Nagayama, K., Yan, J. (2021) Deterministic error compensation for slow tool servo-driven diamond turning of freeform surface with nanometric form accuracy. *Journal of Manufacturing Processes* 64: 45–57.
[25] Yan, J.W., Tamaki, J., Syoji, K., Kuriyagawa, T. (2004) Ductile regime machining of single-crystal CaF2 for aspherical lenses. *Key Engineering Materials* 257–258: 95–100.
[26] Chen, Y., Sun, L., Wu, Y. (2022) Improving the machining performance in single-point diamond turning of curved Zerodur optics by using straight-nosed cutting tools. *Journal of Materials Processing Technology* 310: 117777.
[27] Jasinevicius, R.G., Duduch, J.G., Cirino, G.A., Pizani, P.S. (2013) Diamond turning of small Fresnel lens array in single crystal InSb. *Journal of Micromechanics and Microengineering* 23 (5): 055025.
[28] Gao, X., Xue, C., Chao, Y., Liu, Y. (2021) Optimization method of manufacturing for diamond turning soft-brittle materials' harmonic diffractive optical elements. *Applied Optics* 60 (1): 162–171.
[29] Yan, J., Maekawa, K., Tamaki, J., Kuriyagawa, T. (2005) Micro grooving on single-crystal germanium for infrared Fresnel lenses. *Journal of Micromechanics and Microengineering* 15 (10): 1925–1931.
[30] Yan, J., Syoji, K., Tamaki, J.I. (2003) Crystallographic effects in micro/nanomachining of single-crystal calcium fluoride. *Journal of Vacuum Science & Technology B: Microelectronics and Nanometer Structures Processing, Measurement, and Phenomena* 22 (1): 46–51.
[31] Matsumura, T. (2018) Micro milling for functional surface. In: J. Yan (eds) *Micro and Nano Fabrication Technology*. Micro/Nano Technologies. Springer, Singapore.
[32] Chen, N., Li, H.N., Wu, J., Li, Z., Li, L., Liu, G., He, N. (2021) Advances in micro milling: From tool fabrication to process outcomes. *International Journal of Machine Tools and Manufacture* 160: 103670.
[33] Fang, F.Z., Wu, H., Liu, X.D., Liu, Y.C., Ng, S.T. (2003) Tool geometry study in micromachining. *Journal of Micromechanics and Microengineering* 13 (5): 726–731.
[34] Aurich, J.C., Reichenbach, I.G., Schüler, G.M. (2012) Manufacture and application of ultra-small micro end mills. *CIRP Annals* 61 (1): 83–86.
[35] Ali, M.Y., Ong, A.S. (2006) Fabricating micromilling tool using wire electrodischarge grinding and focused ion beam sputtering. *The International Journal of Advanced Manufacturing Technology* 31 (5): 501–508.
[36] Yan, J., Uchida, K., Yoshihara, N., Kuriyagawa, T. (2009) Fabrication of micro end mills by wire EDM and some micro cutting tests. *Journal of Micromechanics and Microengineering* 19 (2): 025004.
[37] Zhang, Z., Peng, H., Yan, J. (2013) Micro-cutting characteristics of EDM fabricated high-precision polycrystalline diamond tools. *International Journal of Machine Tools and Manufacture* 65: 99–106.
[38] Zong, W.J., Li, Z.Q., Sun, T., Li, D., Cheng, K. (2010) Analysis for the wear resistance anisotropy of diamond cutting tools in theory and experiment. *Journal of Materials Processing Technology* 210 (6): 858–867.
[39] Yan, J., Syoji, K., Kuriyagawa, T. (2000) Ductile-brittle transition at large negative tool rake angles. *Journal of the Japan Society for Precision Engineering* 66 (7): 1130–1134.

[40] Zong, W.J., Sun, T., Li, D., Cheng, K. (2009) Design criterion for crystal orientation of diamond cutting tool. *Diamond and Related Materials* 18 (4): 642–650.
[41] Zhang, S., Zong, W. (2020) Micro defects on diamond tool cutting edge affecting the ductile-mode machining of KDP crystal. *Micromachines* 11 (12): 1102.
[42] Dodmani, A., Subbiah, S. (2022) Accurate measurement of cutting-edge radius on a single-crystal diamond tool. *Manufacturing Letters* 33: 469–478.
[43] Ohta, T., Yan, J., Yajima, S., Takahashi, Y., Horikawa, N., Kuriyagawa, T. (2007) High-efficiency machining of single-crystal germanium using large-radius diamond tools. *International Journal of Surface Science and Engineering* 1 (4): 374–392.
[44] Kubota, A., Nagae, S., Motoyama, S. (2020) High-precision mechanical polishing method for diamond substrate using micron-sized diamond abrasive grains. *Diamond and Related Materials* 101: 107644.
[45] Hird, J.R., Field, J.E. (2004) Diamond polishing. Proceedings of the Royal Society of London. *Series A: Mathematical, Physical and Engineering Sciences* 460 (2052): 3547–3568.
[46] Xin, Y., Lu, J., Li, C., Luo, Q., Li, Z., Ke, C., Xu, X. (2024) Effects of processing parameters and grinding direction on the material removal mechanism of (100) surface single crystal diamond in self-rotating mechanical grinding. *Diamond and Related Materials* 141: 110685.
[47] Furushiro, N., Higuchi, M., Yamaguchi, T., Sugimoto, T., Shimada, S. (2010) Thermo-Chemical polishing of diamond cutting tool based on carbon diffusion (machine elements, design and manufacturing). *Transactions of the Japan Society of Mechanical Engineers Series C* 76 (764): 974–980.
[48] Furushiro, N., Higuchi, M., Yamaguchi, T., Shimada, S., Obata, K. (2009) Polishing of single point diamond tool based on thermo-chemical reaction with copper. *Precision Engineering* 33 (4): 486–491.
[49] Zhang, S.J., Fang, F.Z., Xu, Z.W., Hu, X.T. (2009) Controlled morphology of microtools shaped using focused ion beam milling technique. *Journal of Vacuum Science & Technology B: Microelectronics and Nanometer Structures Processing, Measurement, and Phenomena* 27 (3): 1304–1309.
[50] Miyamoto, I., Taniguchi, N. (1982) Polishing and sharpening of diamond point tools by ion sputter-machining. *Precision Engineering* 4 (4): 191–194.
[51] Ding, X., Lim, G.C., Cheng, C.K., Butler, D.L., Shaw, K.C., Liu, K., Fong, W.S. (2008) Fabrication of a micro-size diamond tool using a focused ion beam. *Journal of Micromechanics and Microengineering* 18 (7): 075017.
[52] Takayama, N., Ishizuka, J., Yan, J. (2018) Microgrooving of a single-crystal diamond tool using a picosecond pulsed laser and some cutting tests. *Precision Engineering* 53: 252–262.
[53] Kitagawa, D., Meshram, T., Yan, J. (2023) Microgrooving of ferrous metals by using cBN tools with laser-modified edges. *Journal of the Japan Society for Abrasive Technology* 67 (6): 341–346.
[54] Yan, J., Uchida, K., Yoshihara, N., Kuriyagawa, T. (2009) Fabrication of micro end mills by wire EDM and some micro cutting tests. *Journal of Micromechanics and Microengineering* 19 (2): 025004.

# 5 Machining Mechanism and Machinability of Ductile Materials

## 5.1 INTRODUCTION

The ease and/or difficulty of machining a certain material to achieve the required surface quality is usually described using the term machinability. Materials with good machinability are easy to obtain high surface quality with low power to cut, high machining efficiency, and little wear on cutting tools. The machinability of materials is varied with the material's properties (mainly mechanical and thermal properties), such as microstructure, micro hardness, fracture toughness, yield strength, thermal conductivity, thermal expansion, melting points, and chemical composition.

Most metals, such as gold (Au), silver (Ag), platinum (Pt), nickel (Ni), copper (Cu), aluminum (Al) and their alloys, low-carbon steel, and certain polymer materials such as polyethylene (PE) and polycarbonate (PC) are ductile materials which exhibit strong plastic deformation under tensile stress. In other words, they have high fracture toughness and yield strength, which contributes to a good machinability in cutting. However, generally speaking, the factors that improve a material's performance often degrade its machinability. As more and more new materials with high performances are being developed, finding new approaches for improving their machinability without harming their performance is a crucial task for manufacturing engineers.

Clarifying the machining mechanism is an essential step to optimize a machining process. As a machining operation has so many variables, it is very difficult to accurately predict the machining characteristics of every material under a specific machining condition. For this reason, a huge number of machining tests are necessary to establish a database of machinability for various materials. For conventional metal cutting, machinability can be characterized by tool life, usable cutting speed, cutting force, cutting temperature, surface roughness, and ease of chip disposal. In ultraprecision machining, as the machining scale is extremely small, the cutting force is small, cutting temperature is low, chip disposal is easy, as well as cutting speed is not too high, only tool wear and surface roughness are regarded as the important factors in evaluating the machinability.

DOI: 10.1201/9781003292388-5

## 5.2 MACHINING MECHANISM

### 5.2.1 Chip Formation

In macro-machining, the depth of cut is far bigger than the edge radius of a cutting tool. The cutting edge is even considered to be ideally sharp without an edge radius, and it is assumed that the cutting tools can completely remove the required depth of material from the surface of workpiece. Figure 5.1 illustrates a schematic of orthogonal cutting, which is widely used to describe a macro-cutting process, and the Merchant's model is usually used for analyzing the cutting forces in orthogonal macro-cutting [1,2].

In micro/nanoscale machining, however, the tool edge roundness cannot be neglected. This is because a commercially available cutting tool has an edge radius also in the micro/nano level (for single-crystal diamond ~50 nm, for tungsten carbide tools ~1 μm), which is comparable to the machining scale. Figure 5.2 presents a schematic illustrating orthogonal micro/nano-cutting. Due to the edge roundness, there is a critical point A in the cutting zone. The upper part of material above point A is piled up to form a chip, while the lower part of material under point A is suppressed downwards in front of the tool, causing significant elastic/

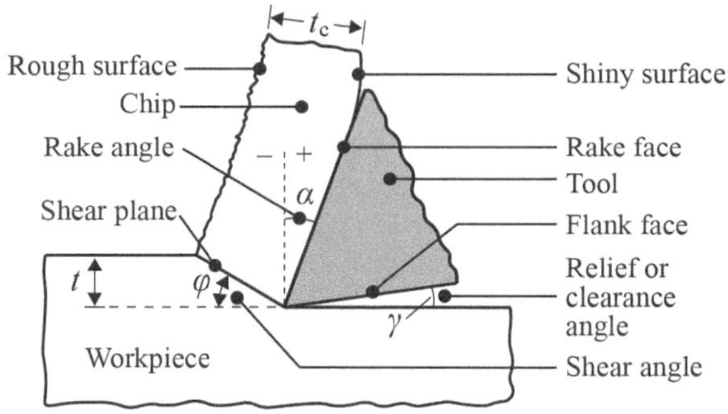


**FIGURE 5.1** Schematic of orthogonal macro-cutting.

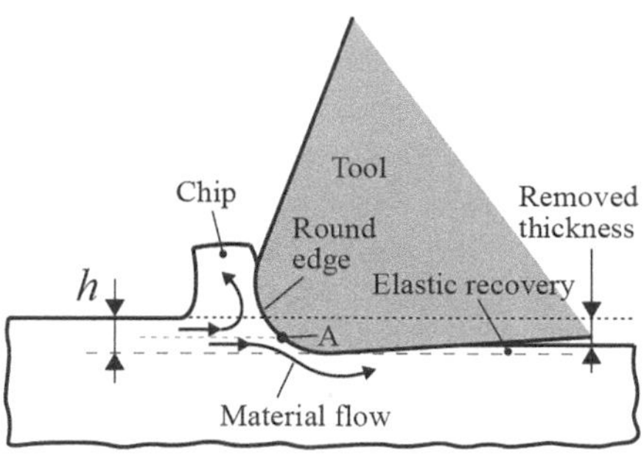


**FIGURE 5.2** Schematic of orthogonal micro/nano-cutting.

plastic deformation. Then after the tool passes, elastic recovery takes place, which causes an error in actual depth of cut, i.e., the actual thickness of the removed material is smaller than the ideal undeformed chip thickness $h$, and the difference between the two is the elastic recovery.

When the undeformed chip thickness becomes extremely small, the elastic recovery may be equal to the ideal $h$, and the actual removal thickness is zero. In other words, chip formation may not occur, as shown in Figure 5.3(a). In this case, a highly negative effective tool rake angle dominates the machining process by rubbing and compressing the workpiece surface instead of cutting [3]. This phenomenon causes surface roughening and elastic recovery. Therefore, a minimum chip thickness $h_{min}$ exists. The undeformed chip thickness must be larger than this critical chip thickness to form a chip. When $h$ increases beyond $h_{min}$, as shown in Figure 5.3(b), although elastic deformation still occurs, chips are formed by shearing or extrusion of material from the workpiece; a better surface finish will be consequently obtained. The elastic deformation decreases as the $h$ increases, approaching a situation where the material within the entire undeformed chip thickness is removed as a chip [4].

In general, the minimum chip thickness is determined by tool edge radius, material property, and lubricating condition. As the minimum chip thickness is critical for the process modeling, planning, and optimization, quantitative measurement of the minimum chip thickness attracted much attention many years ago. Ikawa et al. [5,6] explored measuring the minimum thickness of the cut of copper and aluminum by diamond turning using a well-defined diamond tool and found that at a depth of cut of 1 nm, chip formation was possible. Figure 5.4 shows SEM observations of chips of copper where the stable shear-banded chip is generated even at a nanometric uncut chip thickness. The chips are corrugated and folded due to the intermittent buckling of chip. Judging from the facts that the thickness, lamellar spacing, and folding interval of chips continuously decrease with the uncut chip thickness, the mechanism of nanometric chip removal process seems to be similar to that of conventional cutting. Molecular dynamics simulations revealed that, with perfect machine performance, the minimum thickness of

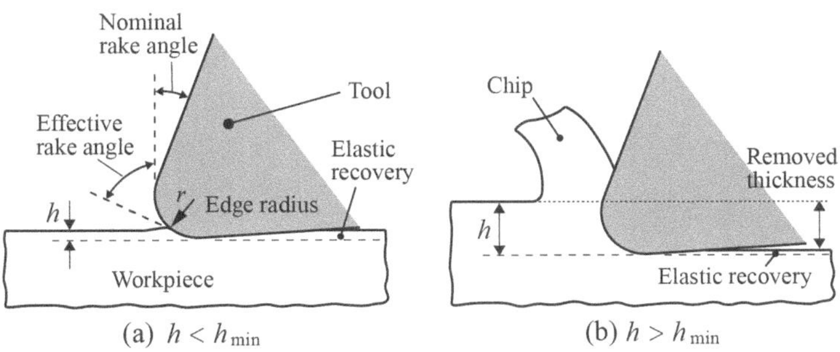


**FIGURE 5.3** Schematic of the minimum chip thickness hmin: (a) $h$ is smaller than $h_{min}$, and (b) $h$ is larger than hmin.

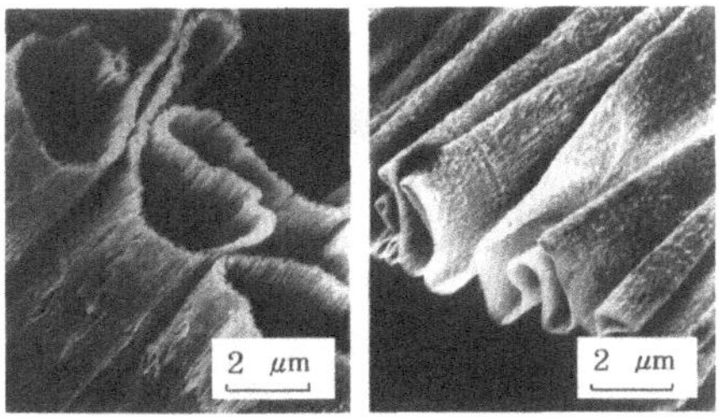


**FIGURE 5.4** SEM micrographs of chips at a nominal thickness of cut of (a) 1 nm; (b) 30 nm [7].

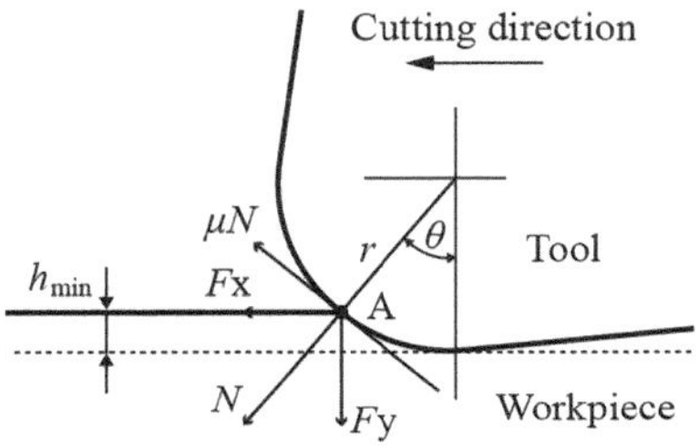


**FIGURE 5.5** The relationship of the minimum cutting thickness to the tool edge radius and cutting forces.

the cut was strongly influenced by the sharpness of the cutting edge. The minimum thickness of the cut could be in the order of 1/20 to 1/10 of the edge radius.

Yuan et al. [8] analyzed the relationship of the minimum cutting thickness to the tool edge radius and cutting forces, as shown in Figure 5.5. If the uncut surface contacts the tool at the critical point A, the value of minimum cutting thickness can be given by:

$$h_{min} = r\left(1 - \cos\theta\right) = r\left(1 - \left(1 + \tan\theta\right)^{-1/2}\right) \tag{5.1}$$

There are two forces acting at point A, i.e., horizontal force $F\mathrm{x}$ and vertical force $F\mathrm{y}$. These two forces can be divided into normal force $N$ and tangential force $\mu N$,

$$N = F\mathrm{y} \cdot \cos\theta + F\mathrm{x} \cdot \sin\theta \tag{5.2}$$

$$\mu N = F\mathrm{x} \cdot \cos\theta - F\mathrm{y} \cdot \sin\theta \tag{5.3}$$

where $\mu$ is the friction coefficient between the cutting tool and the workpiece material.

From Eqs. (5.2) and (5.3),

$$\tan\theta = \frac{Fx - \mu Fy}{\mu Fx + Fy} \tag{5.4}$$

and substituting (5.4) into (5.1), $h_{min}$ can be expressed by:

$$h_{min} = r\left(1 - \frac{Fy + \mu Fx}{\sqrt{\left(Fx^2 + Fy^2\right)\left(1 + \mu^2\right)}}\right) \tag{5.5}$$

It can be seen from Eq. (5.5) that if the radius of the cutting edge $r$ is a given value, the minimum cutting thickness depends on the ratio of $Fy/Fx$ and the friction coefficient $\mu$ between the tool and the workpiece, which are affected by material properties and lubrication conditions.

Liu et al. [9] developed a model to predict normalized minimum chip thickness, which accounts for the effects of thermal softening and strain hardening on the minimum chip thickness, as well as the effects of cutting speed and cutting edge radius on the minimum chip thickness. Various carbon steels with different carbon contents and aluminum alloy Al6082-T6 were used as workpieces, and the machining tests were performed under a range of cutting velocities and tool edge radii. Results showed that the minimum chip thickness increased as the cutting velocity and tool edge radius increased when machining carbon steels. This was attributed to the predominance of the thermal softening effect over the strain hardening effect. The higher the carbon content, the larger the normalized minimum chip thickness. However, the normalized minimum chip thickness was found to stay almost constant over a range of cutting velocities and tool edge radii when machining Al6082-T6. This was because the thermal softening effect and strain hardening effect were equally important and canceled out each other.

### 5.2.2 Cutting Force and Energy

The cutting force characteristics contain a great deal of information about the cutting process. In the past decades, cutting force measurements in micro/nano machining of a variety of materials at different cutting conditions have been performed. Furukawa and Moronuki [10] observed that when the depth of cut decreases, the specific energy (i.e., cutting force per unit area of cut) increases exponentially with the smaller depth of cut, as shown in Figure 5.6. This phenomenon is also known as "size effect", that is, at a small depth of cut, though the cutting force is small, the pressure is high.

The size effect is unaccountable for the conventional sharp-edge cutting model (see Figure 5.1) assuming that the resultant force is affected only by shear along one shear plane and friction at the rake face [1]. In micro-machining, as the depth of cut

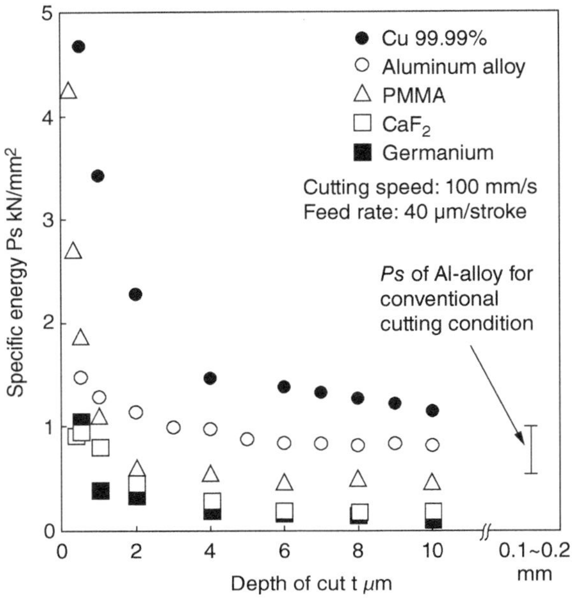


**FIGURE 5.6** Specific energy in micro-cutting of various materials [10].

becomes of the same order as the tool edge radius the assumption that the tool edge is perfectly sharp is no longer valid and micro-machining may involve significant sliding along the flank face of the tool due to the elastic recovery of the workpiece material. In other words, sliding along the flank face of the tool due to the elastic recovery of the workpiece and plowing due to the negative effective rake angle resulting from the tool edge radius become important in micro-machining [11]. This effect in micro-machining may increase tool wear and worse machined surface quality.

Taminiau and Dautzenberg [12] witnessed an increase in specific energy while machining brass with decreased chip thickness. They found that the specific energy depended only on the ratio of the uncut chip thickness to the cutting edge radius when the uncut chip thickness was smaller than the edge radius. Similarly, Lucca et al. [13] pointed out that the specific energy at an extremely small chip thickness will be greatly higher than the total energy required for shearing and for overcoming friction on the rake face of the tool. The significance of plowing and elastic recovery is a major cause of specific energy. When the uncut chip thickness approached the size of the edge radius, the effective rake angle determined the resulting forces and dissipated energy. Then, they investigated the transition from a shearing-dominated process to a plowing-dominated process in diamond turning by studying the angle of the resultant cutting forces in orthogonal cutting [14]. Results indicated that, as shown in Figure 5.7, as the uncut chip thickness decreased, the forces per unit width in the cutting and thrust directions became smaller, and the reduction of the force in the cutting direction was more significant. As a result, the resultant force gradually deflected to the thrust direction,

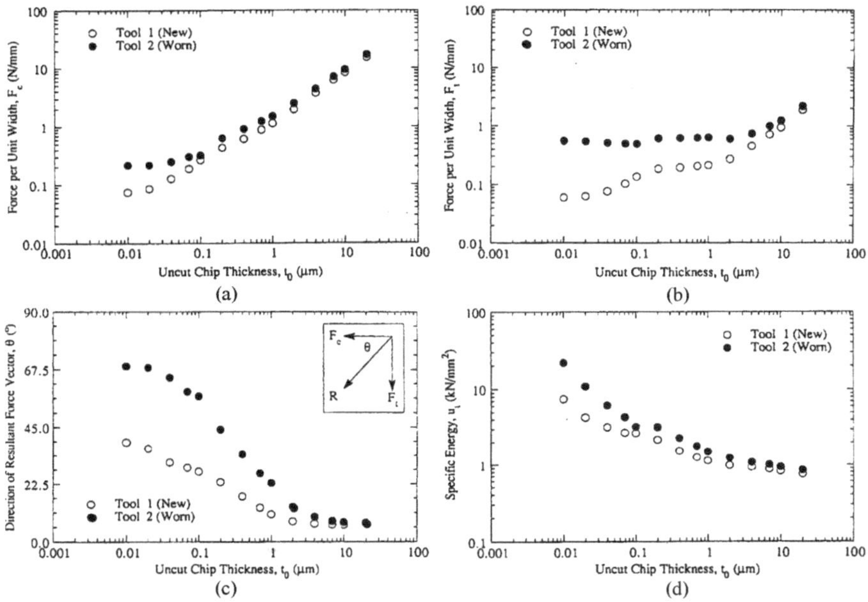


**FIGURE 5.7** Comparison of the (a) force in cutting direction, (b) force in thrust direction, (c) direction of resultant force, and (d) specific energy obtained for new and worn tools with the same overall tool geometry [14].

which means that the force in the cutting direction that contributed to material removal tended to become less dominant. Therefore, for the same volume of material removed, the energy consumed also became greater. In addition, the tool edge condition (tool wear) was found to have a significant effect on the thrust forces when the depth of cut was below the tool edge radius. Thus, a worn tool caused a higher specific energy.

### 5.2.3 Surface Roughness

Surface roughness is one of the major parameters to describe the quality of a produced surface by cutting processes. It depends on the machine tool accuracy and stiffness (which determine the stability of extremely fine tool feed and depth of cut), the tool geometry (nose radius, rake angle, relief angle, edge sharpness, and tool wear), the properties of workpiece material, and the interface between the tool and workpiece (friction coefficient, lubrication consistency).

If a surface is perfectly machined based on the movement provided by the machine tool, the surface roughness can be simplified as a geometric issue in terms of tool-workpiece interference. This theoretical surface roughness is determined by tool geometry and cutting parameters. Here, we consider cutting models with two kinds of simple tool geometries: straight- and round-nosed tools. The tool-workpiece interference models of straight- and round-nosed tools are shown in Figure 5.8 and Figure 5.9, respectively.

As shown in Figure 5.8, for a straight-nosed tool without nose radius, surface roughness is the maximum height (peak to valley, $R_{max}$) of the scallops, and can be given by:

$$R_{max} = \frac{f}{\cot\alpha + \cot\beta} \tag{5.6}$$

where $d$ is the depth of cut, $f$ is the tool feed, $\alpha$ is the major cutting edge angle, and $\beta$ is the minor cutting edge angle.

For a round-nosed tool with a nose radius $R$, as shown in Figure 5.9, the maximum height of the scallops is given by:

$$R_{max} = R - \left(R^2 - \frac{f^2}{4}\right)^{\frac{1}{2}} \approx \frac{f^2}{8R} \tag{5.7}$$

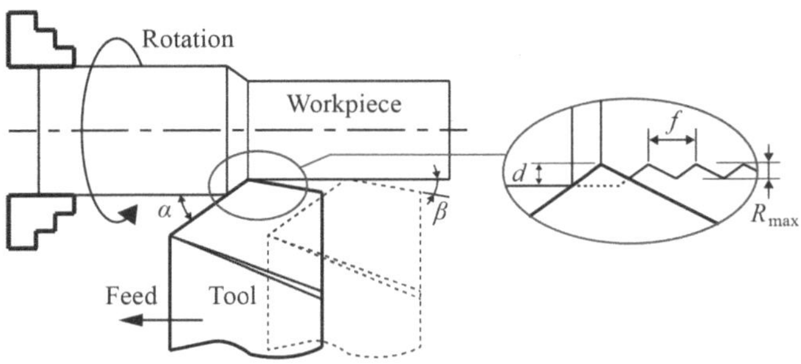


**FIGURE 5.8** Tool-workpiece interference model of straight-nosed tools.

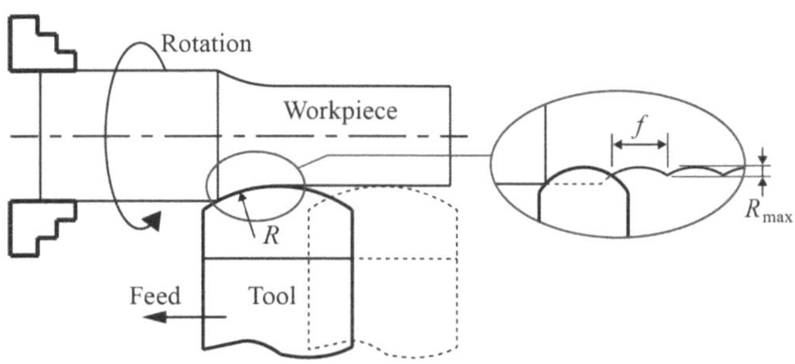


**FIGURE 5.9** Tool-workpiece interference model of round-nosed tools.

For example, when cutting with a straight-nosed tool with major/minor cutting edge angles of $\alpha = \beta = 45°$ at a tool feed of 10 μm/rev, $R_{max}$ will be 5 μm according to Eq. (5.6). In the case of cutting with a round-nosed tool with a nose radius of 1 mm at a tool feed of 10 μm/rev, $R_{max}$ will be 0.0125 μm or 12.5 nm, according to Equation (5.7). Generally speaking, a round-nosed tool with large radius is beneficial in terms of low surface roughness compared to a straight-nosed tool. To obtain a smooth surface with straight-nosed tools, it is important to use sufficiently small cutting-edge angles $\alpha$ and $\beta$ [15].

However, in practical machining processes, the surface finish is also affected by the following factors, causing the actual surface roughness is always higher than the theoretical surface roughness:

a. Dynamic errors of the machine tool, including the instability and vibration of the spindles and slide tables

In ultra-precision machining, the performance of the machine tool directly determines the achievable shape accuracy and surface roughness of the machined parts, which is the fundamental factor affecting the surface generation. Currently, air bearing spindles, pneumatic/hydrostatic slides, laser position feedback with nanometric resolution, and active vibration cancellation system are widely used in ultra-precision machine tools. These techniques provide significant improvements in the performance of the machine tool, enabling high accuracy and low surface roughness [16].

b. Processing parameters, including cutting speed, tool feed, and depth of cut

Cheung and Lee [17–19] performed micro-cutting experiments on aluminum alloy and identified that the machined surface roughness decreases with increasing spindle speed and increases with increasing feed rate, while no evident relationships were found between depth of cut and surface roughness, as shown in Figure 5.10(a)–(c). It is worth noting that the machined surface roughness is large due to the formation of build-up edge (BUE) at a low cutting speed, while at high cutting speeds, no BUE is generated and the surface roughness is small. According to Eqs. (5.6) and (5.7), the smaller the feed rate, the smaller the theoretical surface roughness. However, if the feed rate is reduced so that the undeformed chip thickness is lower than the minimum chip thickness, the machined surface roughness increases. Although the depth of cut is not directly related to surface roughness, it has a quasi-linear proportion with cutting force, which indirectly influences surface roughness via heat generation, elastic-plastic deformation, vibration, and so on. In addition, the depth of cut also determines the chip thickness, thus if it is smaller than a critical value, the actual chip thickness becomes less than the minimum chip thickness, which causes plowing starts to make a principal contribution to material removal, and consequently, the machined surface roughness increases reversely [7,8].

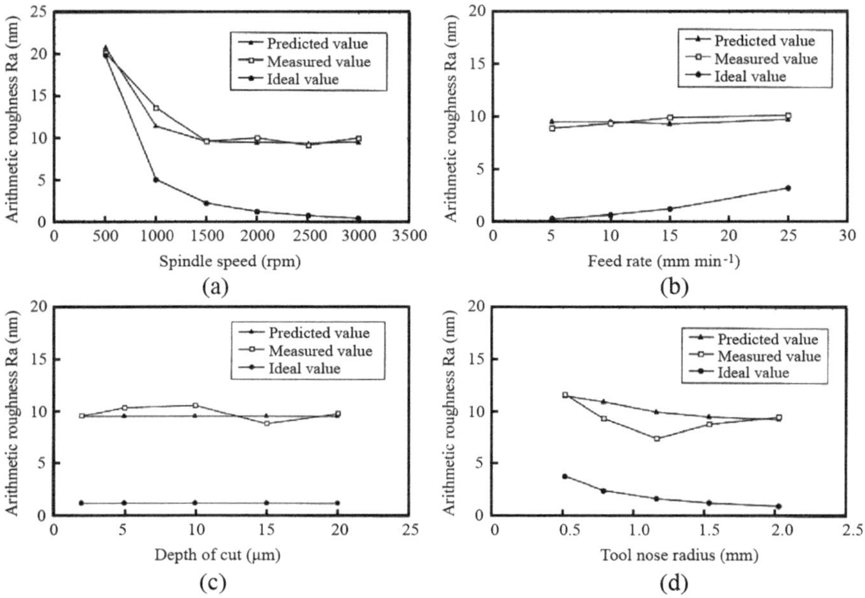


**FIGURE 5.10** Arithmetic roughness of the machined aluminum alloy surface at different (a) spindle speed, (b) feed rate, (c) depth of cut and (d) nose radius [17].

c. Cutting tool geometrical conditions, including initial geometry and tool wear

In ultraprecision machining, single-crystal diamond tools are considered ideal cutting tools. The geometrical parameters, such as tool nose radius, profile error, edge radius, rake angle and clearance angle, have a significant influence upon chip formation, heat generation, and tool wear which in turn influence surface roughness [20]. According to Eq. (5.7), the larger the tool nose radius, the smaller the theoretical surface roughness. However, Cheung and Lee [17] pointed out that the surface roughness decreases with increasing tool nose radius within a certain range, and then increases as the tool nose radius further increases, as shown in Figure 5.10(d). Tool edge radius means the sharpness of a tool. According to Eq. (5.5), the smaller edge radius, the smaller the minimum chip thickness. Therefore, the rubbing and plowing effects are less likely to occur when using a sharp tool, resulting in a better surface finish [21]. Tool rake angle is another important feature parameter influencing machined surface roughness and material removal. For machining brittle materials, a negative rake angle is preferred to create hydrostatics pressure in the material and prevent surface cracking. For machining ductile materials, however, a positive rake angle is preferred to achieve better surface quality, because a positive rake angle tool facilitates the reduction of cutting force [22] and cutting temperature [23], and increase of the shear angle [24]. Tool clearance angle (also known as tool relief angle) determines the contact of machined surface with the tool flank face. An excessively large relief angle can lead to tool chipping, whereas an overly small angle may cause increased friction due to rubbing between the tool flank surface and the machined surface, resulting in excessive flank wear.

Micro/nanoscale wear of the cutting-edge tools also has a great impact on surface roughness. Tool wear is inevitable and dynamically changes in the cutting process, and the wear patterns are dependent on material hardness and cutting distance. The major types of tool wear include abrasive wear, which includes crater wear, flank wear, and edge chippings. Depending on the types of the wear pattern, the worn tool may produce a blunter edge radius (Figure 5.11(a)), larger contact area between the tool flank face and machined surface (Figure 5.11(b)), or unstable tool-workpiece contact at the cutting area (Figure 5.11(c)), which roughens the surface. Tool wear can affect surface roughness either directly or indirectly through other factors. For example, tool wear causes cutting forces to increase, and those forces promote vibration in the cutting process, which further worsens surface roughness [16].

d. Workpiece material properties, such as mechanical and physical properties and material heterogeneity

For different isotropic materials, the machined surface roughness depends on the mechanical properties of the materials. For example, as shown in Figure 5.12, copper alloy had a higher surface roughness than aluminum alloy due to material swelling and recovery.

For anisotropic materials, such as single-crystal materials, the chip morphology, shear angle, cutting forces, and lamellae spacing greatly vary with crystallographic orientation. As a result, the anisotropy of materials induces the variation of surface roughness. For example, as shown in Figure 5.13, the surface roughness of an anisotropic aluminum single crystal varies with cutting direction and exhibits fourfold symmetry due to the crystal symmetry, while isotropic aluminum alloy 6061 shows a constant surface roughness along various cutting directions.

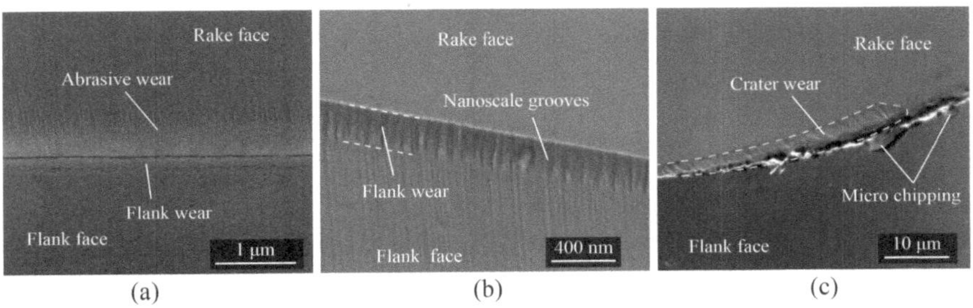


**FIGURE 5.11** SEM images of diamond tool wear after cutting (a) ZnSe [25], (b) SiC [26], and (c) $CaF_2$ [27].

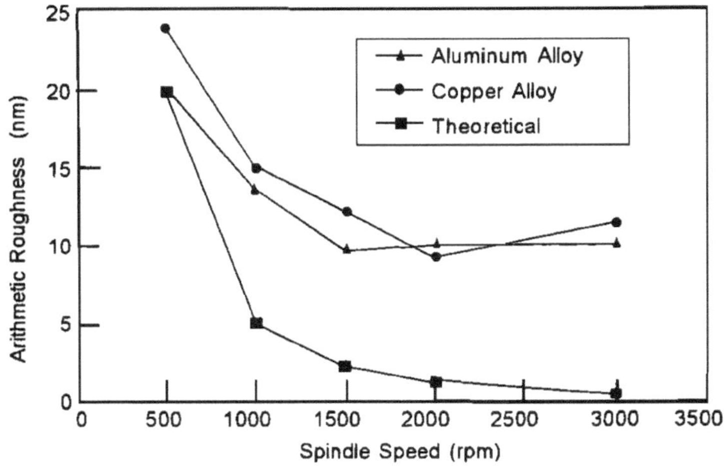


**FIGURE 5.12** Comparison of surface roughness of Al alloy and copper alloy at various spindle speeds [18].

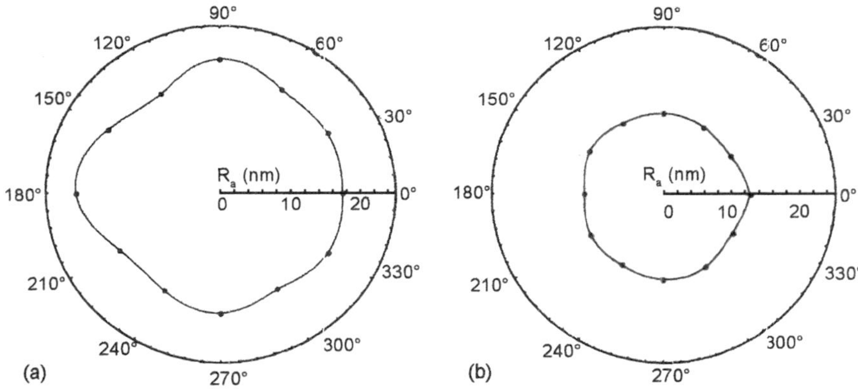


**FIGURE 5.13** Variation of roughness for diamond-turned surface of (a) aluminum single crystal (001) and (b) aluminum alloy 6061 [18].

For polycrystalline materials, grain boundary step (i.e., a height difference between two adjoining crystal grains appearing at their common border due to the crystallographic effect in the material property) is a critical factor of surface roughness. For example, in oxygen-free high conductivity (OFHC) copper, as different grains have different relative orientations and the elastic recovery of a grain after cutting depends on the grain's orientation, small steps with step heights of a few nanometers appeared at the grain boundaries of a diamond machined surface, as shown in Figure 5.14. Besides, the material heterogeneity, such as impurities, inclusions, and preexisting defects also affects the machine's surface roughness.

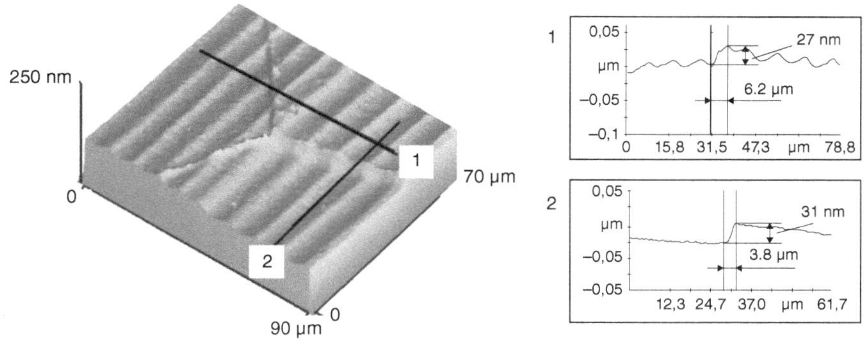


**FIGURE 5.14** AFM image of a diamond-turned OFHC copper surface exhibiting small steps at grain boundaries [28].

e. Tool-workpiece interfacial conditions, including cooling, lubrication, and chemical effects

Any variation in the tool-workpiece interfacial conditions would deteriorate surface quality due to the relative motion between the tool and the workpiece. The disturbance sources can be cutting temperature change, friction coefficient variation, and physical/chemical reactions. Moriwaki et al [29] systematically investigated the effect of cutting heat-induced thermal deformation of cutting tool and workpiece on the machining performance in ultra-precision diamond cutting of aluminum alloy and oxygen-free copper. The temperature rise of the tool due to the cutting heat can reach up to the order of 10°C and the machining error due to the thermal expansion reaches to the order of a few µm, while the temperature rise of the workpiece is relatively low. The temperature rise and the machining error can be reduced by spreading the kerosene over the workpiece surface prior to cutting and they are further reduced by spraying the kerosene mist to the cutting point during cutting. The temperature rise of the tool in the case of cutting of oxygen-free copper is much lower than that of aluminum alloy. To et al. [30] analyzed cutting forces in the cutting of Zn–Al alloy and found that the coolant is effective for decreasing the high-frequency vibration. In addition, the changes in the microstructure become more obvious on samples machined with a coolant than without a coolant. As a result, the coolant by refining the microstructure improves the roughness of the machined surfaces.

f. Vibration of workpiece and cutting tool, and machine tool chatter

Vibration occurs in ultra-precision machining even when the machine is supported by air mounts to prevent the vibration from the floor. In general, any vibration of unknown origin that ultimately causes unstable contact between tool and workpiece is called machine tool vibration, which is characterized by uncontrollability. The vibration can be caused by workpiece, tool tip, spindle rotation, and machine tool chatter [16]. Takasu [31] pointed out that vibration with small amplitude and low frequency usually remains in precision machine tools even supported by air mounts, and inevitably deteriorates the surface roughness. They estimated the surface roughness as the function of both the ratio of vibration amplitude to geometrical roughness, and the phase shift of the vibration to workpeice rotation, and clarified that the roughness can be reduced to 20-30% of the sum of the geometrical roughness and the vibration amplitude at optimal cutting conditions.

Kim et al. [32] investigated the effect of the frequency ratio between the vibration frequency $f_0$ and spindle speed $n$ on surface topography. The frequency ratio is defined as follows to describe the radial profile:

$$f_0 n = a + b \tag{5.8}$$

where $a$ is an integer, and $b$ is the fraction in the range of –0.5 to 0.5. The simulated surface topographies at different frequency ratios are shown in Figure 5.15.

### 5.2.4 Burr Formation

A burr is a raised edge or small piece of material (usually unwanted) that remains attached to a workpiece after a machining process. Burr formation was found to be a critical problem in micro grooving of ductile materials, especially in dry cutting or when the cutting fluid is not sufficiently provided into the cutting region

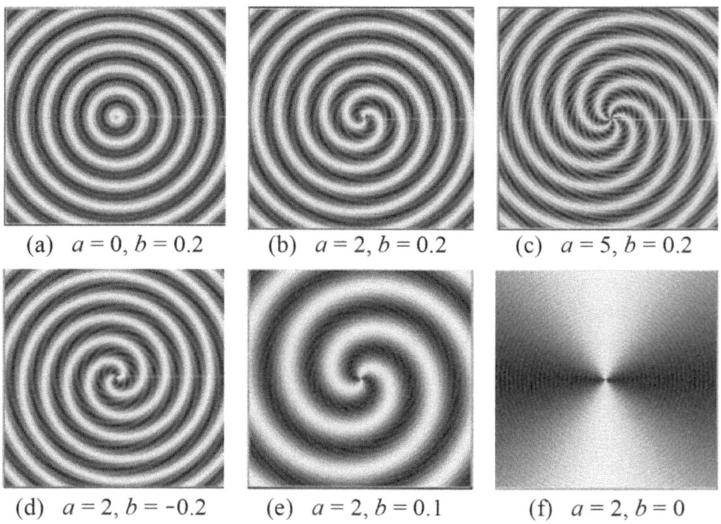


**FIGURE 5.15** Simulated surface topographies at different frequency ratios: (a) $n$ = 10 rev/s, $f_0$ = 2 Hz; (b) $n$ = 10 rev/s, $f_0$ = 22 Hz; (c) $n$ = 10 rev/s, $f_0$ = 52 Hz; (d) $n$ = 10 rev/s, $f_0$ = 18 Hz; (e) $n$ = 10 rev/s, $f_0$ = 21 Hz; and (f) $n$ = 10 rev/s, $f_0$ = 20 Hz [32].

[33]. Figure 5.16 shows SEM photographs of microgrooves cut by single cuts at a depth of 10 μm with a micro-chamfered 60° V-shaped tool. Burr formation at the side edges of the microgroove is very significant, although the inside surface of the groove is smooth.

The burr formation in Figure 5.16 is presumably caused by the side flow of material in the cutting region. As shown in Figure 5.17(a), the cross-section of the uncut chip is triangular, which is different from normal orthogonal cutting conditions where uncut chip thickness is uniform along the cutting edge. Therefore, during the cutting the flow of material generated by the two side edges is directed toward the center of the tool and interferes with each other, leading to a high-pressure material stagnation zone. The material stagnation at the tool center will cause significant side flows of material to form side burrs.

To suppress the burr formation in micro grooving, two-step cutting is proposed. The first cut is done as a rough cut at a large depth, and the second cut is done as a finish cut at a small chip thickness. In this case, the cutting model of the finish cut is shown schematically in Figure 5.17(b). The cutting conditions at both side edges are in the orthogonal cutting state. This way, the material flow will be fluent, which contributes to burr-free surface formation.

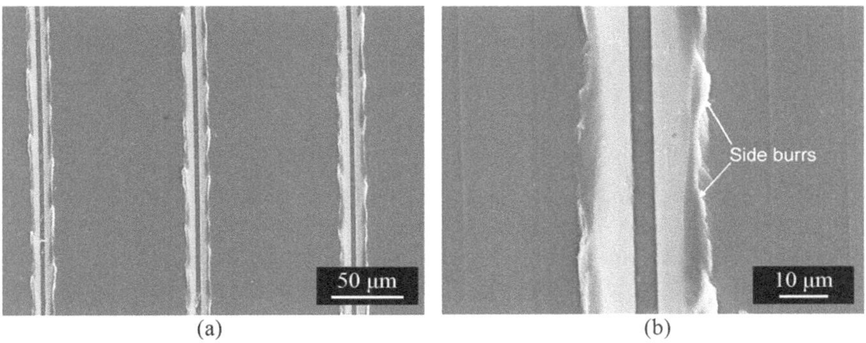


**FIGURE 5.16** SEM photographs at different magnifications showing microgrooves with side burrs [33].

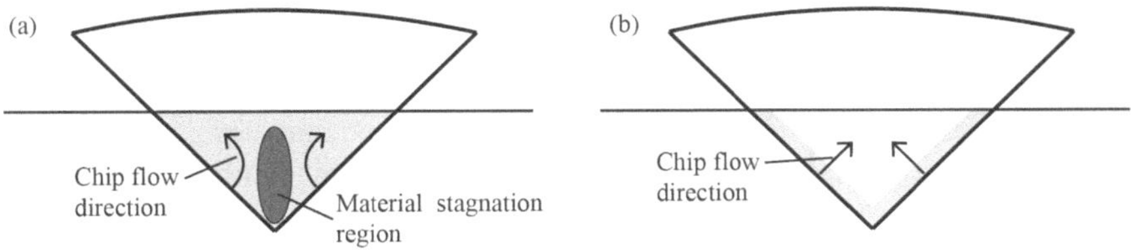


**FIGURE 5.17** Cross-sectional models for (a) one-step grooving process showing a material stagnation region, and (b) the second cut of the two-step microgrooving process [33].

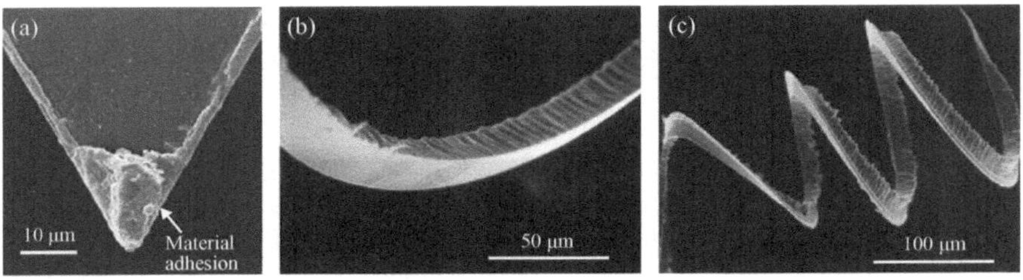


**FIGURE 5.18** SEM photographs of (a) the tool tip showing material adhesion, and the cutting chips in (b) the first cut, and (c) the second cut [33].

Figure 5.18(a) shows an SEM photograph of the cutting tool after one-step grooving process. The material adhesion was observed on the tool tip, indicating the occurrence of material stagnation. The material stagnation causes a highly negative effective rake angle, and in turn, increases cutting forces. The phenomenon of material stagnation can be partially prevented by applying cutting fluid effectively into the cutting region, but it is difficult to eliminate it completely. Figure 5.18(b) and (c) show geometry of the chip generated by the first cut and second cut of the two-step cutting. As evident, the chip with triangular cross-section is produced by the first cut, whereas the chip with a V-shaped cross-section and a uniform thickness is produced by the second cut, which is consistent with the model illustrated in Figure 5.17.

## 5.3 MACHINABILITY OF METAL MATERIALS

Metal materials compatible with diamond turning include copper (Cu), aluminum (Al), gold (Au), silver (Ag), platinum (Pt), electroless plated nickel (NiP), beryllium (Be), lead (Pb), and so on. However, ferrous (iron) metals, such as carbon steels and alloy steels containing Cr, Ti, Ni, etc. elements, are normally considered non-machinable, because chemical reaction occurs between the workpiece material and diamond tool causing excessive tool wear [34]. In this section, the machinability of several metal materials will be introduced, and a comparison will be made.

### 5.3.1 Copper and Its Alloys

Copper is a chemical element with atomic number 29. Copper has higher thermal and electrical conductivity than most other metals. It also has a high reflection ratio, especially for infrared light. Pure copper is soft and malleable, and it is usually used in the form of an alloy. In this section, the machinability of two copper materials, oxygen-free copper and brass, will be taken as examples.

Oxygen-free copper (OFC) refers to a group of wrought copper alloys that have been electrolytically refined to reduce the level of oxygen to 0.001% or below. Oxygen-free copper is valued for its chemical purity and is widely used in the applications where an environment with no oxygen is required to avoid undesirable chemical reactions with other materials in the environment. For example, OFC is widely used as reflecting mirrors for high-power $CO_2$ lasers, where a high-standard surface quality (in terms of surface roughness and residual stress condition) is required to guarantee the quality of reflectors. It is also used in vacuum devices such as particle accelerators, and plasma deposition processes used for the manufacture of semiconductors and superconductor components.

Moriwaki [35] investigated the chip formation, cutting forces, and surface roughness in micro/nanoscale cutting of OFC (purity 99.99%) with single-crystal diamond tools under various uncut chip thicknesses from 3.0 μm to 2.5 nm. He confirmed continuous chip formed even at the uncut chip thickness of 2.5 nm, but the lowest surface roughness was obtained when the uncut chip thickness was 25 nm. As the OFC contains crystal grains in average diameter of 10–100 μm, grain boundary step (a height of a few tens of nanometers) is a major cause of surface roughness. Figure 5.19 shows an SEM micrograph of a diamond-turned OFC surface at various uncut chip thicknesses. Boundary steps are observed on all the surfaces, and are more significant at larger uncut chip thicknesses. The boundary steps were suppressed by optimizing the cutting conditions, such as reducing the uncut chip thickness, to induce a kind of "burnishing effects", as shown in Figure 5.20. By minimizing the grain boundary step, it was possible to obtain a surface roughness of 10–20 nm Ry on oxygen-free copper.

Surface roughness can also be improved by optimizing the raw material fabrication process. Zhang et al. [36] found that the degree of crystal nonuniformity has a great influence on the finished surface quality. The higher the degree of crystal nonuniformity, the worse the finished surface quality. Samples subjected to cold deformation-recrystallization annealing are of fine and uniform crystal structure and show excellent surface quality. Sugano et al. [37] performed diamond turning experiments on various kinds of OFC and found that the raw material produced by hot/cold rolling followed by heat treatment was the most suitable to improve the surface integrity, compared to the forged, hot rolled, and hot/cold

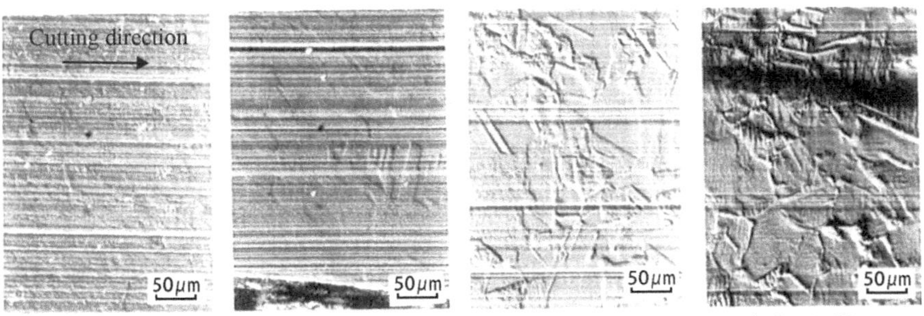


(a) $h = 0.005$ μm (b) $h = 0.082$ μm (c) $h = 0.41$ μm (d) $h = 1.67$ μm

**FIGURE 5.19** SEM photographs of annealed oxygen-free copper surfaces machined by orthogonal fly cutting at various uncut chip thickness $h$ (cutting speed 754 m/min) [35].

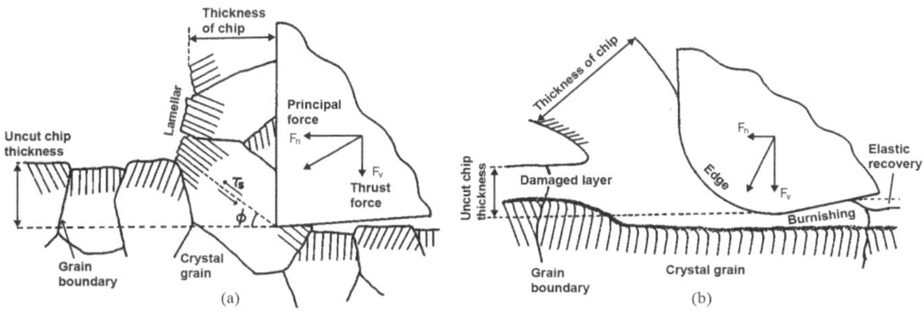


**FIGURE 5.20** Cutting models of polycrystal material in diamond cutting: (a) at a large uncut chip thickness, (b) at a small uncut chip thickness [35].

rolled material without heat treatment. The surface roughness was stable at ~10 nm Rz in a cutting distance range of 500 km. As for tool wear, crater wear was remarkably large, while flank wear was negligible. They also found that the cutting forces of OFC were 50% larger than those of aluminum.

Brass is an alloy of copper and zinc; the proportions of zinc and copper can be varied to create a range of brasses with varying properties. For example, β-brass is a copper alloy with 45–50% zinc content, which can only be hot-worked or cast. Ueda et al. [38] investigated the mechanism of diamond turning of single crystalline β-brass using a special in-situ micro cutting system equipped in an SEM. They found that the chip formation was largely affected by the workpiece rotation angle, that is, the cutting direction with respect to the crystalline orientations of the workpiece. Figure 5.21(a) shows the SEM images of the chip formation of specimen A (001), which exhibits a typical lamellar slip structure at the rotation angles of +25° and +45°, but the chip changes become discontinuous at the rotation angles of −25°, −15°, and 0°. In contrast, Figure 5.21(b) shows the SEM images of cutting chips in cutting specimen B (101). As the rotation angle of the workpiece is changed, mostly the chips formed are continuously characterized by lamellar slip structure. However, the chip pattern transforms noticeably from the lamellar type to a discontinuous type accompanied by micro-cracking in the shear band when the crystal is machined at a rotation angle of +90°. The more detailed examination of the morphology of chips shows that there appears a transition in the lamellar slip structure from fine coarse around the rotation angle of +30°. The variation in the chip morphology described above has a periodicity of about 90° and 180° for specimen A (001) and specimen B (101), respectively. These periods coincide well with those of the crystallographic system of each specimen.

### 5.3.2 Aluminum and Its Alloys

Aluminum is a member of the boron group of chemical elements, the atomic number of which is 13. Like other single-crystal metals, single-crystal aluminum is highly anisotropic in physical and mechanical properties. Yuan et al. [39]

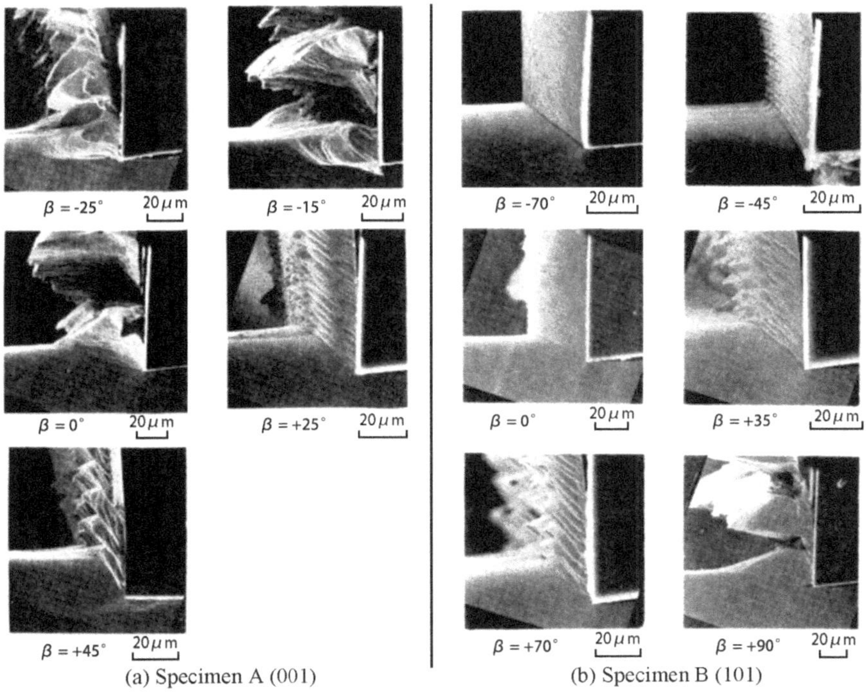


**FIGURE 5.21** Typical SEM images showing the variation in the chip formation pattern with the rotation angle of $\beta$-brass: (a) specimen A (100) and (b) specimen B (101) [38].

investigated the effect of crystallographic orientation on cutting forces in diamond cutting of single-crystal aluminum. When face turning is performed, the cutting is along different crystallographic directions on the same crystal plane; the variation of cutting forces with the cutting directions is shown in Figure 5.22(a). In the case of cutting on (110) plane, the value of cutting force is the biggest at [1–10] direction while it reaches a minimum at [001] direction. The magnitude of the fluctuation in cutting forces with crystallographic orientations on (111) plane is smaller than that on (110) plane. When cylindrical turning is performed, cutting is performed along various directions on different crystal planes; the relationship between the cutting forces and cutting directions is shown in Figure 5.22(b). In the case of cutting around (100) axis, four peak values were found during a revolution of workpiece. The magnitude of the fluctuation in cutting forces with crystallographic orientations around (111) axis is smaller than that on (100) axis.

Lee et al. [40] found that the pattern of periodic fluctuation of the cutting forces in diamond turning of single-crystal aluminum was due to not only the crystallographic orientation of the materials but also the friction between the chip and the tool rake face. The frictional effect was found to cause a change in the pattern and the periodicity of the fluctuation of the cutting force. It should be pointed out that

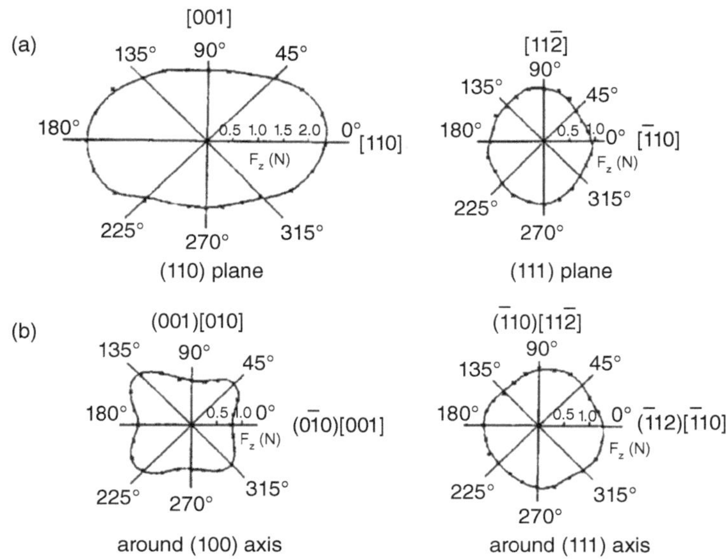


**FIGURE 5.22** Variation of cutting forces with crystal orientations in (a) face turning (b) cylindrical turning of single crystal aluminum [39].

single-crystal aluminum is not widely used in industry, because it is too chemically reactive to exist in nature as a free metal. Therefore, most aluminum used today is alloyed with other elements, such as aluminum–magnesium (Al–Mg) alloy, which has a low density, relatively high strength, and ability to resist corrosion due to the phenomenon of passivation. Aluminum alloy parts with high surface quality are widely used in modern industries, such as polygon mirrors, hard discs, exposure drums, and optical mirrors.

In the cutting of aluminum alloys, material adhesion to the tool rake face is a serious problem. This might be due to the softening and melting of aluminum alloys in the cutting region. Especially when the cutting speed is high and lubrication is insufficient, the highest temperature in the cutting region may exceed the melting point of the material. The material adhesion to the surface of the cutting tool may also lead to built-up edge formation, which causes scratches and protrusions on the finished surface. In this case, the cooling and lubrication of the cutting tool is essential. To prevent chips from adhering, Sugihara and Enomoto [41] used diamond-like carbon (DLC) coated tools with extremely low friction and fabricated functional textures on the tool surface. After cutting an aluminum alloy for 1800 m, the tool rake faces were examined to evaluate the tool surface adhesion. As seen from the surface topographies, SEM images and EDX analyses in Figure 5.23, the nano/micro texture with the DLC film significantly improved anti-adhesion in wet cutting compared with the conventional DLC-coated tool in wet cutting. However, in dry cutting, as shown in Figure 5.24, the textured coated tool was less effective in preventing chip adhesion. From this result, it is presumable that the

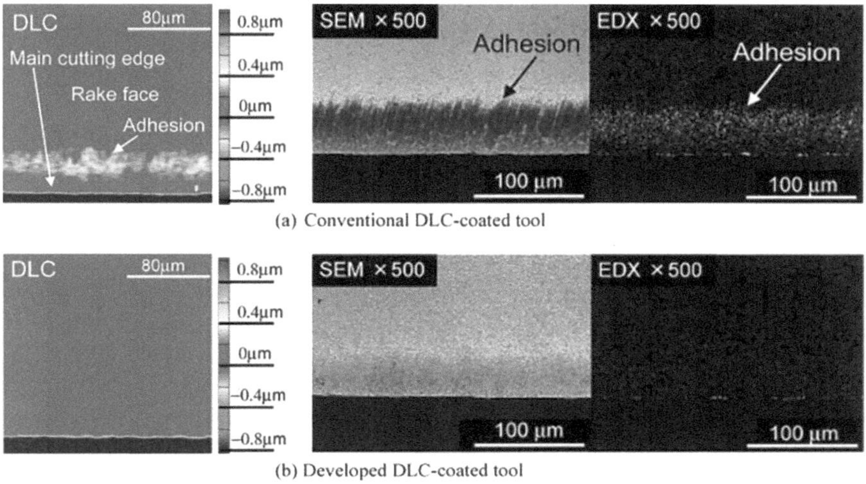


**FIGURE 5.23** Surface topography, SEM images, and EDX analyses of rake face of DLC-coated cutting tools in wet cutting [41].

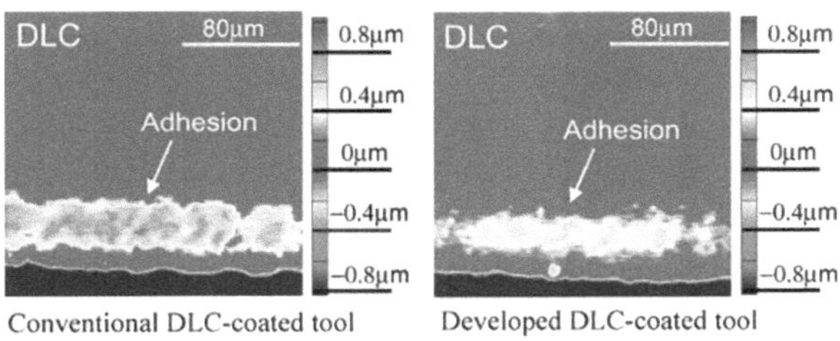


**FIGURE 5.24** Surface topography of rake face of DLC-coated cutting tools in dry cutting [41].

effect of improving the anti-adhesive properties of the nano/micro texture was exerted by retaining cutting fluid in the grooves and preventing lubricant film on the rake face from being broken by a severe sliding contact between tool and chip.

Sugano et al. [42] investigated the effects of the wear of diamond tools on the surface roughness and the residual stress in diamond turning of aluminum for a cutting distance of 900 km. They found that despite tool wear, the roughness of diamond-turned surface remains almost constant due to the "burnishing" effects. The surface roughness was varied by tool feed rate, nose radius, and thrust forces.

### 5.3.3 Nickel Alloy

Nickel (Ni) is a metal with atomic number 28 and the face-centered cubic crystalline structure. Nickel is a corrosion-resistant material and has found many uses in alloys. However, the machinability of pure nickel is very poor due to severe tool wear. The tool wear is presumably caused by the chemical reaction between the pure nickel and diamond [43] Compared to pure nickel, electroless nickel plating with the addition of a small mount (10–12%) of phosphorus (P) significantly improves the machinability for diamond turning.

Electroless nickel plating has many applications. One of them is to use electroless nickel-plated molding dies in optical component manufacturing. Electroless nickel plating is an auto-catalytic chemical technique used to deposit a layer of nickel-phosphorus alloy on a substrate. The process relies on the presence of a reducing agent, for example, hydrated sodium hypophosphite, which reacts with the metal ions to deposit metal. It is possible to produce alloys with varying percentages of phosphorus, ranging from 2 to 14, and the metallurgical properties of these alloys depend on the percentage of phosphorus. If the added amount of phosphorus is high (>10%), electroless nickel plating results in an amorphous structure. Consequently, no boundary steps occur on the machined surface, and a surface finish of a few nanometers can be achieved. However, if the added amount of phosphorus is low (<8%) or high-temperature heat treatment is performed after the plating process, crystalline nickel grains form within the material, leading to distinct grain boundary steps on the machined surface, causing a higher surface roughness.

The lubrication/cooling conditions significantly affect the chip formation mechanisms, and consequently affect the machined surface quality. Figure 5.25(a) and (b) show the machined surface of electroless nickel with a straight-nosed tool under dry cutting and wet cutting, respectively. On the dry-cut surface, although a section of the surface appears smooth with periodical grooves, some protrusions can be observed on the surface due to chip adhesion. These protrusions subsequently led to a significantly high surface roughness and form error. In contrast, on the wet-cut surface, periodic grooves were generated at the nanometric level without the formation of protrusions. The surface roughness is 0.015 μm Ry, which is close to the theoretical value.

As shown in Figure 5.26(a1), the chips generated under dry cutting are in the form of narrow fibers with widths significantly smaller than the nominal width of the cut. In contrast, the chips generated under wet cutting (Figure 5.26(a2)) are in the form of flow type ribbons with widths similar to the nominal width of the cut.

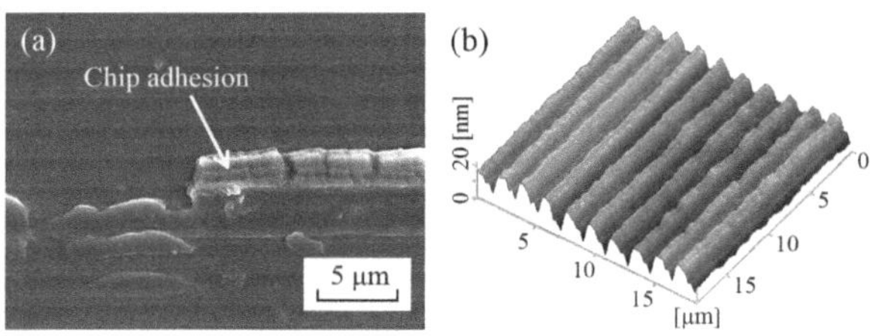


**FIGURE 5.25** Machined surfaces of electroless nickel at an undeformed chip thickness of 0.02 μm: (a) SEM image of the surface by dry cutting, (b): AFM image of the surface by wet cutting [44].

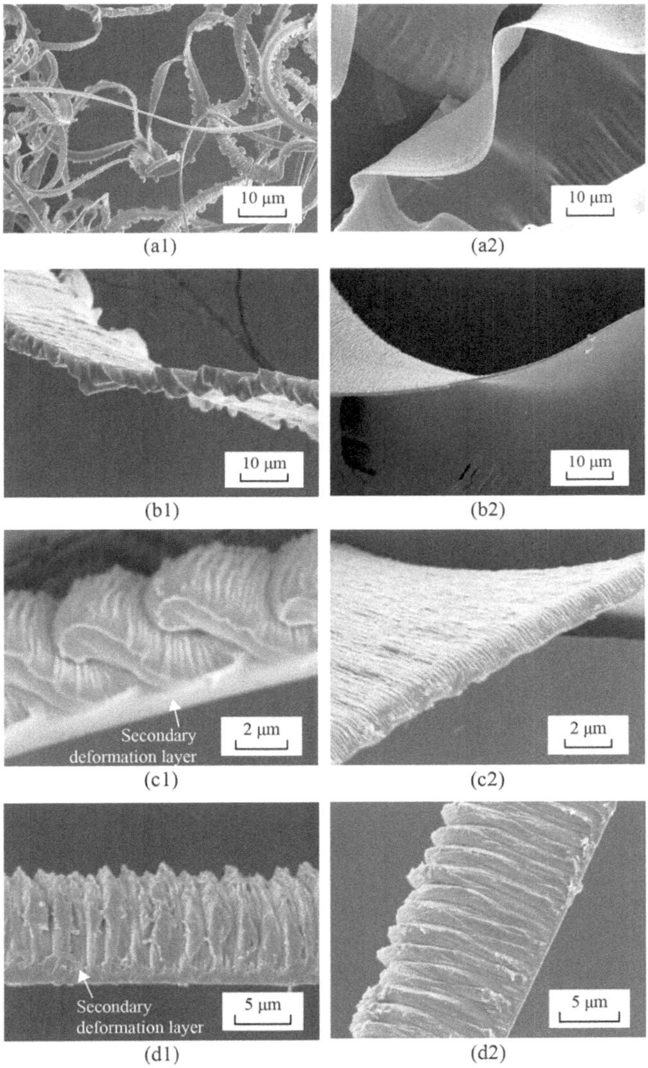


**FIGURE 5.26** SEM images of the chips at different undeformed chip thicknesses: (a) 0.02 μm, (b) 0.1 μm, (c) 0.5 μm, and (d) 5 μm. (a1–d1) dry cutting; (a2–d2) wet cutting [44].

This implies that the chip has been split into many sections along the cutting edge under dry cutting. The split of chips may be caused by the fact that in dry cutting, the high temperature in the cutting region results in increased surface energy for both the newly generated chip and the workpiece. As a result, these surfaces become physically and chemically active, making them prone to adhesion. This adhesion becomes particularly significant when the undeformed chip thickness is extremely small, leading to a high negative effective rake angle due to the edge radius. This situation hinders chip flow and gives rise to built-up edges. Consequently, the actual cutting conditions vary along the cutting edge, leading to the formation of narrow fibers as the chips are split. These small, narrow chips are easily recirculated into the cutting region and adhere to the workpiece surface due to the squeezing action of the tool, as shown in Figure 5.25(a).

As the undeformed chip thickness increases to 0.5 μm, as shown in Figure 5.26(b1,2), the split of cutting chips is suppressed; instead, surface waviness is formed by the stick-slips resulting from chip folding. Furthermore, when the undeformed chip thickness exceeds 1 μm, as shown in Figure 5.26(c1,2) and (d1,2), a thick secondary deformation layer is formed within the chip near the chip-tool contact boundary, indicating severe friction occurring under dry cutting conditions. However, under wet cutting conditions, very smooth workpiece surfaces were produced with yielding flow and shear-type ribbon chips. These results indicate that the cutting fluid plays an important role in the ultra-precision cutting of electroless nickel.

In diamond cutting of electroless nickel plating, tool wear is suppressed significantly compared with that in cutting pure nickel, which is attributed to the phosphorous addition reducing the erosion of diamond and carbon diffusion into Ni [45]. Nevertheless, compared to cutting other metal materials, such as aluminum and copper, the cutting of electroless nickel still leads to a more severe flank wear. Figure 5.27 shows images of a diamond tool after cutting for 3.18 km distance on electroless nickel. A long and narrow wear land was formed on the flank face with uniform and nearly symmetrical about the tool tip. Additionally, extremely fine microgrooves in the direction parallel to the cutting direction can be observed on the wear land. The shape of these grooves was regular and uniform. The groove pitch was 2 μm, which matched with the tool feed rate exactly.

### 5.3.4 Gold and Other Rare Metals

Gold is a chemical element with the symbol Au and atomic number 79. It is a soft metal and is considered the most ductile metal on the periodic table. On the other hand, gold is a precious metal, thus parts are typically not made entirely of gold;

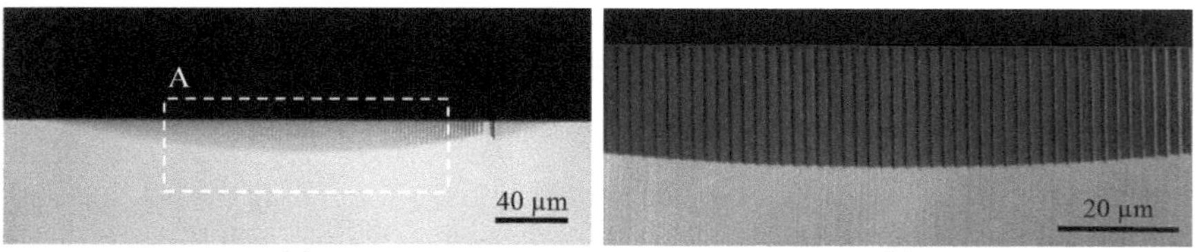


(a) (b)

**FIGURE 5.27** Flank face of the worn diamond tool with a 5 mm nose radius after machining electroless nickel: (a) general view, (b) close-up view of area A in (a). (Cutting direction is from left to right in (a) with a feed rate of 2 μm/rev) [46].

instead, only some key elements are made of gold, for example, three-dimensional implementation of large-scale integrated (LSI) chips, which are usually stacked and pressure-bonded via micro gold bumps electroplated on silicon wafers. To create gold plating with a specific shape on silicon wafers, photoresist masks must be developed first. Then, the gold and photoresist are planarized together by cutting, and finally, the photoresist is removed. Thus, the cutting process can be regarded as cutting a photoresist-gold composite material.

Figure 5.28 shows SEM images of the surfaces of the photoresist-gold composite after diamond turning with various cutting parameters. In Figure 5.28(a), the photoresist surface displays numerous microfractures, and burrs are evident surrounding the gold bumps. In contrast, in Figure 5.28(b), the photoresist surface exhibits fewer fractures, and there are no apparent burrs surrounding the gold bumps. This suggests a strong correlation between the cutting characteristics of the two materials.

After diamond turning, the photoresist was removed with acetone to observe burr formation on the gold bumps. Figure 5.29(a) and (b) present the bumps machined by diamond tools with edge radii of 200 and 50 nm, respectively. The observation was performed from the direction perpendicular to the feed direction. The size of the exit burr became smaller with decreasing tool edge radius. Figure 5.29(c) and (d) show the bumps machined at depth of cut of 16 and 2 μm, respectively. In this case, the observation was performed from the direction perpendicular to the cutting direction. A big side burr was formed at the larger depth of cut, while no obvious burr was formed at the smaller depth of cut. Therefore, it is evident that the formation of exit burrs depends on the tool edge radius, while side burr formation is influenced by the depth of cut. To prevent burr formation on the gold bumps, it is preferred to use a sharp tool with a shallow depth of cut and a low tool feed rate. Figure 5.29(e) and (f) show general views of bumps before and after cutting under such conditions. Before cutting, the bump surface was very rough, but it became distinctly smoother after the cutting process, with a surface roughness of 18 nm Ry and a flatness of 43 nm.

Another example of rare metals is niobium. Niobium (Nb) is a chemical element with atomic number 41. Single-phase niobium is utilized in superconducting accelerator components within particle accelerating facilities, where the

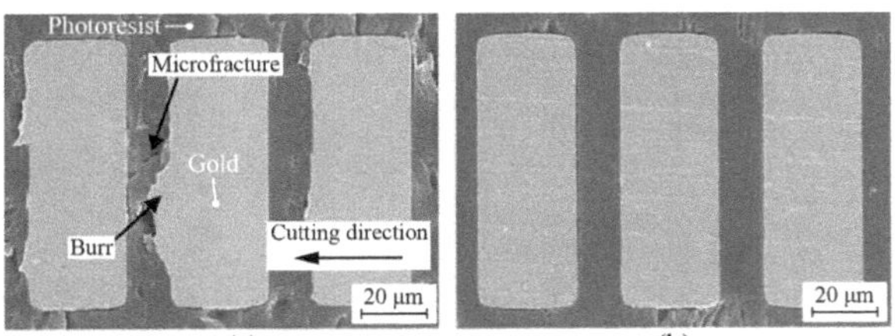


(a) (b)

**FIGURE 5.28** SEM images of diamond-turned surfaces of the photoresist-gold composite, showing correlation between bump burr formation and photoresist microfracture [47].

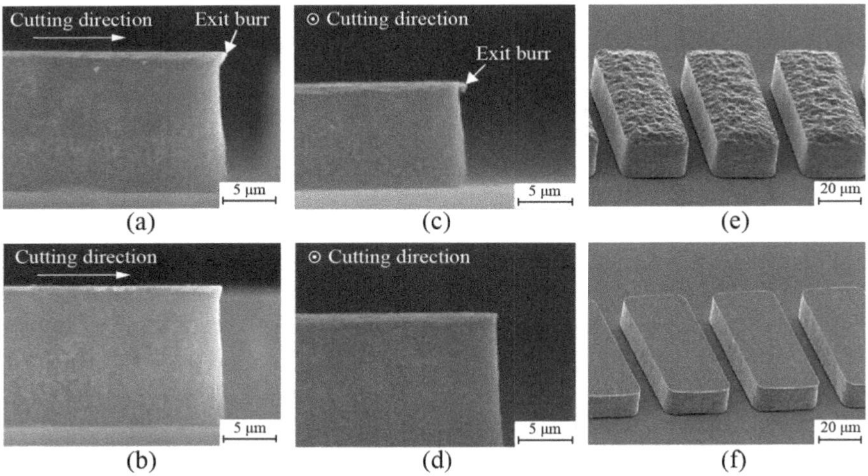


**FIGURE 5.29** SEM images of diamond-turned gold bumps: exit burr formation when using diamond tools with edge radius of (a) 200 nm and (b) 50 nm; side burr formation when cutting at depth of cut of (c) 16 μm and (d) 2 μm; the general appearance of gold bumps (e) before and (f) after diamond turning [47].

functionality of these components requires high surface quality. Single-phase Nb, as opposed to its alloys, is a highly ductile metal. During a cutting process, the high ductility may result in surface defects including chip adhesion, built-up edges, side flow, and burr formation.

The machinability of Nb was investigated by using various kinds of tools. It was found that CBN and single-crystal diamond tools were not suitable for machining Nb because they exhibited immediate tool failure [48]. PCD tools and TiAlN-TiSiN nanolaminate (TH1000) coated cemented carbide tools achieved the smallest surface roughness at low feed rates, while uncoated and CVD diamond-coated cemented carbide tools offered poor surface quality, as shown in Figure 5.30(a). Although the surface roughness was minimized to 0.46 μm Ra by PCD tools, as shown in Figure 5.30(b), this tooling experiences rapid wear and leads to significant subsurface deformation. The TH1000 coated cemented carbide tools could achieve less subsurface deformation, but the machined surface exhibited a larger roughness of Ra = 0.75 μm, as shown in Figure 5.30(c).

### 5.3.5 Ferrous Metals

Industrially used ferrous metals include alloy steel, carbon steel, cast iron, and wrought iron. These metals are prized for their tensile strength and durability. Carbon steel, also known as structure steel, is a staple in the construction industry and is used in skyscrapers and bridges. Ferrous metals are also used in shipping containers, industrial piping, automobiles, railroad tracks, molding dies, etc. As

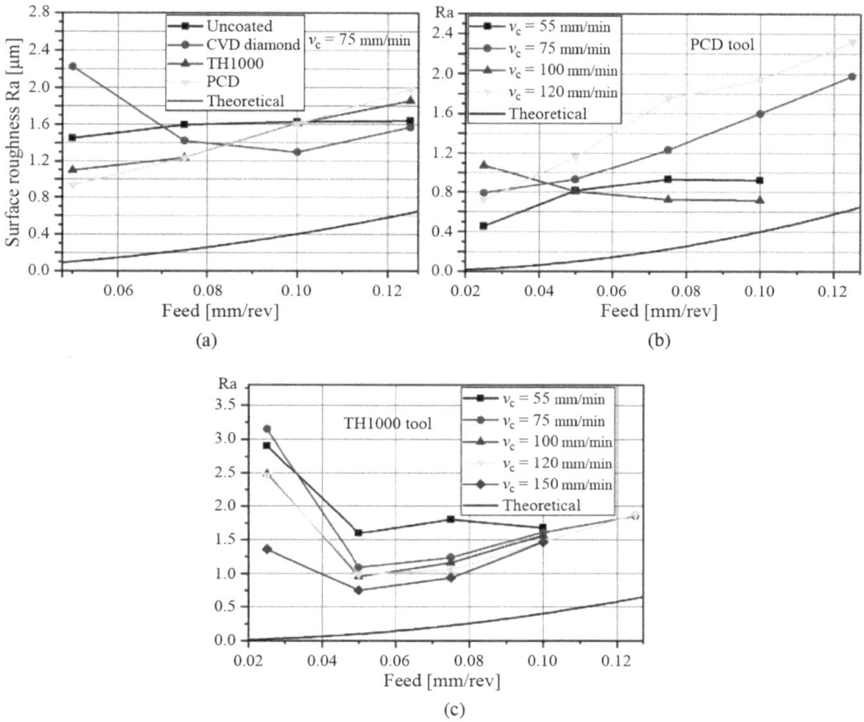


**FIGURE 5.30** Surface roughness of Nb for (a) different cutting tools at $v_c$ = 75 mm/min, (b) PCD tool at different cutting speeds and feeds, and (c) TH1000 tool at different cutting speeds and feeds [48].

mentioned in Chapter 4, ferrous materials are difficult to machine using diamond tools. Due to the chemical reaction between diamond and ferrous elements, the diamond tool wear is very severe, and a durable cutting process is impossible. For example, the wear rate of a diamond tool in turning mild steel is $10^4$ times greater than that when turning brass of comparable hardness. It is accepted that the tool wear proceeds by the graphitization of the diamond. Due to the catalytic properties of steel and high cutting temperature, the surface layer of a diamond tool transforms to graphite and is continually removed either by the abrasion of the steel or by solution in the steel [49].

To reduce diamond tool wear when cutting ferrous metals, various efforts have been made to use special cutting environment. For example, Casstevens tested [50] $CO_2$ gas and $CH_4$ gas in the atmosphere when cutting hardening steel with diamond tools. In the $CH_4$ gas environment, a significant decrease in tool wear was confirmed compared to that in air and $CO_2$. A diamond-turned surface on annealed AISI 1090 steel showed a low surface roughness of 45 nm Ry. Furthermore, it has been observed that the absence of water vapor or oxygen enhances the adhesion of metals to diamond. Therefore, when cutting in the gas environment like argon or nitrogen, the gas effects on diamond tool wear may be detrimental. This was confirmed by the study comparing the impact of oxygen, air, and argon gas environments on the suppression of diamond tool wear during ultrasonic vibration cutting of tungsten, as shown in Figure 5.31 [51]. Oxygen shielding led to a reduced tool wear rate and a better machined surface finish compared to air and argon shielding. This phenomenon is attributed to that oxygen increases the growth rate and thickness of ferrous metal oxides, which, in turn, suppresses the chemical reaction between the ferrous metal and the diamond tool.

Since severe tool wear in diamond cutting of ferrous materials is primarily attributed to chemical reactions, and the rate of all chemical reactions increases exponentially with temperature, cryogenic machining is an effective way to suppress the tool wear. Evans and Bryan [52] developed a cryogenic machining system comprising tool post and chuck with constrained liquid nitrogen flows. With this system, both the diffusion and catalyzed graphitization of the diamond tool were significantly slowed at low temperatures, which achieved a surface finish of 25 nm Ra on stainless steel.

Furthermore, as the catalytic reaction significantly depends on the surface electric potential of the catalyst, the metal catalytic activity can be either increased or decreased by changing the surface electric potential. Based on this concept, Zhang [53] proposed an electric field-assisted machining method that involves electrically charging the workpiece within an electric field during diamond machining to adjust the workpiece surface to a desired electric potential. This surface electric potential may inhibit the physical and chemical adsorption of diamond carbon onto the workpiece surface, and thus the diamond tool life can be extended.

Instead of using diamond tools, selecting materials that do not chemically react with ferrous metals as tools is the simplest and most effective approach. Cubic

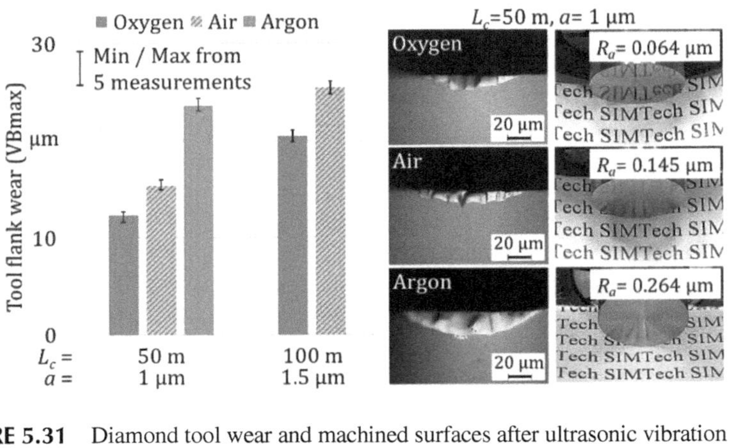


**FIGURE 5.31** Diamond tool wear and machined surfaces after ultrasonic vibration cutting of tungsten with different shielding gases (feed rate 15 µm/rev, cutting speed 1.25 m/min) [51].

boron nitride (cBN) is such a tool material for cutting ferrous materials. Fujisaki et al. [54] developed a new tool using ultra-fine-grain (<100 nm) binderless cBN material, which was fabricated by transforming hexagonal boron nitride into cBN through sintering under an ultra-high pressure of 10 GPa at 1800°C. In this way, the cutting edges of the newly developed cBN tool can be made as sharp as those of single-crystal diamond tools. For electroless nickel (NiP) workpiece, both diamond and cBN tools can produce very smooth surfaces without any visible scratches or rubbed patterns, as shown in Figure 5.32(a) and (b). For a heat-treated stainless steel (SUS420J2) workpiece, the surface machined with the diamond tool was covered with scars, and the tool itself had a large crack at the cutting edge. In contrast, the cBN tool produced a significantly better surface quality, as

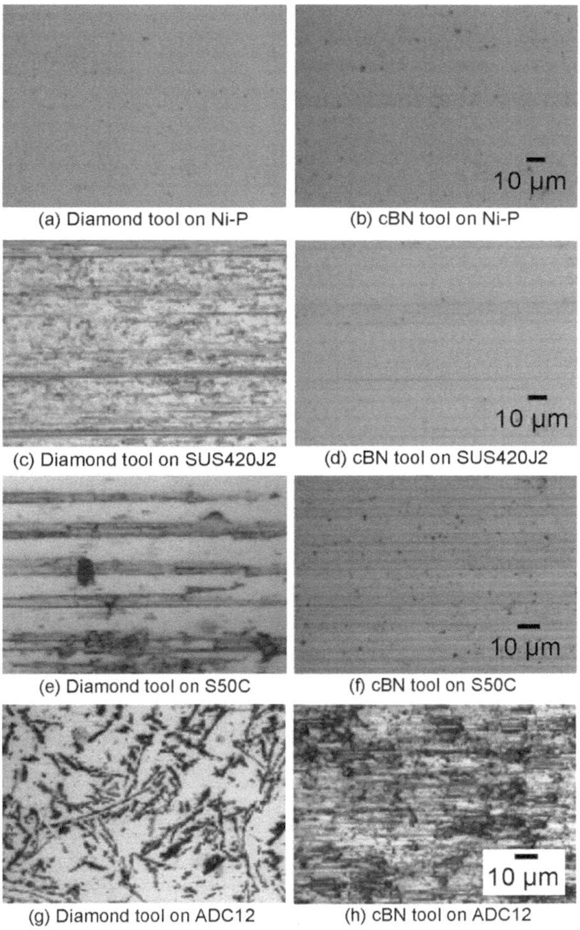


**FIGURE 5.32** Comparison of machined surfaces of ferrous metals using diamond tools and cBN tools [54].

shown in Figure 5.32(c) and (d). For medium carbon steel (S50C), the surface cut by the cBN tool was very similar to that of the heat-treated stainless steel. When S50C was cut with the diamond tool, as shown in Figure 5.32(e) and (f), the surface exhibited some scratched paths, although not as significant as those of the heat-treated stainless surface cut with the diamond tool. For the aluminum cast alloy (ADC12) workpiece, the surface cut by the diamond tool exhibited a metallographic structure, whereas the surface cut by the cBN tool resembled an abrasion surface, as shown in Figure 5.32(g) and (h). The cBN tool showed the formation of build-up edge after cutting the ADC12 specimens, which made it difficult to obtain a mirror surface.

### 5.3.6 Titanium/Magnesium Alloys

Alloying has long been applied to impart desirable properties to materials. Typically, it involves the addition of relatively small amounts of secondary elements to a primary element. Some popular alloys are titanium alloys and magnesium alloys.

Titanium alloys are composed of a titanium base metal and additional elements, such as aluminum, manganese, zirconium, chromium, and cobalt. While pure titanium is already one of the strongest elements, titanium alloys are characterized by their high strength, biocompatibility, excellent corrosion resistance, and good toughness, particularly at room or elevated temperatures. Consequently, there is a growing demand for these lightweight metallic materials in aerospace, medical, and automotive industries. Machining titanium alloys presents challenges due to their low thermal conductivity, which leads to heat accumulation in the cutting region. This, in turn, results in elevated temperatures, rapid tool wear, and reduced tool life [55].

Narutaki et al. [56] used natural diamond tools, which are characterized by the highest thermal conductivity among all available tool materials, and applied abundant water-soluble coolant to the cutting Ti6Al4V. It was found that the natural diamond tool exhibits an excellent cutting performance in machining titanium alloys, and the cutting speed could be increased up to 3.33 m/s when applying abundant coolant. In addition, the quality of surface machined with the natural diamond tool was found to be better than those cut with the other tools. Yip and To [57] found that, in the diamond cutting of Ti6Al4V, the assistance of magnetic field could increase the friction coefficient at the tool/workpiece interface, resulting in a reduction in the minimum chip thickness. This, in turn, contributes to an improved machined surface roughness. The chip generated at the depth of cut of 1 μm under magnetic field was continuous and smooth with a consistent shape (Figure 5.33(c)), while that without magnetic field was discontinuous (Figure 5.33(a)). Without magnetic field, when increasing the depth of cut to 2 μm, the chip formed continuously but the chip edges exhibited a sawtooth shape (Figure 5.33(b)), and the chip thickness looked thinner and narrower than the chip formed with the magnetic field (Figure 5.33(d)).

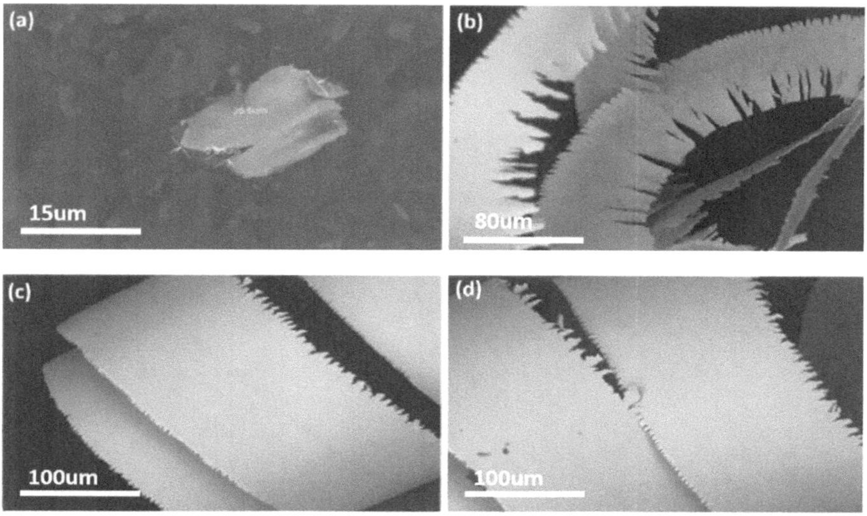


**FIGURE 5.33** Chip formation at a depth of cut of (a) 1 μm, (b) 2 μm without magnetic field, (c) 1 μm and (d) 2 μm with magnetic field [57].

Magnesium alloys are well-known for being the lightest structural alloys, consisting of mixtures of magnesium with other metals, including aluminum, zinc, manganese, silicon, copper, rare earths, and zirconium. Cast magnesium alloys are used for many components of automobiles and have been used in some high-performance vehicles; die-cast magnesium is also used for camera bodies and components in lenses.

In the cutting of magnesium alloys, cutting speed is a very important parameter because the ignition point of magnesium is low and flammability is high. During dry machining of magnesium alloys, fires are likely to occur when the melting point of material (400–600°C) is exceeded. Additionally, increasing the cutting speed may lead to significant flank build-up due to adhesion between the cutting tool and the workpiece [58]. Under dry conditions, the machined surface tends to exhibit more scratches compared to the surface obtained under wet conditions, as shown in Figure 5.34. These scratches are caused by material adhesions on the rake/flank faces of the cutting tool and hard inclusions in the workpiece. The material adhesion can be suppressed by using suitable cutting fluids.

### 5.3.7 High-Entropy Alloys

In recent decades, a new alloying strategy has been in vogue that involves the combination of multiple principal elements in high concentrations to create novel materials called high-entropy alloys (HEAs). The quinary CrMnFeCoNi alloy was one of the first equi-atomic HEAs reported to crystallize as a single-phase FCC solid solution. CrMnFeCoNi alloy exhibits outstanding strength-ductility

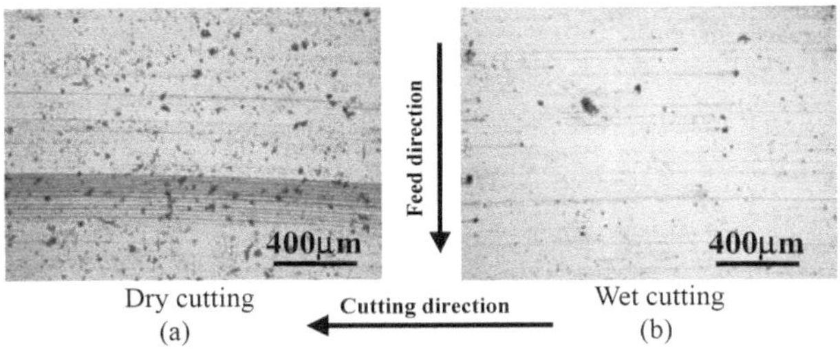


**FIGURE 5.34** Microscope images of the surfaces of AZ31 alloy machined under (a) dry and (b) wet conditions [59].

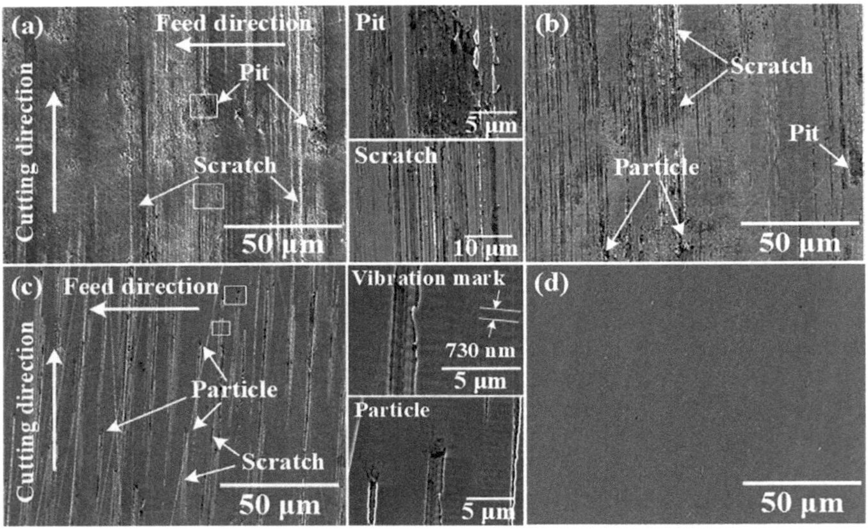


**FIGURE 5.35** SEM images of CrMnFeCoNi alloy surfaces machined by conventional cutting at tool feed rates of (a) 10 and (b) 1 μm/rev, and ultrasonic vibration cutting at (c) 10 and (d) 1 μm/rev [60].

trade-off, especially at cryogenic temperature, and thus has potential applications in space technology, and so on. However, as all the elements are ferrous transition metals having a strong chemical affinity with diamond, conventional diamond cutting of FeCrCoMnNi alloy has been considered impossible. Nevertheless, ultrasonic vibration cutting has been validated as an effective method to suppress the chemical wear of diamond tools when cutting FeCrCoMnNi alloy. Figure 5.35 shows SEM images of the surfaces machined with conventional and vibration-assisted diamond turning. For conventional diamond turning, numerous micro pits, particle adhesions, and scratches were observed. In contrast, the surface defects were significantly reduced in the surface produced through vibration-assisted diamond turning. Although the scratches along the cutting direction and the tool marks perpendicular to the cutting direction were present when the tool

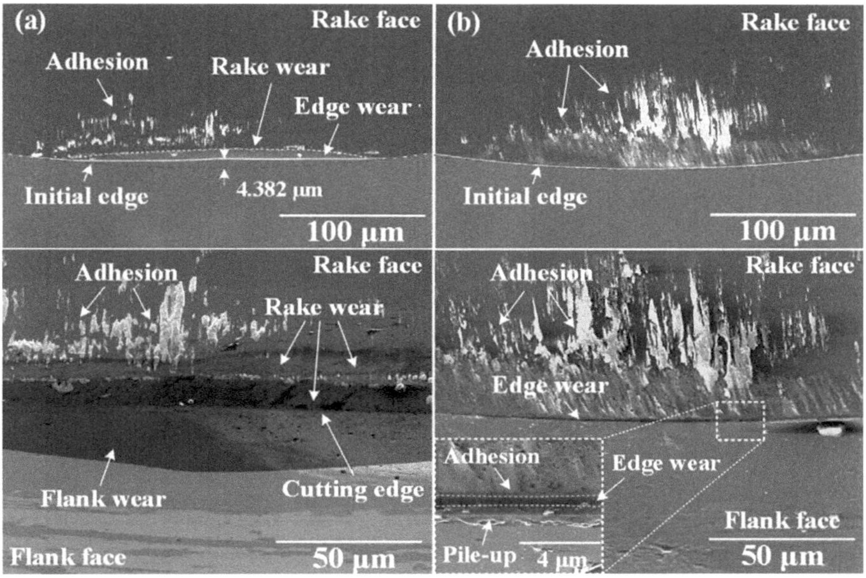


**FIGURE 5.36** SEM images of edges of diamond tool after (a) conventional cutting, and (b) ultrasonic vibration cutting of CrMnFeCoNi alloys [60].

feed rate exceeded 4 μm/rev, a smooth surface was achieved when the tool feed rate was less than 4 μm/rev, and surface defects were completely eliminated.

Figure 5.36 compares the cutting edges of the tools after conventional cutting and ultrasonic vibration cutting for a cutting distance of 88 m. As seen in Figure 5.36(a), severe flank wear was found for conventional cutting, causing a 4.3 μm recession of the cutting edge. In addition, rake wear and material adhesion occurred. For the ultrasonic vibration, as shown in Figure 5.36(b), a distinctly small tool wear with width <1 μm was formed on the flank face, which is attributed to the reduction in effective contact time between the tool and the workpiece induced by vibration. Additionally, in the inset, a uniform thin layer of material adhesion covering the rake surface was identified. This thin adhesion layer might be a result of interfacial ultrasonic impacts and could have served to protect the rake face from erosion by a fresh chip surface.

## 5.4 MACHINABILITY OF POLYMETRIC MATERIALS

A polymer is a substance or material consisting of very large molecules, often composed of repeating structural units, formed through the process of polymerization. Their consequently large molecular mass, relative to small molecule compounds, produces unique physical properties including toughness, elasticity, viscoelasticity, and a tendency to form amorphous and semicrystalline structures rather than crystals. Additionally, the softening points of commonly used polymers are much lower than metals. In the machining of polymers, the alterations in material properties induced by cutting heat can significantly affect the machined surface quality and form errors. This section introduces the machinability of several polymer materials.

### 5.4.1 Photoresist

Photoresist is a light-sensitive material employed in photolithography and photoengraving to create a patterned coating on a surface. It is applied as a liquid and spin-coated to ensure uniform thickness. While photoresist is typically removed by a developer liquid through a chemical reaction, there are instances where mechanical machining of photoresist becomes necessary. For example, in the bump planarization application discussed in Section 5.3.4, to prevent contamination, the photoresist used as a mask in the bump electroplating process needs to be flattened together with the bumps and subsequently removed after the planarization step. As the planarization quality of the bumps is strongly influenced by that of the photoresist, it is essential to understand the cutting characteristics of the photoresist.

Figure 5.37 shows Nomarski micrographs of the machined surface at different radial positions on a photoresist-coated silicon wafer. In the inner region, where the cutting speed is low, the photoresist surface is free from fractures, whereas in the outer region, where the cutting speed is high, numerous fractures are observed. This indicates that the cutting speed is a significant factor influencing the surface quality of the photoresist. Figure 5.38(a) further compares the percentage of the fractured area of the surface machined with various cutting speeds ($V_c$). Microfractures begin to form significantly when the cutting speed exceeds 300 m/min. Figure 5.38(b) illustrates the trend of cutting temperature changes with cutting speed. As the cutting speed increases, the cutting temperature gradually rises. Figure 5.38(c) and (d) plot variations in principal and thrust cutting forces with respect to cutting speed. With an increase in cutting speed, both forces exhibit a decrease.

High cutting speed might cause two effects in a photoresist, thermally induced softening/melting and high strain rate induced embrittlement. These two effects

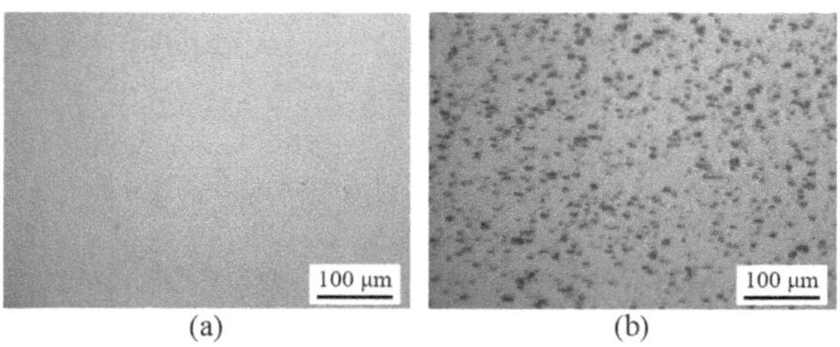


**FIGURE 5.37** Nomarski micrographs of a machined surface at different radial positions of the photoresist workpiece: (a) inner region, (b) outer region [47].

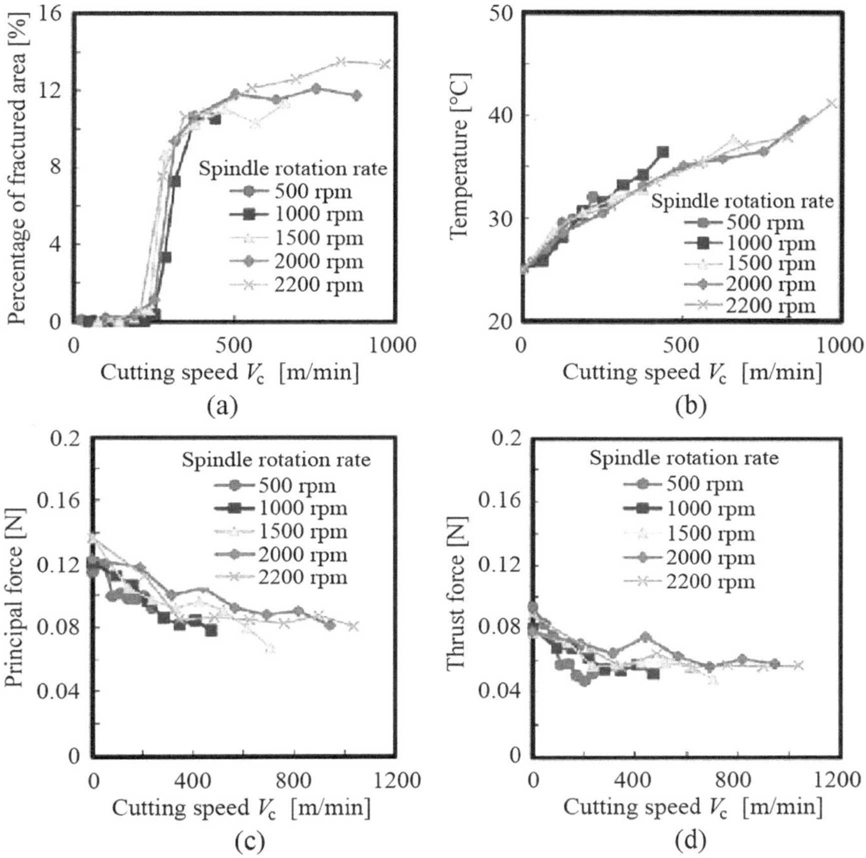


**FIGURE 5.38** Effects of cutting speed on (a) percentage of fractured area, (b) cutting temperature, (c) principal force, and (d) thrust force in photoresist cutting [47].

may coexist, and the predominance of either depends on the composition and properties of the photoresist. Melting of the photoresist occurred during cutting due to the rise of temperature, as indicated by the chip adhesion observed on the tool rake face. Therefore, thermally induced softening/melting might be predominant in the cutting process, and this material softening resulted in a decrease in cutting forces. After cutting, however, the photoresist becomes brittle due to cooling; meanwhile, the photoresist undergoes shrinking and hardening. Thus, tensile stress is induced, leading to the generation of microfractures.

Figure 5.39(a) depicts the changing trend of fracture area percentage with the depth of cut. As the depth of cut increases, microfracture generation in the photoresist becomes significant, even at a low cutting speed. Figure 5.39(b) shows results for various tool feed rates. The general trend of the fracture area percentage change with feed rate is similar to that with depth of cut. Figure 5.39(c) plots

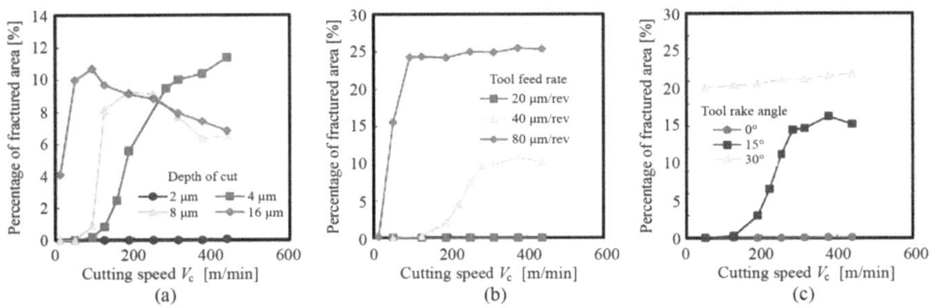


**FIGURE 5.39** Effect of (a) depth of cut, (b) feed rate, and (c) tool rake angle on percentage of fractured area in photoresist cutting [47].

the percentage of fractured area with different tool rake angles. The percentage increases dramatically with the tool rake angle. A high rake angle enhances tensile stress in the cutting region, which causes a brittle material to fracture; this demonstrates again that the photoresist has a brittle nature. Therefore, microfractures generated at a high rake angle, a large depth of cut, and a high tool feed rate are due to non-thermal reasons, which are distinctly different from microfracture generation at a high cutting speed. From the above results, we can conclude that photoresist is a material having double-face properties: significant thermal plasticity and high brittleness. These two aspects must be considered in the precision machining of photoresist.

## 5.4.2 PMMA

Precision plastic components, such as optical lenses and prisms, are typically mass-produced through plastic injection molding. However, in specific instances, such as the prototyping of camera lenses and contact lenses, direct diamond turning of plastic becomes essential. The machining characteristic of plastic significantly depends on the chemical composition and material properties of the plastic workpiece. For example, polymethyl methacrylate (PMMA), polycarbonate (PC), polypropylene (PP), and high-density polyethylene (HDPE) exhibit relatively good micro machinability, whereas low-density polyethylene (LDPE), polytetrafluoroethylene (FTFE), and epoxy resin (EP) show poor machinability [61]. When machining LDPE and PTFE, tearing phenomena were observed, while EP cutting resulted in the generation of fractured chips. Figure 5.40 compares the surfaces of PMMA, PC, and HDPE machined at different feed rates (*f*), where clear differences between these machined surfaces can be seen. When the feed rate decreases to less than 100 μm/rev, the machined surfaces of PC and HDPE exhibit less prominent feed marks due to the deteriorating surface quality. In contrast, the machined surface of PMMA achieves a cracks-free surface, and fine-pitched patterns perpendicular to the cutting direction are diminished. An optical-grade surface roughness of about 5 nm Ra can be achieved for PMMA at a cutting speed of 730 m/min,

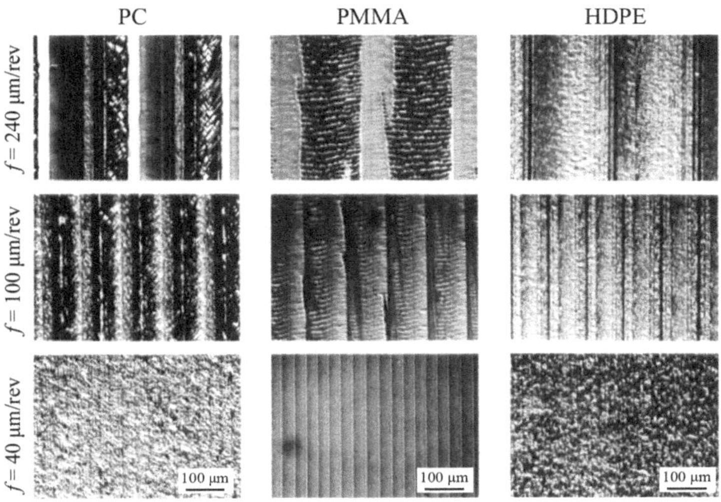


**FIGURE 5.40** Comparison of surfaces of PMMA, PC, and HDPE machined at different feed rates (*f*) with a cutting speed of 300 m/min and a depth of cut of 10 μm [61].

a depth of cut of 10 μm, and a feed rate of 3 μm/rev. Therefore, PMMA is the most suitable plastic for optical parts using a cutting process.

The machining characteristics of PMMA show differences in continuous diamond turning (with workpiece fixed at spindle center) and intermittent cutting (with workpiece fixed at the spindle outer edge). Figure 5.41 shows the results of micro grooving experiments for intermittent and continuous cutting, respectively, at the same cutting speed (45 m/min). Continuous cutting produced a smooth surface under all feed rates; in contrast, fine scratch marks were observed during intermittent cutting. The reason for this phenomenon is that in intermittent cutting, the temperature rise of the PMMA surface is small due to the cooling effect of air cuts, the chips removed from the edge region of the workpiece are attached to the tool tip and scratch the groove surface. In continuous cutting, however, there are no such chip attachments and surface scratching does not occur.

Moreover, the width of the grooves is different, indicating that the actual depth of the grooves is also different, although the designed depth of the groove is the same (10 μm). The comparison of the residual depth of the grooves is plotted in Figure 5.42. The groove depth achieved by continuous cutting was deeper than that by intermittent cutting across all feed rates. In continuous cutting, the groove depth remained stable at about 9.7 μm with minimal variation with feed rate, while in intermittent cutting, the groove depth decreased as the feed rate was reduced. This suggests that continuous cutting provides greater stability than intermittent cutting, even when varying the tool feed rate, and that a groove depth closely approximating the theoretical depth of cut can be obtained.

**FIGURE 5.41** Surface morphology of grooves cut under various feed rates in depth direction: (a), (d) $f = 0.1$ μm/rev, (b), (e) $f = 0.2$ μm/rev, and (c), (f) $f = 0.5$ μm/rev. (a)–(c) are intermittent cutting. (d)–(f) are continuous cutting.

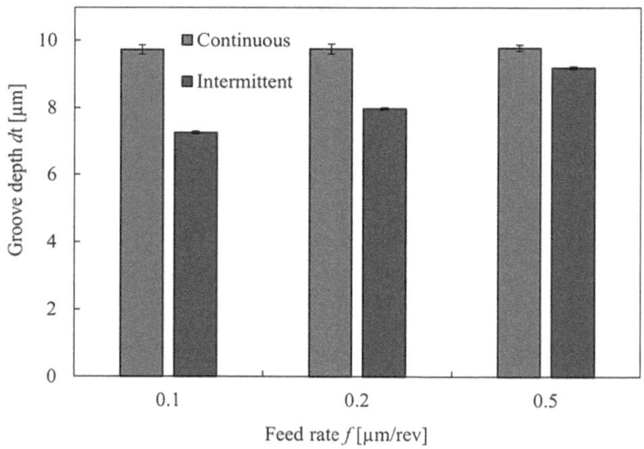


**FIGURE 5.42** Comparison of the depth of the grooves cut by continuous cutting and intermittent cutting at various feed rates.

Figure 5.43 shows the relationship between cutting force and tool feed rate for continuous and intermittent cutting. Both the principal force and thrust force were greater in continuous cutting across all feed rates. The different machining characteristics observed during the cutting of PMMA are attributed to the different heat generation and cooling situations during intermittent and continuous cutting

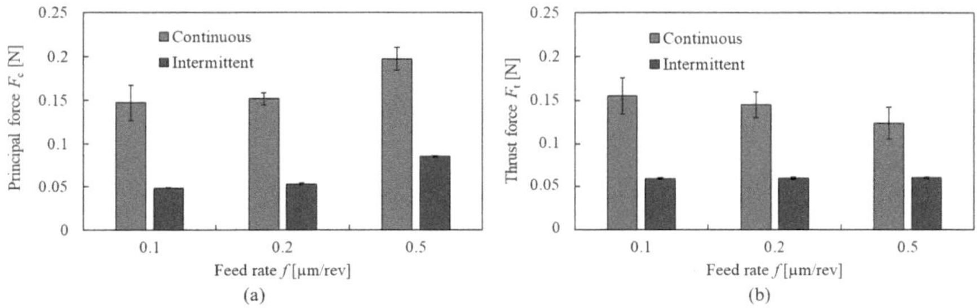


**FIGURE 5.43** Comparison of (a) principal force and (b) thrust force during continuous cutting and intermittent cutting at various feed rates.

processes. PMMA has a relatively lower softening point compared to metals and, thus is sensitive to the presence/absence of cooling.

Plastics machining is challenging because of the lower melting point and poor thermal conductivity compared to metals. The cutting heat is accumulated in cutting zones, which degrades the tool life and locally melts the plastic. Once the plastic melts, it flows around the cutting edge instead of being removed. The material stagnation not only leads to an increase in surface roughness, but also causes increase in cutting force. The machinability of plastics can be enhanced by utilizing a high-lubricity coolant to maintain the tool edge free of chip build-up. Additionally, the coolant helps cool the material, preventing it from viscously deforming or adhering to the cutting edge. Elastic recovery is another factor that influences the surface quality and form accuracy in machining plastics. To suppress the elastic deformation of the workpiece, it is essential to use a diamond tool with a sharp edge to reduce the cutting force, especially the thrust force.

### 5.4.3 Cellulose Nanofiber-Reinforced Plastics

Cellulose nanofiber (CeNF) is an attractive material for a variety of industrial applications, owing to its excellent properties such as high rigidity, light weight, low coefficient of thermal expansion, and optical transparency. Mixing polymer with CeNF can improve the mechanical strength of the polymer without compromising its transparency. In recent years, various CeNF-reinforced plastics have been developed, for example, CeNF-reinforced homogeneous polypropylene (homo PP), which has wide potential applications in the manufacturing of optical and mechanical parts.

Figure 5.44 compares the surfaces of pure homo PP and CeNF-reinforced homo PP (CeNF-PP) machined at various tool feed rates and cutting speeds. For pure PP, the machined surfaces were smooth at a low cutting speed ($V$ = 0–30 m/min) regardless of the tool feed rate, as shown in Figure 5.44(a1) and (b1). In contrast, surface defects caused by material tearing are observed on the surfaces at a high cutting speed ($V$ = 180–220 m/min), as shown in Figure 5.44(a2) and (b2). For CeNF-PP, submicron-scale tearings are generated even at the low cutting

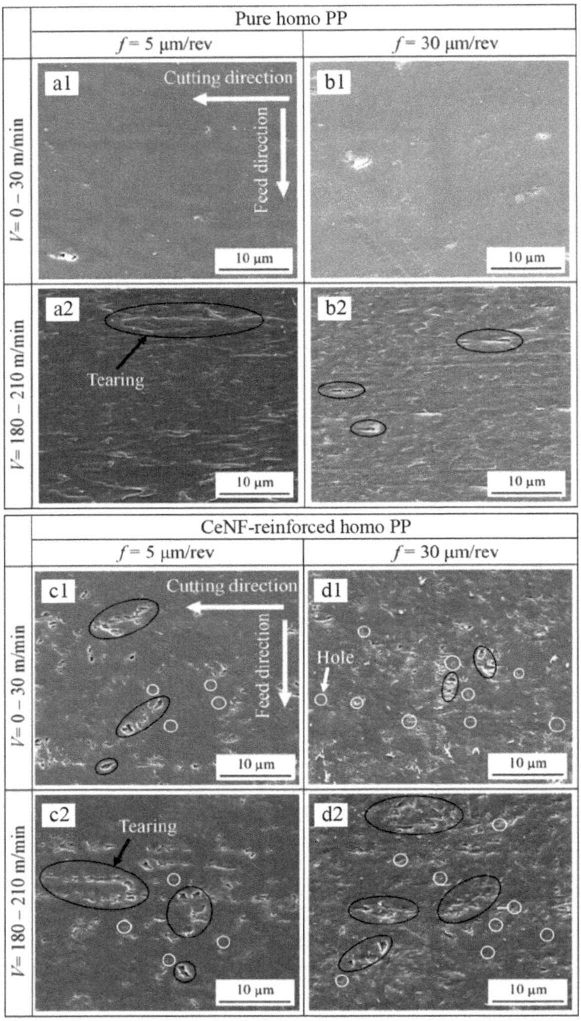


**FIGURE 5.44** SEM images of machined surfaces of pure homo PP and CeNF-reinforced homo PP at various tool feed rates and cutting speeds [62].

speed, which are different from the results obtained in pure PP. This might be due to that during the cutting process, some CeNFs were pulled out with cutting chips, instead of being cut off. In addition, the flatness of the machined surfaces under a small feed rate ($f$ = 5 μm/rev) shown in Figure 5.44(c1,2) are better than those at a large feed rate ($f$ = 30 μm/rev) shown in Figure 5.44(d1,2). The cutting speed has less effect on the surface quality of PP-CeNF compared with the tool feed rate. This might be because the tensile strength of PP-CeNF is much larger than that of pure PP. Consequently, the viscoelastic induced tearings can be suppressed under the high cutting speed.

Figure 5.45 shows SEM images of cutting chips of pure PP and PP-CeNF. From a general view, ribbon-like continuous cutting chips were generated in each cutting condition. This indicates that both pure PP and PP-CeNF were removed in ductile mode. However, these cutting chips exhibit different features at the high-magnification view. For pure PP, the surface of cutting chip is smooth without any holes even at the large feed rate, as shown in Figure 5.45(a). The cutting chips generated at the small feed rate are similar to those at the large feed rate, although not shown. In contrast, for PP-CeNF, at the large feed rate, tearing-induced surface defects are dominant on the chip surface, as shown in Figure 5.45(b). In addition, protruding CeNFs, which were pulled out from the polypropylene matrix, are directly observed on the chip surface. When the feed rate is small, the chip surface is smoother than that at the large feed rate, as shown in Figure 5.45(c). It is worth noting that there are more holes on the chip surface, indicating that at the small feed rate, CeNFs were more likely to be pulled out.

Figure 5.46 shows the surface roughness of the machined pure PP and PP-CeNF workpieces. The figure also compares the surface roughness at different cutting speeds. In all cases, the surface roughness at a small feed rate is lower than that at a large feed rate under the same cutting speed. For pure PP, the surface roughness abruptly increases at high cutting speeds for both feed rates of 5 and 30 μm/rev, while it remains at the same level for other cutting speeds. This indicates that there is a critical cutting speed of around 150–180 m/min, beyond which a large number of viscoelastic-induced tearings are generated. In contrast, for PP-CeNF, under the large feed rate, the surface roughness gradually increases with the cutting

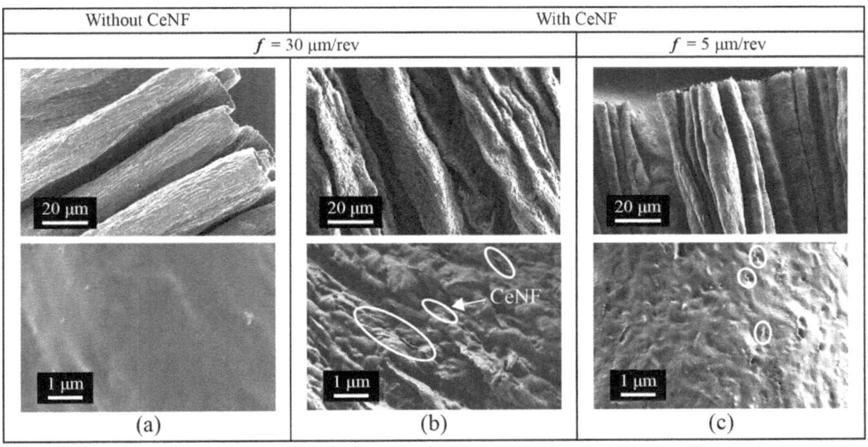


**FIGURE 5.45** SEM images of cutting chips: (a) pure homo PP at feed rate (*f*) = 30 μm/rev; CeNF-reinforced homo PP at (b) *f* = 30 μm/rev, and (c) *f* = 5 μm/rev [62].

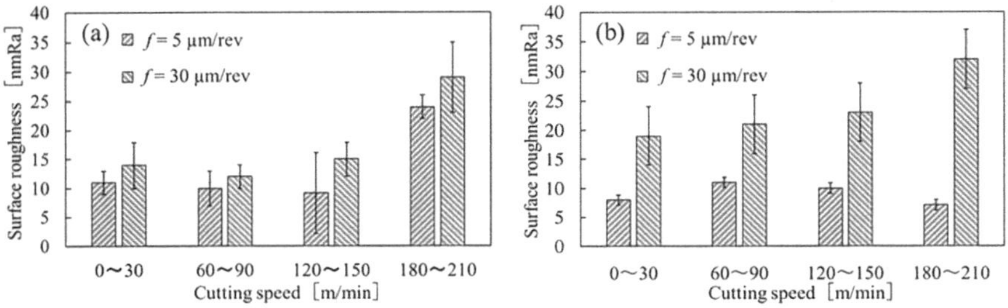


**FIGURE 5.46** Surface roughness of the machined workpieces: (a) pure homo PP; (b) CeNF-reinforced homo PP [62].

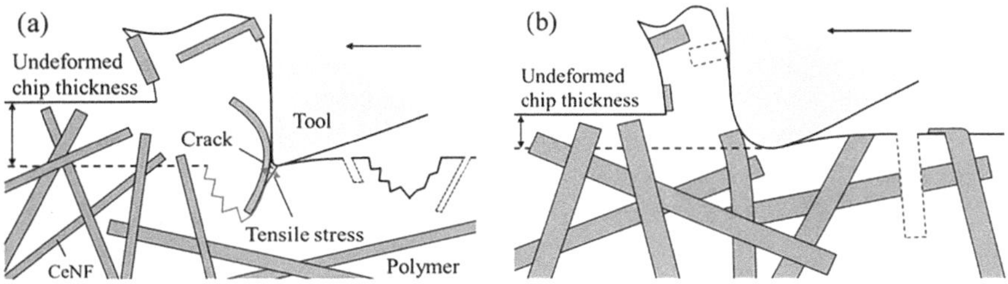


**FIGURE 5.47** Schematic models of cutting PP-CeNF: (a) at a large undeformed chip thickness; (b) at a small undeformed chip thickness [62].

speed, while under the small feed rate, the surface roughness is hardly affected by the cutting speed.

At the large feed rate, a violent tensile stress occurs below the tool tip due to the large undeformed chip thickness, leading to crack initiation. Meanwhile, the polypropylene matrix surrounding the tool tip softens due to high temperature. Therefore, if the length of fiber inside the matrix is long, the fibers will be pulled out by strong friction force and take away a lot of matrix material. As a result, tearings are generated on the machined surface of PP-CeNF, as illustrated in Figure 5.47(a). Whereas, at the small feed rate, the undeformed chip thickness is also small. Thus, the short length of fibers inside the chips makes them easily pulled out, resulting in the generation of holes on the cutting chips, as illustrated in Figure 5.47(b). Since the tensile stress below the tool tip is suppressed at a small undeformed chip thickness, cracks are less likely to occur. In addition, the polypropylene matrix surrounding the tool tip is relatively hard. Consequently, the fibers are primarily cut by the tool edge through strong shear stress or embedded inside the matrix under compressive stress, leading to significant suppression of the pulling-out of fibers and induced tearings. Therefore, the machined surface of the workpiece becomes smooth with only a few tearings and holes.

## REFERENCES

[1] Merchant, M.E. (1945) Mechanics of the metal cutting process. I. Orthogonal cutting and a type 2 chip. *Journal of Applied Physics* 16 (5): 267–275.
[2] Merchant, M.E. (1945) Mechanics of the metal cutting process. II. Plasticity conditions in orthogonal cutting. *Journal of Applied Physics* 16 (6): 318–324.
[3] Komanduri, R. (1971) Some aspects of machining with negative rake tools simulating grinding. *International Journal of Machine Tool Design and Research* 11 (3): 223–233.
[4] Chae, J., Park, S.S., Freiheit, T. (2006) Investigation of micro-cutting operations. *International Journal of Machine Tools and Manufacture* 46 (3): 313–332.
[5] Ikawa, N., Shimada, S., Tanaka, H., Ohmori, G. (1991) An Atomistic Analysis of Nanometric Chip Removal as Affected by Tool-Work Interaction in Diamond Turning. *CIRP Annals* 40 (1): 551–554.
[6] Shimada, S., Ikawa, N., Tanaka, H., Ohmori, G., Uchikoshi, J., Yoshinaga, H. (1993) Feasibility study on ultimate accuracy in microcutting using molecular dynamics simulation. *CIRP Annals* 42 (1): 91–94.
[7] Ikawa, N., Shimada, S., Tanaka, H. (1992) Minimum thickness of cut in micromachining. *Nanotechnology* 3 (1): 6.
[8] Yuan, Z.J., Zhou, M., Dong, S. (1996) Effect of diamond tool sharpness on minimum cutting thickness and cutting surface integrity in ultraprecision machining. *Journal of Materials Processing Technology* 62 (4): 327–330.
[9] Liu, X., DeVor, R.E., Kapoor, S.G. (2005) An analytical model for the prediction of minimum chip thickness in micromachining. *Journal of Manufacturing Science and Engineering* 128 (2): 474–481.
[10] Furukawa, Y., Moronuki, N. (1988) Effect of material properties on ultra precise cutting processes. *CIRP Annals* 37 (1): 113–116.
[11] Kim, J., Kim, D.S. (1995) Theoretical analysis of micro-cutting characteristics in ultra-precision machining. *Journal of Materials Processing Technology* 49 (3): 387–398.
[12] Taminiau, D.A., Dautzenberg, J.H. (1991) Bluntness of the tool and process forces in high-precision cutting. *CIRP Annals* 40 (1): 65–68.
[13] Lucca, D.A., Rhorer, R.L., Komanduri, R. (1991) Energy dissipation in the ultraprecision machining of copper. *CIRP Annals* 40 (1): 69–72.
[14] Lucca, D.A., Seo, Y.W., Komanduri, R. (1993) Effect of tool edge geometry on energy dissipation in ultraprecision machining. *CIRP Annals* 42 (1): 83–86.
[15] Yan, J., Syoji, K., Kuriyagawa, T., Suzuki, H. (2002) Ductile regime turning at large tool feed. *Journal of Materials Processing Technology* 121 (2): 363–372.
[16] Zhang, S.J., To, S., Wang, S.J., Zhu, Z.W. (2015) A review of surface roughness generation in ultra-precision machining. *International Journal of Machine Tools and Manufacture* 91: 76–95.
[17] Cheung, C.F., Lee, W.B. (2000) A theoretical and experimental investigation of surface roughness formation in ultra-precision diamond turning. *International Journal of Machine Tools and Manufacture* 40 (7): 979–1002.
[18] Cheung, C.F., Lee, W.B. (2000) Study of factors affecting the surface quality in ultra-precision diamond turning. *Materials and Manufacturing Processes* 15 (4): 481–502.
[19] Cheung, C.F., Lee, W.B. (2001) Characterisation of nanosurface generation in single-point diamond turning. *International Journal of Machine Tools and Manufacture* 41 (6): 851–875.

[20] Dogra, M., Sharma, V.S., Dureja, J. (2011) Effect of tool geometry variation on finish turning - A review. *Journal of Engineering Science and Technology Review* 4 (1): 1–13.
[21] Malayath, G., Mote, R.G. (2022) A review of cutting tools for ultra-precision machining. *Machining Science and Technology* 26 (6): 923–976.
[22] Günay, M., Korkut, O., Aslan, E., Şeker, U. (2005) Experimental investigation of the effect of cutting tool rake angle on main cutting force. *Journal of Materials Processing Technology* 166 (1): 44–49.
[23] Li, A., Zang, J., Zhao, J. (2020) Effect of cutting parameters and tool rake angle on the chip formation and adiabatic shear characteristics in machining Ti-6Al-4V titanium alloy. *The International Journal of Advanced Manufacturing Technology* 107 (7): 3077–3091.
[24] Zhang, S., Guo, P., Xiong, Z., To, S. (2020) Cyclic shear angle for lamellar chip formation in ultra-precision machining. *Proceedings of the Institution of Mechanical Engineers, Part C: Journal of Mechanical Engineering Science* 234 (13): 2673–2680.
[25] Huang, W., Yan, J. (2020) Surface formation mechanism in ultraprecision diamond turning of coarse-grained polycrystalline ZnSe. *International Journal of Machine Tools and Manufacture* 153: 103554.
[26] Huang, W., Yan, J. (2023) Mechanisms of tool-workpiece interaction in ultraprecision diamond turning of single-crystal SiC for curved microstructures. *International Journal of Machine Tools and Manufacture* 191: 104063.
[27] Huang, W., Nagayama, K., Yan, J. (2023) Fabrication of microlens arrays on single-crystal CaF2 by ultraprecision diamond turning. *Journal of Materials Processing Technology* 321: 118133.
[28] Brinksmeier, E., Preuss, W., Riemer, O., Rentsch, R. (2017) Cutting forces, tool wear and surface finish in high speed diamond machining. *Precision Engineering* 49: 293–304.
[29] Moriwaki, T., Horiuchi, A., Okuda, K. (1990) Effect of cutting heat on machining accuracy in ultra-precision diamond turning. *CIRP Annals* 39 (1): 81–84.
[30] To, S., Chan, C.Y., Zhang, S.J., Cheung, C.F., Zhu, Y.H., Lee, W.B. (2009) Microstructural characterization of an ultra-precision-machined surface of a Zn–Al alloy. *Journal of Micromechanics and Microengineering* 19 (5): 054005.
[31] Takasu, S., Masuda, M., Nishiguchi, T., Kobayashi, A. (1985) Influence of study vibration with small amplitude upon surface roughness in diamond machining. *CIRP Annals* 34 (1): 463–467.
[32] Kim, D., Chang, I., Kim, S. (2002) Microscopic topographical analysis of tool vibration effects on diamond turned optical surfaces. *Precision Engineering* 26 (2): 168–174.
[33] Yan, J., Oowada, T., Zhou, T., Kuriyagawa, T. (2009) Precision machining of microstructures on electroless-plated NiP surface for molding glass components. *Journal of Materials Processing Technology* 209 (10): 4802–4808.
[34] Guo, J., Zhang, J., Pan, Y., Kang, R., Namba, Y., Shore, P., Yue, X., Wang, B., Guo, D. (2020) A critical review on the chemical wear and wear suppression of diamond tools in diamond cutting of ferrous metals. *International Journal of Extreme Manufacturing* 2 (1): 012001.
[35] Moriwaki, T. (1989) Machinability of copper in ultra-precision micro diamond cutting. *CIRP Annals* 38 (1): 115–118.
[36] Zhang, X., Zhang, Y. (1997) Study on the surface quality of a diamond-turned oxygen-free high-conductance copper reflector used in a high-power CO2 laser. *Optical Engineering* 36 (3): 825–830.

[37] Sugano, T., Takeuchi, K., Yoshida, Y. (1993) Diamond turning of oxygen-free copper for mirrors. *JSME international journal. Ser. C, Dynamics, control, robotics, design and manufacturing* 36 (4): 549–555.
[38] Ueda, K., Iwata, K., Nakayama, K. (1980) Chip formation mechanism in single crystal cutting of β-brass. *CIRP Annals* 29 (1): 41–46.
[39] Yuan, Z.J., Lee, W.B., Yao, Y.X., Zhou, M. (1994) Effect of crystallographic orientation on cutting forces and surface quality in diamond cutting of single crystal. *CIRP Annals* 43 (1): 39–42.
[40] Lee, W.B., Cheung, C.F., To, S. (2003) Friction-induced fluctuation of cutting forces in the diamond turning of aluminium single crystals. *Proceedings of the Institution of Mechanical Engineers, Part B: Journal of Engineering Manufacture* 217 (5): 615–631.
[41] Sugihara, T., Enomoto, T. (2012) Improving anti-adhesion in aluminum alloy cutting by micro stripe texture. *Precision Engineering* 36 (2): 229–237.
[42] Sugano, T., Takeuchi, K., Goto, T., Yoshida, Y., Ikawa, N. (1987) Diamond turning of an aluminum alloy for mirror. *CIRP Annals* 36 (1): 17–20.
[43] Uemura, M. (2004) An analysis of the catalysis of Fe, Ni or Co on the wear of diamonds. *Tribology International* 37 (11): 887–892.
[44] Yan, J.W., Sasaki, T., Tamaki, J., Kubo, A., Sugino, T. (2004) Chip formation behaviour in ultra-precision cutting of electroless nickel plated mold substrates. *Key Engineering Materials* 257-258: 3–8.
[45] Furushiro, N., Tanaka, H., Higuchi, M., Yamaguchi, T., Shimada, S. (2010) Suppression mechanism of tool wear by phosphorous addition in diamond turning of electroless nickel deposits. *CIRP Annals* 59 (1): 105–108.
[46] Kwon, S.C., Hideaki, T., Yoshiharu, N. (2014) Wear inspection of a single-crystal diamond tool used in electroless nickel turning. *Optical Engineering* 53 (3): 034102.
[47] Yan, J. (2011) Ultraprecision cutting of photoresist/gold composite microstructures. *CIRP Annals* 60 (1): 133–136.
[48] Olsson, M., Persson, H., Bushlya, V., Ståhl, J. (2018) Surface roughness and sub-surface deformation measurements in machining of niobium. *Procedia CIRP* 71: 413–417.
[49] Thornton, A.G., Wilks, J. (1978) Clean surface reactions between diamond and steel. *Nature* 274 (5673): 792–793.
[50] Casstevens, J.M. (1983) Diamond turning of steel in carbon-saturated atmospheres. *Precision Engineering* 5 (1): 9–15.
[51] Zhang, X., Deng, H., Liu, K. (2019) Oxygen-shielded ultrasonic vibration cutting to suppress the chemical wear of diamond tools. *CIRP Annals* 68 (1): 69–72.
[52] Evans, C., Bryan, J.B. (1991) Cryogenic diamond turning of stainless steel. *CIRP Annals* 40 (1): 571–575.
[53] Zhang, G. (2007) Method for extending diamond tool life in diamond machining of materials that chemically react with diamond U.S. Patent No. 7198043B1, filed August 23, 2006, and issued April 3, 2007.
[54] Fujisaki, K., Yokota, H., Furushiro, N., Yamagata, Y., Taniguchi, T., Himeno, R., Makinouchi, A., Higuchi, T. (2009) Development of ultra-fine-grain binderless cBN tool for precision cutting of ferrous materials. *Journal of Materials Processing Technology* 209 (15): 5646–5652.
[55] Ezugwu, E.O., Wang, Z.M. (1997) Titanium alloys and their machinability—A review. *Journal of Materials Processing Technology* 68 (3): 262–274.
[56] Narutaki, N., Murakoshi, A., Motonishi, S., Takeyama, H. (1983) Study on Machining of Titanium Alloys. *CIRP Annals* 32 (1): 65–69.
[57] Yip, W.S., To, S. (2019) Reduction of minimum cutting thickness of titanium alloys in micro cutting by a magnetic field assistance. *IEEE Access* 7: 152034–152041.

[58] Fang, F.Z., Lee, L.C., Liu, X.D. (2005) Mean flank temperature measurement in high speed dry cutting of magnesium alloy. *Journal of Materials Processing Technology* 167 (1): 119–123.

[59] Okuda, K., Takeno, Y. (2004) Ultra-precision cutting of magnesium alloys using a single crystal diamond tool. *Journal of Japan Institute of Light Metals* 54 (11): 538–543.

[60] Zhang, L., Hashimoto, T., Yan, J. (2021) Machinability exploration for high-entropy alloy FeCrCoMnNi by ultrasonic vibration-assisted diamond turning. *CIRP Annals* 70 (1): 37–40.

[61] Kobayashi, A., Hirakawa, K. (1984) Ultraprecision Machining of Plastics. Part 1. Polymethyl Methacrylate. *Polymer-Plastics Technology and Engineering* 22 (1): 15–25.

[62] Kamada, Y., Yan, J. (2021) Machinability investigation for cellulose nanofiber-reinforced polymer composite by ultraprecision diamond turning. *International Journal of Automation Technology* 15 (4): 475–482.

# 6 Machining Mechanism and Machinability of Brittle Materials

## 6.1 INTRODUCTION

Most non-metallic single- and polycrystalline materials, such as silicon (Si), germanium (Ge), calcium fluoride ($CaF_2$), and zinc selenide (ZnSe), are brittle materials. Various glass materials, such as borosilicate glass and soda-lime glass, and ceramics, such as reaction-bonded silicon carbide (RB-SiC) and tungsten carbide (WC), are also brittle materials. These materials are widely used in the fields of semiconductor, optics, biomedicine, aerospace, and so on, and thus have become tremendously important in our society. To obtain better performance of parts, the requirements for surface quality and form accuracy are getting higher and higher. However, it is challenging to machine such brittle materials with high surface integrity because they have very low fracture toughness and easily fracture, causing surface and subsurface cracks and other types of defects. For this reason, in the past several decades, precision machining of brittle materials has been a subject of concentrated research interests from both the academic and industrial communities.

It has been conventionally considered that brittle materials cannot be machined precisely like ductile materials, and a smooth surface can only be obtained on brittle materials by loose abrasive finishing processes such as lapping and polishing under extremely low load conditions. However, this situation has changed since the 1980s when it was demonstrated that a nominally brittle material can be machined in a ductile mode by ultraprecision cutting and grinding without generation of microfractures. This technology was later termed "ductile regime machining" or simply "ductile machining". Compared with loose abrasive finishing, ductile machining has remarkably high productivity, numerical controllability, and flexibility for fabricating complex shapes or small components. In this chapter, the fundamentals of ductile machining technology will be introduced and the state-of-the-art research and developments on ductile machining technology will be reviewed. The ductile machinability of various brittle materials will be compared in detail based on the available literature.

DOI: 10.1201/9781003292388-6

## 6.2 MACHINING MECHANISM

### 6.2.1 Brittle–Ductile Transition

"Ductile-brittle transition" is a technical term used in metallurgy. At low temperatures some metals that would be ductile at room temperature become brittle. This is known as a "ductile-to-brittle transition". This transition is essential for the understanding of fracture processes of materials. Experiments indicate that solids are brittle at very low temperatures, but their toughness becomes much higher at an elevated temperature. The transition temperature is a critical measure for structural material. Besides temperature, strain rate also influences the fracture toughness. The strain rate, alternatively, defines the speed imposed on the deformation of a material. A faster strain rate likely induces brittle fracture. With other conditions fixed, a slower loading rate leads to ductile fracture.

The term "brittle–ductile transition" used here has a different meaning. It is used to describe a change in the machining mode from a "brittle mode" where microfractures occur in the tool-workpiece contact region to a "ductile mode" where plastic deformation dominates the material removal. This transition takes place as one or multiple machining conditions, such as depth of cut, cutting speed, tool rake angle, and environmental temperature, are strictly limited to some specific ranges in which brittle fractures are suppressed.

All brittle materials may exhibit a certain amount of ductility although the extents of ductility are normally extremely low. Therefore, if a brittle material is machined under similar conditions to metal machining, i.e., a large undeformed chip thickness and a positive tool rake angle, as schematically shown in Figure 6.1, there will be a tensile stress concentration in the vicinity of the cutting edge of the tool. Due to the low fracture toughness of brittle materials, as the tool advances, cracks will be immediately initiated in the stress concentration region before plastic deformation occurs. This results in brittle mode material removal.

However, a number of studies have demonstrated a fundamental fact that a variety of brittle materials can be machined without brittle fracture at a sufficiently small machining scale and/or a negative tool rake angle. In other words, at an extremely small undeformed chip thickness, even a brittle material can be machined via plastic

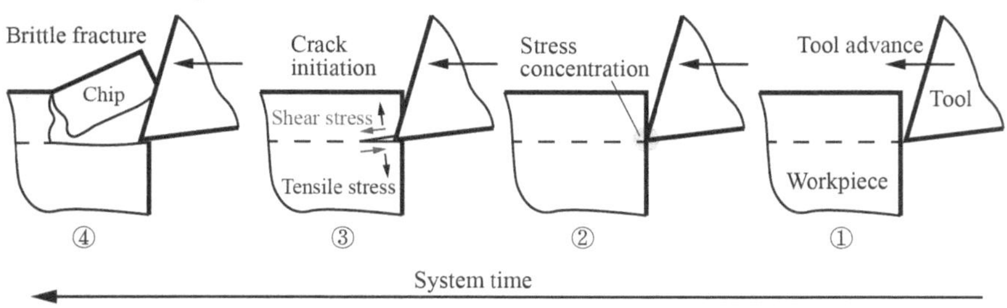


**FIGURE 6.1** Schematic of chip formation in cutting a brittle material under a large undeformed chip thickness with a positive rake angle tool.

flow, like ductile metal machining, leaving a crack-free surface. This is a kind of geometrically induced brittle–ductile transition [1–6], although it is completely different from the brittle–ductile transition of a metal material caused by a temperature rise.

The phenomena of brittle–ductile transition in machining have been confirmed for several decades; however, the scientific reason for this brittle–ductile transition has remained unclear and many hypotheses related to this transition are still controversial. To date, there are three major hypotheses on the origin of brittle–ductile transition.

#### 6.2.1.1 Defect Density Hypothesis

Nakasuji et al. [2] proposed that the brittle/ductile behavior is affected by the density of defects included in the workpiece material. Supposing the density of the defects which act as the nuclei of cleavage fractures is homogeneous in the material, the critical value of fracture occurrence depends on the size of the stress field. Cleavage fracture takes place if numerous defects exist in a large stress field, while plastic deformation occurs in a smaller stress field where the defect number is fewer. Figure 6.2 shows a model of chip removal with a "size effect" in terms of defects distribution. When undeformed chip thickness is small, the size of the critical stress field is small enough to avoid cleavage initiated by the defects, as shown in Figure 6.2(a); while when undeformed chip thickness is large, the critical stress field is extended and cracks would occur because there are enough defects in the stress field to cause fracture, as shown in Figure 6.2(b). Consequently, the transition of material removal from ductile to brittle takes place depending on the undeformed chip thickness. According to this hypothesis, to achieve material removal in ductile mode instead of brittle mode, it is necessary to control the stress field to be small and/or defects density low.

This hypothesis could explain the brittle–ductile transition in machining a material with numerous defects. However, it could not clearly explain the transition in the machining of a material without such internal defects. For example, commercially available high-grade single-crystal silicon is a nearly perfect crystalline material without detectable crystal defects. However, microfractures occur on the surface in brittle-mode machining of such silicon crystals.

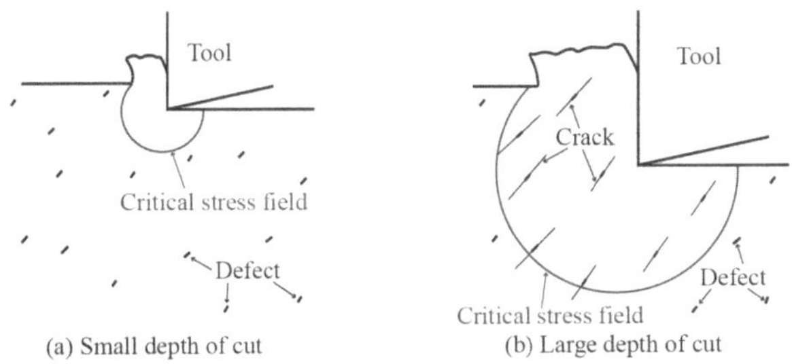


**FIGURE 6.2** A model of chip removal with a size effect in terms of defect distribution [2].

#### 6.2.1.2 Energy Scaling Hypothesis

Puttick [7] found that the relative scaling of the strain energy release rate at a growing crack and its surface energy requirement is an important factor in the relationship between specimen size and brittle fracture initiation. Specifically, the critical condition for brittle fracture is that the elastic tensile strain energy $U$ stored in a stressed volume shall be sufficient to supply the surface work $S$ required by the area $A$ of a growing crack, which can be expressed as follows [3]:

$$\frac{\mathrm{d}U}{\mathrm{d}A} \geq \frac{\mathrm{d}S}{\mathrm{d}A} \tag{6.1}$$

Similarly, Bifano et al. [5] argued the transition from brittle to ductile material removal at smaller cutting depths in the grinding process from the consideration of material removal energy. In their discussion, the energy ($E_\mathrm{p}$) required to plastically deform a specified volume of material ($V_\mathrm{p}$) was written as:

$$E_\mathrm{p} = \sigma_\mathrm{y} V_\mathrm{p} \tag{6.2}$$

where $\sigma_\mathrm{y}$ is the yield stress that characterizes the resistance of a material to plastic flow. The energy ($E_\mathrm{f}$) required for fracture is a function of the area ($A_\mathrm{f}$) of new surface created by crack propagation. Thus:

$$E_\mathrm{f} = G A_\mathrm{f} \tag{6.3}$$

where $G$ is the Griffith crack propagation parameter characterizing the resistance of a material to fracture.

Bifano et al. [5] assumed that for a machining depth $d$, the order of magnitude of both $V_\mathrm{p}$ and $A_\mathrm{f}$ are determined by $d$:

$$V_\mathrm{p} \propto d^3 \tag{6.4}$$

$$A_\mathrm{f} \propto d^2 \tag{6.5}$$

Therefore, the ratio of material removal energies can be given by:

$$\frac{\text{Plastic flow energy}}{\text{Fracture energy}} = \frac{E_\mathrm{p}}{E_\mathrm{f}} \propto d \tag{6.6}$$

Consequently, as the scale of the machining decreases, plastic flow becomes a more likely mechanism for material removal than fracture.

The energy scaling hypothesis could explain the brittle–ductile transition even for the machining of a defect-free brittle material. However, only the machining depth $d$ was involved in the energy model while other factors, such as tool geometry, were not considered. As a result, this hypothesis could not explain the effect of tool rake angle, which takes an important role in brittle material machining.

#### 6.2.1.3 Stress State Transition Hypothesis

In 1995, Morries et al. [8] found that the ductile response of some semiconductor materials in machining originated primarily from the high hydrostatic pressure-induced phase transformation in the cutting region. Many other researchers demonstrated that pressure-induced ductile response of brittle materials also exists beneath a sharp indenter tip during micro/nanoscale indentation tests. Based on this pressure-related research, Yan et al. [9] proposed a stress state transition hypothesis for brittle–ductile transition in ultraprecision machining. As schematically shown in Figure 6.3(a), when undeformed chip thickness is large, there will be a tensile stress concentration region in the vicinity of the cutting edge, whereas most of the upper region is under a low-stress state. Due to the low fracture toughness of brittle materials, cracks will be immediately initiated in the stress concentration region as the tool advances, before the occurrence of material's plastic deformation. This stress state results in brittle mode material removal. On the other hand, as the undeformed chip thickness decreases, the stress in the cutting region will become compressive. Since the effective rake angle induced by the edge radius becomes negative, the material in front of the cutting edge will be under high hydrostatic pressure, as schematically shown in Figure 6.3(b). In this case, plastic deformation will become preferable to crack generation, and ductile mode material removal can be achieved. Yan et al. [10] estimated the hydrostatic pressure in cutting silicon to be ~15 GPa by FEM simulation. This pressure is high enough to cause a phase transformation of silicon to a metallic phase which provides the ductile machining behavior [11,12].

The stress state transition hypothesis can be used to explain the brittle–ductile transition in brittle material machining by considering the effects of tool rake angle. That is, when a negative tool rake angle is used, high hydrostatic pressure can be induced, which enhances the ductile response of materials. This hypothesis is more useful than others from the viewpoint of machining process optimization.

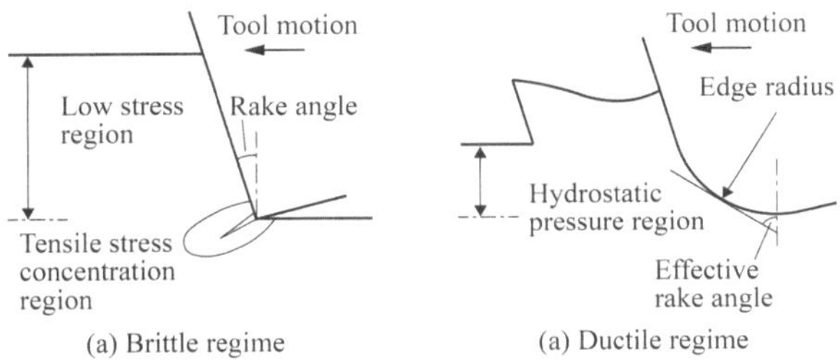


**FIGURE 6.3** Schematic illustration of the stress state transition in (a) brittle regime and (b) ductile regime [9].

### 6.2.2 Ductile Cutting Methods

#### 6.2.2.1 Round-Nosed Tool Cutting

Diamond turning is a typical method for machining a large area surface. When diamond turning of brittle materials is performed using a round-nosed tool, the machining model is schematically shown in Figure 6.4(a). The effective chip thickness $t_{\text{eff}}$ (also known as undeformed chip thickness $h$) varies along the cutting edge and can be calculated according to Eq. (6.7) [4].

$$t_{\text{eff}} = R - R\sqrt{1 - 2\frac{f}{R}\sin\theta} \tag{6.7}$$

where $f$ is the tool feed rate, $R$ is the tool nose radius and $\theta$ is the angle indicating the location of the cutting point.

The maximum effective chip thickness ($t_m$) is around the top of the uncut shoulder, as indicated in Figure 6.4(a). To make the entire chip formed in a ductile mode, the $t_m$ must be smaller than the critical chip thickness ($t_c$) for brittle–ductile transition. Therefore, for a given diamond tool, the tool feed rate $f$ must be kept sufficiently small to make the $t_m$ smaller than $t_c$. In the late 1980s, Blake and Scattergood [13] proposed a model of ductile-regime machining, as illustrated in Figure 6.4(b). The brittle fractures around the top of the uncut shoulder would be removed by the following tool passes and a crack-free surface can be fabricated. Even if the cracks are generated during the cutting process, they will not extend down to the cut surface plane if the tool feed rate is sufficiently small.

However, a low tool feed rate is the main factor limiting processing efficiency. To achieve a small $t_m$ without reducing the tool feed rate, another way is to use cutting tools having large nose radii. According to Eq. (6.7), when machining with a large nose radius tool, the entire effective chip thickness ($t_{\text{eff}}$) would be reduced, which causes the position where $t_{\text{eff}} = t_c$ to move toward the top of the uncut

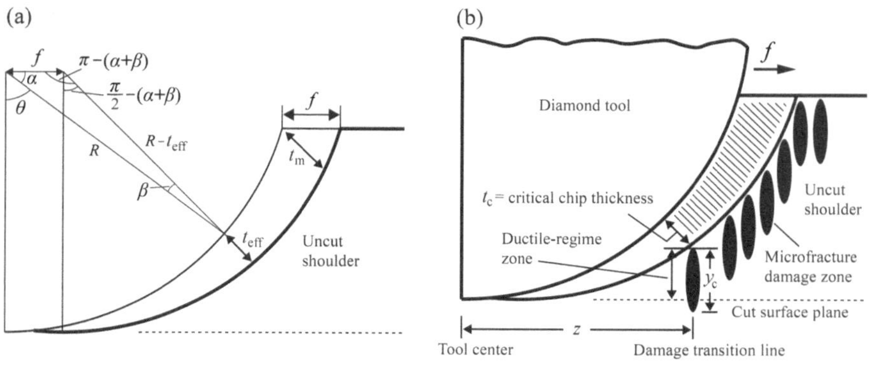


**FIGURE 6.4** Schematics of ductile cutting using a round-nosed tool: (a) calculation model of effective chip thickness and (b) distribution model of brittle fractures.

shoulder. Consequently, a wider ductile-regime zone is obtained, given other parameters, i.e., depth of cut and feed rate, are the same [14,15]. Large radii diamond tools have been used for high-efficiency ductile machining for silicon aspherical lenses in industry [16].

#### 6.2.2.2 Straight-Nosed Tool Cutting

An alternative method for applying a large tool feed is diamond turning using a straight-nosed tool, which was proposed by Yan et al. [9]. The machining model of a straight-nosed tool is presented in Figure 6.5. Compared to the round-nosed tool (see Figure 6.4), the undeformed chip thickness ($h$) is constant along almost the entire main cutting edge. Thus, ductile mode machining can be achieved when $h$ is smaller than the critical value ($h_c$) for brittle–ductile transition. The $h$ can be described by the following equation:

$$h = f \cdot \sin \kappa \tag{6.8}$$

where $f$ is the feed rate, and $\kappa$ is the cutting-edge angle.

According to Eq. (6.8), when using the straight-nosed tool, an extremely small $h$ can still be achieved even at a large tool feed by setting a small cutting edge angle $\kappa$, which ensures higher efficiency of ductile machining compared to the round-nosed tool [9]. Moreover, the uniform thickness of cutting chips over a wide cutting width $b$ provides a plane strain condition. This condition will promote the maintenance of hydrostatic pressure in the cutting region, which is necessary for ductile material removal.

A Nomarski micrograph of a silicon surface machined using a straight-nosed tool under a continuously changed tool feed is shown in Figure 6.6. As tool feed is increased from left to right, the cutting mode transits from ductile to brittle. Microfractures begin to occur at a critical tool feed $f_c$. The value of critical undeformed chip thickness $h_c$ can be then calculated from $f_c$ from the relationship shown in Eq. (6.8). It should be pointed out that, for single-crystal silicon, the critical undeformed chip thickness $h_c$ ranges from a few tens of nanometers to a few hundreds of nanometers, depending on tool geometries and workpiece material orientations [9].

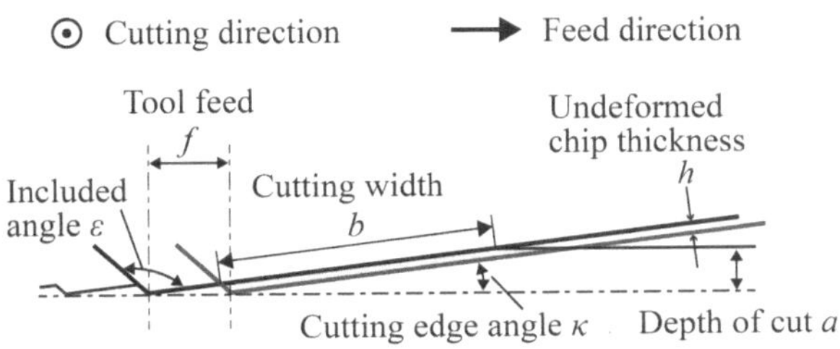


**FIGURE 6.5** Ductile machining using a straight-nosed diamond tool.

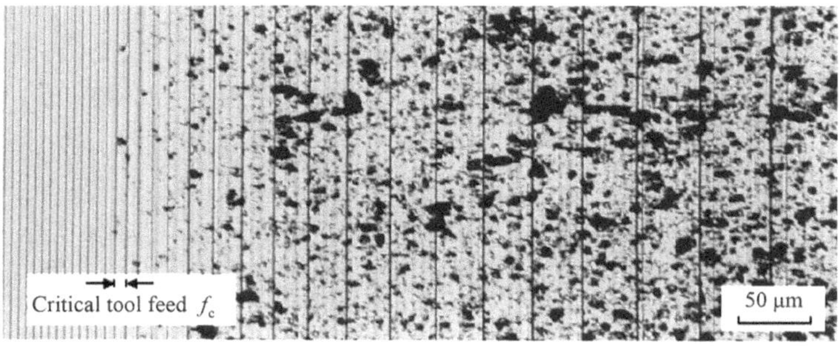


**FIGURE 6.6** Nomarski micrograph of the machined surface under continuously varied tool feed for measuring critical chip thickness of single-crystal silicon (machining conditions: [−211] orientation on (111) silicon, rake angle = −40°, $\kappa$ = 1°) [9].

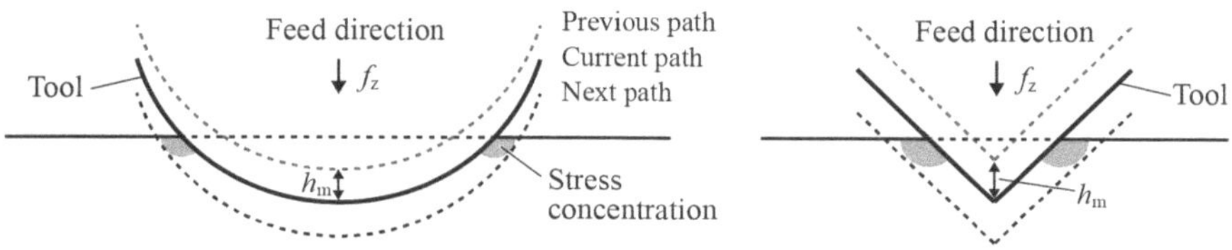


**FIGURE 6.7** Ductile mode grooving with round-nosed and V-shaped tools.

### 6.2.2.3 Grooving

In grooving processes, there is no continuous cross-feed motion of tools, but the tools feed along the direction of the depth of cut. The machining models of grooving with a round-nosed tool and a V-shaped tool are illustrated in Figure 6.7. When the tool edge profile is determined, the undeformed chip thickness is only determined by the tool feed rate $f_z$, and thus the feed rate must be controlled precisely to achieve ductile machining. Besides, when machining grooves, high-stress concentration occurs at the apex of the groove edge. Owing to the brittleness of the material, cracks are likely to initiate at this region, causing material spalling. In addition, the anisotropic effect may be more significant in grooving than in face turning, especially when using V-shaped tools, because the left- and right-side surfaces of a groove have different relative positions with respect to the crystal orientation [17].

The critical undeformed chip thickness for ductile mode grooving of a brittle material can be measured by plunge grooving experiments [18]. Figure 6.8 shows a schematic model of a microgroove plunge-cutting with a round-nosed tool. The groove is created by a single pass of tool; thus, the critical undeformed chip thickness is equal to the critical depth of cut, and the brittle–ductile transition can be observed within the machined microgroove as the depth of cut increases. For example, Figure 6.9 is a micrograph of a plunge-cut microgroove on silicon. From the right to the left of the figure, as the depth of cut increases, the microgroove is

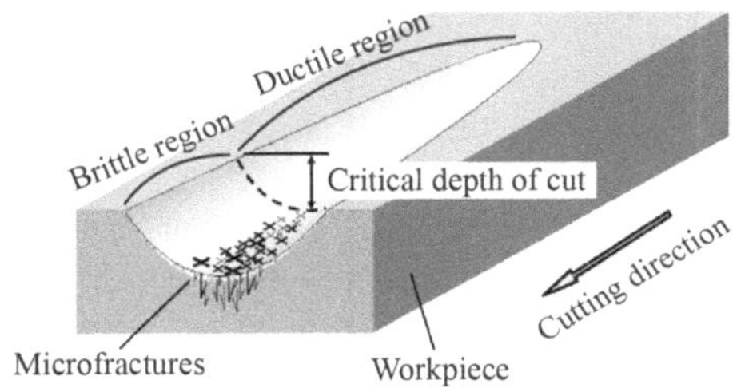


**FIGURE 6.8** Schematic model of a microgroove plunge-cut with a round-nosed tool [18].

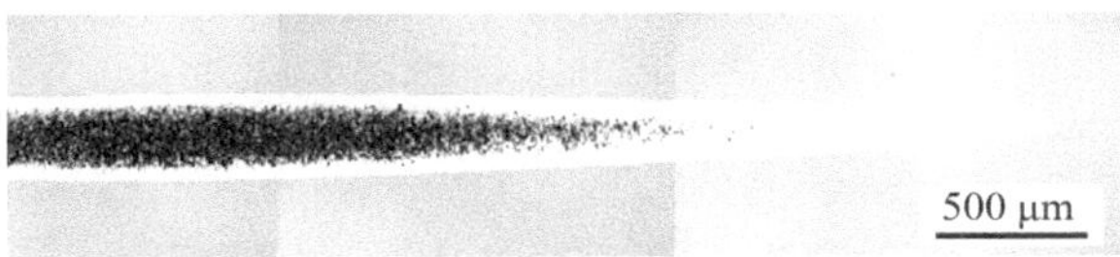


**FIGURE 6.9** Micrograph of a plunge-cut microgroove on a silicon wafer [18].

gradually broadened. Numerous microfractures have occurred at the deep region of the microgroove on the left side, whereas the right side and the fringe of the microgroove where the local depth of cut is shallow remain very smooth. A clear brittle–ductile transition boundary can be identified.

It should be pointed out that the value of the critical undeformed chip thickness measured by the plunge-cut test is slightly different from that measured by the diamond turning. This is because that diamond turning is a continuous process where a certain point on the workpiece surface is cut repetitively by multiple tool passes due to the cross-tool feed; thus, with the exception of the first cut, all subsequent cuts are made upon a damaged or deformed material layer and not the initial defect-free material [13]. On the other hand, in plunge cutting, the cutting is performed on the initial material without preexisting damages caused by previous cuts. Generally speaking, the pre-existing defects will cause a decrease in the threshold depth of brittle–ductile transition.

### 6.2.3 Effect of Tool Edge Geometry

#### 6.2.3.1 Tool Rake Angle

The machining mechanism of brittle materials is significantly affected by tool geometry. Tool rake angle, relief angle, and edge radius are three important factors of tool geometry. Generally speaking, a negative rake angle is advantageous for machining brittle materials, because the high hydrostatic pressure induced by the negative rake angle facilitates a nominally brittle material to deform in a plastic manner and be removed in a ductile mode.

For example, Fuchs et al. [19] reported that using a –45° rake angle tool in the diamond turning of KDP crystal could generate minimal subsurface damage.

Blake and Scattergood [13] conducted diamond turning tests of single-crystal germanium and silicon with diamond tools with rake angles varying from 0 to –30°. They determined the critical chip thickness $t_c$ by measuring the distance $z$ between the tool center line and the microfracture damage transition line in the uncut shoulder region (see Figure 6.4(b)), and found that the critical chip thickness increases sharply at a negative rake angle of –30°. Nakasuji et al. [2] investigated diamond turning of single-crystal germanium, silicon, and $LiNbO_3$ with diamond tools having 0 and –25° rake angles. In all cases, the negative rake angle tool enabled a larger critical chip thickness.

To reveal the effect of tool rake angle on the machining performance of brittle materials, Yan et al. [20] conducted a series of diamond turning experiments on single-crystal silicon using several specially fabricated straight-nosed diamond tools with the same clearance angle but different rake angles from 0 to –80°. Figure 6.10 shows the variation of critical chip thickness $h_c$ with the tool rake angle. The $h_c$ increases from 70 to 175 nm as the negative rake angle is changed from 0 to –40°. Further change of the negative rake angle to –80° results in a decrease in the $h_c$ to ~40 nm. The above result implies that there exists an optimum negative rake angle for maximizing $h_c$. The optimum tool rake angle is approximately –40° for cutting single-crystal silicon.

However, it should be pointed out that a highly negative rake angle may also be harmful to ductile machining from the point of view of subsurface damage. A highly negative rake angle could cause severe subsurface damage and side flows of materials. Yan et al. [20] found that at a –70° rake angle or higher, chips were formed not only in front of the rake face of the tool, but also behind the flank face. This fact indicates that the fracture occurred in the tensile stress field trailing behind the tool after the tool has passed. As the rake angle is highly negative, microfractures in front of the tool will be significantly suppressed by the high hydrostatic pressure. However, chip flow along the rake face in front of the tool

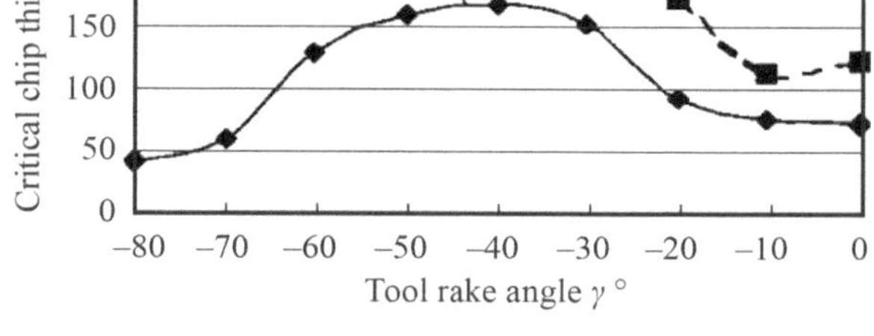


**FIGURE 6.10** The critical chip thickness of silicon under various tool rake angles.

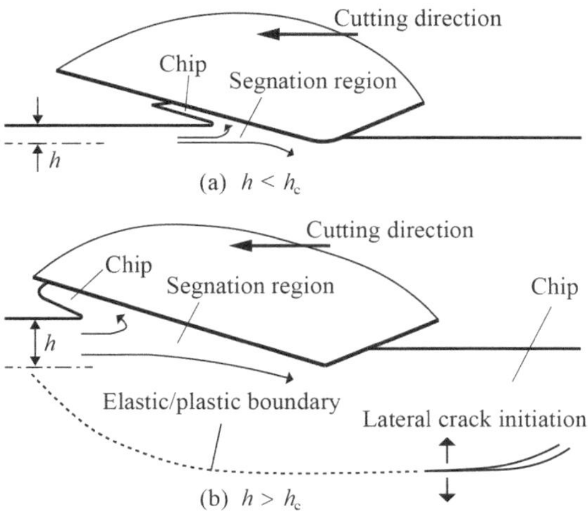


**FIGURE 6.11** Schematic of material removal mechanism at a highly negative rake angle.

will be obstructed and the downward flow will occur, which is similar to metal cutting under highly negative rake angles [21]. When the chip thickness $h$ is smaller than the critical value $h_c$, the volume of downward flow of material is small, a smaller residual stress in the subsurface region and a crack-free surface may be achieved, as shown in Figure 6.11(a). When $h > h_c$, the increased downward flow of material will lead to a large elastic/plastic deformation layer beneath the tool. After the tool passes, the residual tensile stress causes lateral crack initiation along the plastic-elastic boundary, as schematically shown in Figure 6.11(b). This fracture mechanism is similar to the lateral crack generation in indentation tests during unloading [22]. This type of fracture corresponds to a large cutting force due to the formation of a thick plastically deformed layer (subsurface damage layer) and increased surface roughness.

### 6.2.3.2 Edge Radius

A commercially available diamond tool usually has an edge radius of some tens of nanometers or larger [23]. In the cutting of brittle materials, to achieve ductile-mode material removal, the undeformed chip thickness has to be reduced to the same order as the edge radius. Therefore, the edge radius cannot be ignored like in the conventional cutting process, because the effective rake angle is no longer equal to the nominal rake angle and varies with the ratio of undeformed chip thickness ($h$) to edge radius ($r$), as shown in Figure 6.12. The effective rake angle ($\gamma_e$) of the cutting tool can be approximated with the following equation:

$$\gamma_e = -\frac{\pi}{2} + \frac{1}{2}\cos^{-1}\left(1 - \frac{h}{r}\right) \tag{6.9}$$

where $h/r$ is known as relative tool sharpness (RTS).

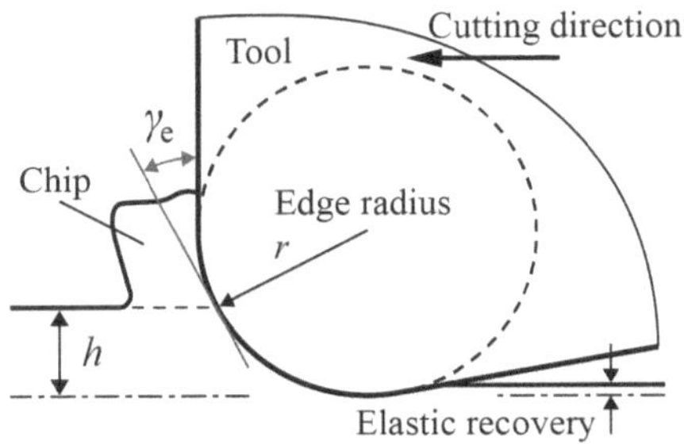


**FIGURE 6.12** Schematic of effective rake angle ($\gamma_e$) of a cutting tool.

With the variation of RTS value, the chip formation mechanisms exhibit four distinct behaviors, including shearing, extruding, plowing, and rubbing, which are independent of one another [24,25]. It is commonly accepted that there is a stagnation point S (threshold) in nanometric cutting. When the RTS is below the threshold, as shown in Figure 6.13(a), the material is not removed from the workpiece but deforms elastic-plastically and passes the lowest cutting edge point L; after the material has passed through the tool, the elastic portion $D_e$ springs back and the plastic deformed portion $\Delta$ leads to a permanent deformation behind the point P (if the material is compressible. If not, $\Delta$ should be 0). When the RTS is greater than the threshold, as shown in Figure 6.13(b), the material above the threshold will be removed and chip formation will take place.

Although these models can qualitatively describe the material removal mechanism when the undeformed chip thickness is at nanometric scale, judging the threshold for a specific material is a challenge, because the real-time observation of chip formation at such a small size scale is very difficult. Therefore, finite element method (FEM) and molecular dynamics (MD) simulations are common methods to investigate the effects of edge radius [10,26].

For example, in the diamond cutting of silicon, Liu et al. [28] applied MD simulations to understand the influence of edge radius, in which three tools with

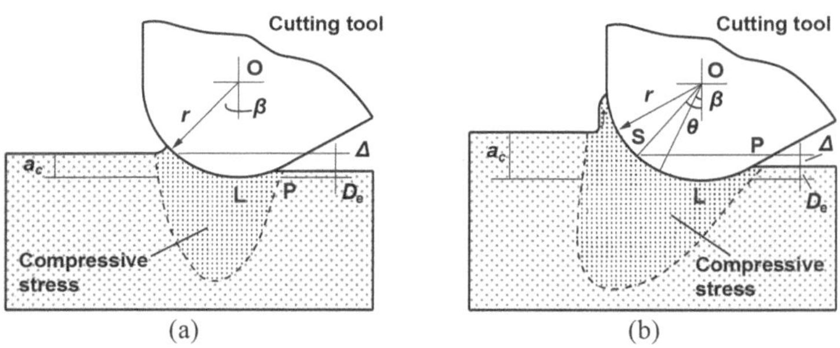


**FIGURE 6.13** Schematics of cutting models at the relative tool sharpness (a) less than the threshold and (b) greater than the threshold [27].

different edge radii (10, 20, and 40 nm) were used, and the undeformed chip thickness $h$ was set to 20 nm. In the simulated results, a stagnation region is clearly observed in all cases, as shown in Figure 6.14, indicating that there is a minimum undeformed chip thickness $h_{min}$ during cutting, and if $h$ is less than $h_{min}$, only elastic-plastic deformation occurs without chip formation. In addition, the $h_{min}$ increases with the increase of the tool edge radius; the material removal behavior changes from shear to extrusion; the larger the tool edge radius, the larger cutting forces. Ameli Kalkhoran et al. [29] conducted an MD simulation of silicon cutting considering the ratio of undeformed chip thickness and tool edge radius. It is found that, for the same RTS ratio, the larger the tool edge radius, the lower the elastic recovery rate of the material and the thicker the subsurface damage layer. Yan et al. [10] conducted FEM simulations of silicon cutting at the practically important submicron scale and investigated machining characteristics in the cutting region when the tool edge radius was changed. Simulation results showed that as the edge radius increases, the region under high equivalent stress expands and moves downward and forward away from the tool tip, as shown in Figure 6.15. Also, as the edge radius increases, the primary high-temperature zone shifts from the rake face side to the flank face side, which will cause a change in the tool wear pattern from crater wear to flank wear.

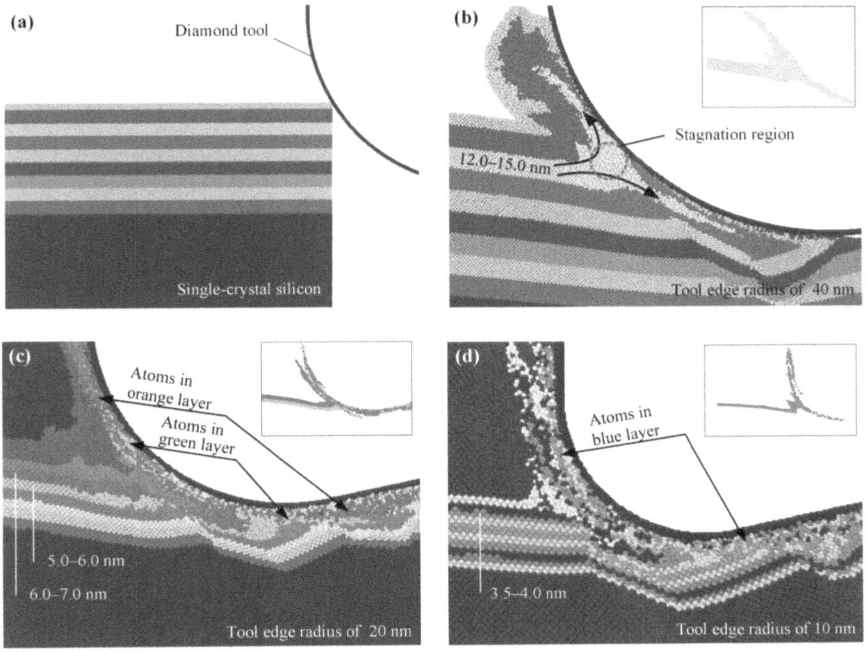


**FIGURE 6.14** Atoms flow during nanometric cutting of silicon with different edge radii: (a) initial model of atom labeling, (b) tool edge radius of 40 nm, (c) tool edge radius of 20 nm, (d) tool edge radius of 10 nm [28].

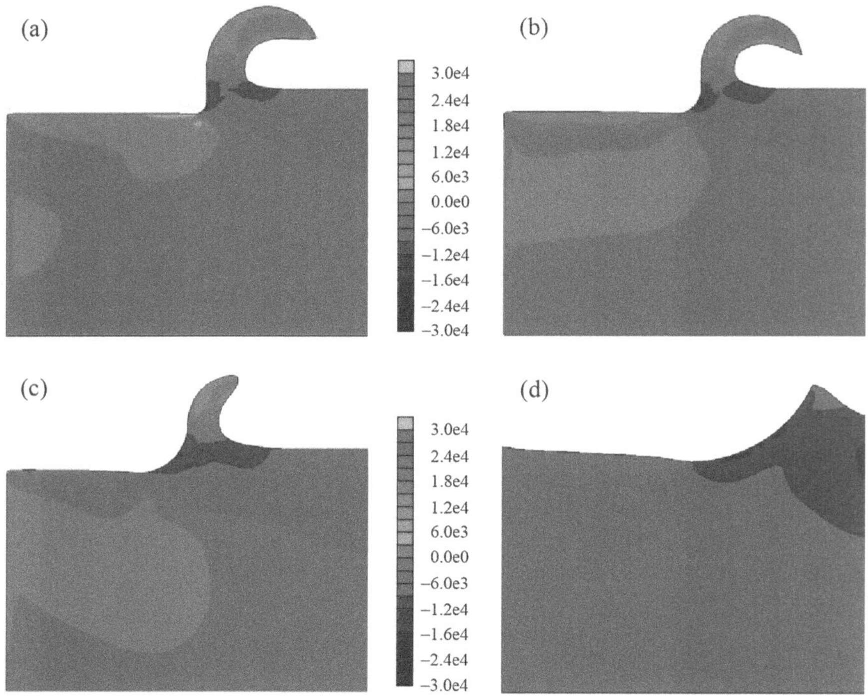


**FIGURE 6.15** Hydrostatic stress distribution in the cutting region at different tool edge radii: (a) 50 nm, (b) 100 nm, (c) 200 nm, and (d) 500 nm [10].

### 6.2.3.3 Tool Relief Angle

From geometrical point of view, the relief angle of tools will affect the contact area between the tool flank face and machined surface, and the strength of the tool. Consequently, the relief angle of the tool would affect the machining characteristics of brittle materials. Blake and Scattergood [13] experimentally investigated the critical chip thickness $t_c$ of silicon and germanium machined using diamond tools with a relief angle of 6° and 16°. For both silicon and germanium, the larger relief angle tool produced a greater $t_c$. Dai et al. [30] performed MD simulations of nano-cutting of silicon considering the effect of relief angle. Results indicated that a larger relief angle tool can produce a slightly better subsurface quality. The tool with a larger relief angle could result in a lower temperature rise in workpiece and smaller cutting forces. In the diamond cutting of KDP crystal, Zong et al. [31] indicated that both the principal and thrust forces decrease with the increase of relief angle, especially for the thrust force, due to a decreasing contact area between the machined surface and the tool flank face. The maximum contact pressure shows a trend of decreasing first and then increasing, which comes to be the lowest at the relief angle of 8°. According to the statistics for machined surface roughness, it is believed that both the smallest friction and burnishing effects at a relief angle of 8°

contribute to the best surface quality of KDP crystal. In the simulation of cutting of tungsten, Zong et al. [32] found that the quality of machined surface became better as the relief angle increased. This may be because the large tool clearance angle reduces the friction between the tool flank surface and the machined surface, and consequently reduces the fluctuation of cutting force, leading to a stable cutting process. However, the results also showed that the surface elastic recovery rate increased with increasing tool relief angle. This phenomenon may cause cracks to be generated on the machined surface of some materials after the tool has passed. For example, in cutting ZnSe, which has low hardness and high brittleness, a large relief angle tool caused submicron-scale pits to form on the machined surface, which were not found when using a small relief angle tool [15].

### 6.2.4 Effects of Crystal Orientation

The critical undeformed chip thickness for brittle–ductile transition depends on the intrinsic properties of the workpiece material. This thickness value is an important indicator of "ductile machinability" of this material. It should be pointed out that the ductile machinability of a crystalline material varies with its crystal orientation because of the crystallographic anisotropy effect in material properties.

During the turning process, cutting direction always changes with respect to the crystal orientations as the workpiece rotates. The direction of the principal cutting force changes with respect to the orientations of slipping systems and cleavage planes of the crystal. Supposing that the high pressure-induced phase transformation of the material is not significantly affected by crystalline orientation, the resolved tensile/shear stresses acting on the cleavage/slipping planes change during machining and the behavior of these planes determines whether brittle fracture or plastic deformation occurs.

For example, the critical undeformed chip thickness $h_c$ of single-crystal silicon changes from a few tens of nanometers to a few hundred of nanometers by changing the crystal orientations [9, 33]. The polar plot of $h_c$ versus crystallographic orientation of silicon (111) wafer is shown in Figure 6.16. Symmetrical peak-valley distributions of $h_c$ are observed along the circumferential orientations. Three large valleys and three small valleys appear during one revolution of the wafer. Similar crystallographic orientation effects were confirmed in diamond turning of single-crystalline germanium (Ge) [34]. The crystallographic orientation effect becomes much more significant in the machining of $CaF_2$ [35]. Crystal orientation effects may cause anisotropic surface roughness under rough machining conditions, leading to non-uniform surface integrity; moreover, it could cause circumferential waviness for the workpiece and reduce its form accuracy.

The crystal orientation dependence of the ductile machinability is a critical problem for optical fabrication applications where the finished surface quality is required to be uniform. The crystallographic effects can be eliminated significantly by using an undeformed chip thickness smaller than the critical undeformed chip thickness $h_{cmin}$ for the most difficult-to-cut orientations. This effect is attributed to the effect of tool edge roundness. As the undeformed chip thickness is

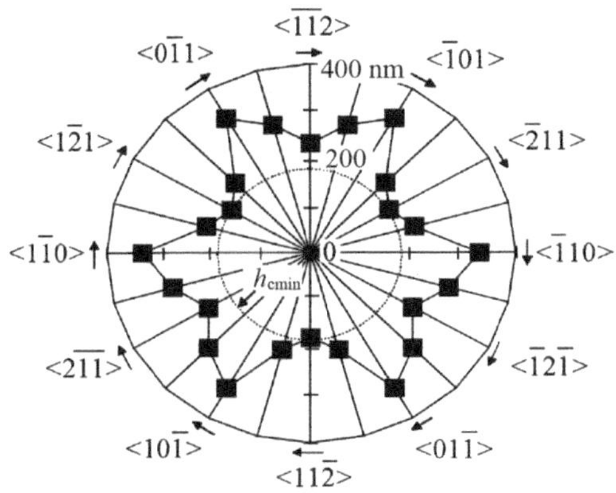


**FIGURE 6.16** Crystal orientation dependence of critical undeformed chip thickness on single-crystal silicon (111) plane [9].

reduced to the same level as the tool edge radius, the tool edge roundness decreases stress concentration and yields a relatively uniform stress field around the tool tip. As shown in Figure 6.17(a), the tool tip acts as a radially expanding core, exerting a uniform hydrostatic pressure on its surroundings, which is similar to the situation of indentation as shown in Figure 6.17(b) [11]. Therefore, compared to cutting at a larger undeformed chip thickness, the high-pressure phase transformation will dominate the material removal despite the change in crystal orientations. The deviatoric stresses causing slips (dislocations) and cleavage fractures are weak and directed to indefinite directions so that the crystallographic effects in machining phenomenon become insignificant. From this viewpoint, we can also say that the hydrostatic pressure is essential for a ductile machining process and this stress state may be achieved by optimizing the undeformed chip thickness and edge radius of the cutting tool.

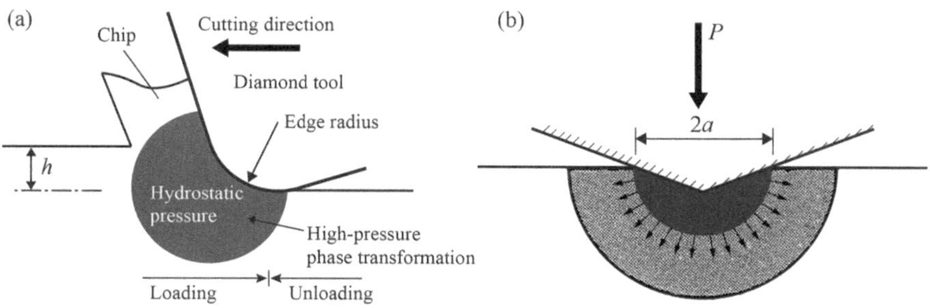


**FIGURE 6.17** Schematics of expanding cores of hydrostatic pressure in (a) cutting and (b) elastic-plastic indentation. The dark regions denote "hydrostatic core", the shaded region "plastic zone" and the surrounding region "elastic matrix" [11].

## 6.3 MACHINABILITY OF CRYSTALLINE BRITTLE MATERIALS

### 6.3.1 SILICON

Single-crystal silicon is not only a dominant semiconductor material for the fabrication of microelectronic and micromechanical components but also an important infrared optical material The ultraprecision machining of single-crystal silicon has tremendous importance. Major physical and mechanical properties of silicon are listed in Table 6.1. Silicon has a strong directional covalent bonding with a diamond structure (4-fold coordination) with a predominant slip system <110>{111}. Pure edge dislocations generally do not form in silicon; instead, pure screw dislocations form with a Burgers vector at 60° to the dislocation line. However, at room temperature, the dislocations are relatively immobile in silicon, and hence, silicon responds in a brittle manner.

Figures 6.18(a) and (b) show a machined surface and the corresponding cutting chips in machining silicon (111) plane at an undeformed chip thickness of 928 nm [9]. The machined surface has significant microfractures across the entire surface. The chips are irregular particles having sharp ends, with an appearance similar to crushed rock. These kinds of surface microfractures and particle-like chips imply that brittle fracture is predominant in material removal whereas almost no plastic deformation occurs. Figures 6.18(c) and (d) show the machined surface and the corresponding cutting chips at an undeformed chip thickness of 44 nm. The machined surface is extremely smooth without any fractures across the entire surface. Also, the chips produced are long and continuous, similar to those produced

**TABLE 6.1**
**Properties of Single-Crystal Silicon [36]**

| Parameters | Value |
|---|---|
| Crystal structure | Diamond |
| Density | 2.329 g/cm$^{-3}$ |
| Lattice constant | 5.431 Å |
| Optical phonon energy | 0.063 eV |
| Hardness | 7 on the Mohs scale |
| Surface microhardness (using Knoop's pyramid test) | 1150 kg/mm$^2$ |
| Cleavage plane | {111} |
| Bulk modulus (compressibility $^{-1}$) | $Bs = 9.8 \times 10^{11}$ dyn/cm$^2$ |
| Shear modulus | $C' = 5.2 \times 10^{11}$ dyn/cm$^2$ |
| Young's modulus [111] | $Y_{111} = 18.5 \times 10^{11}$ dyn/cm$^2$ |
| Young's modulus [110] | $Y_{110} = 17.0 \times 10^{11}$ dyn/cm$^2$ |
| Young's modulus [100] | $Y_{100} = 13.0 \times 10^{11}$ dyn/cm$^2$ |
| Poisson ratio [100] | $\sigma_{100} = 0.28$ |
| Poisson ratio [111] | $\sigma_{111} = 0.26$ |

when machining ductile materials such as aluminum and copper. Atomic force microscope (AFM) measurement of the machined surface showed typical feed marks generated in machining and nanometric surface roughness without any fracture. These experimental results demonstrate that although silicon is a highly brittle material, it can be machined in a completely ductile manner, yielding extremely high-quality surfaces. Figure 6.18(e) shows a single-crystal silicon

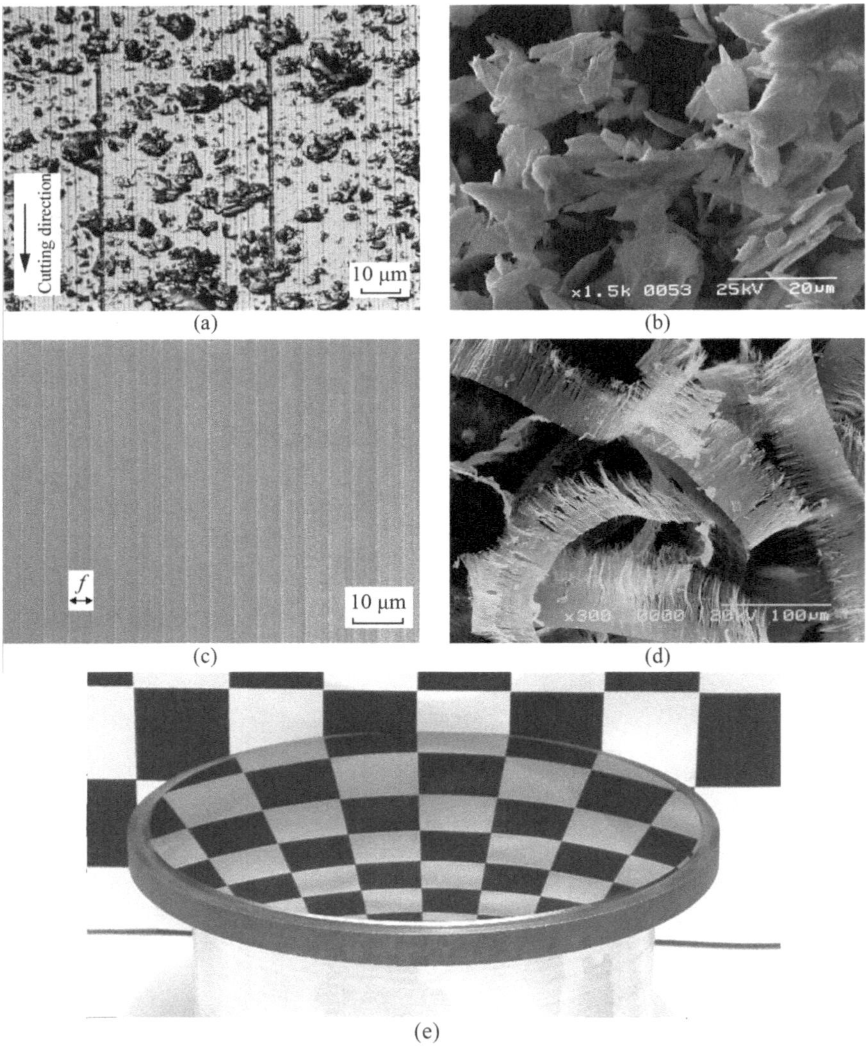


**FIGURE 6.18** (a) Micrograph of machined silicon surface in brittle mode, (b) particle-type chips, (c) machined surface in ductile mode, (d) ribbon-type chips, (e) a ductile-cut silicon aspherical lens [9].

aspherical lens fabricated by diamond turning using a straight-nosed tool in a ductile cutting mode. The lens surface is very smooth, looking like a mirror.

Phase transformation and dislocation are two possible mechanisms for silicon plastic deformation, where the mobility of dislocation in silicon is still controversial. In the 1990s, it was reported that in nanoindentations of silicon at room temperature, no dislocations occurred, and instead, amorphous phase formed in the indent; while dislocations occurred after the temperature increased to 750°C [37]. However, it was found later that under a high pressure, dislocation mobility can be activated at room temperature, resulting in plastic deformation of silicon without phase transformation during in-situ nanoindentation [38]. However, the speed of dislocation motion is by far lower than the cutting speed. Thus, although dislocation is an aspect of subsurface damage in silicon, it does not contribute significantly to the ductile mode removal.

Therefore, the ductile machining behavior of silicon is more likely to be facilitated by high-pressure-induced phase transformations. As an abundance of literature has demonstrated, silicon undergoes phase transformation in situations where high hydrostatic pressure exists, like indentation tests [39, 40]. It is generally accepted that, as a result of the high pressure (10–13 GPa), a structural change from diamond cubic (Si-I) to a metallic state $\beta$-Sn (Si-II) occurs under the indenter during loading. The material around the indenter would then become ductile enough to sustain plastic flow. Measurements of electrical conductivity during indentation close to the indenter on silicon showed a significant increase in conductivity, from semiconducting to highly conducting [41, 42], which strongly supports the idea that a transformation into the metallic state occurs underneath the indenter.

On the hand, the subsurface phase transformation of silicon in machining was confirmed by cross-sectional TEM observations [18]. Figure 6.19(a) shows a TEM micrograph of the uncut region of a silicon wafer; no subsurface defects can be identified. Figure 6.19(b) is a TEM micrograph of the region cut at a depth of 30 nm with a −30° rake angle tool. Clearly, a thin layer of amorphous silicon (the gray layer) was found beneath the cut surface. The average thickness of the layer

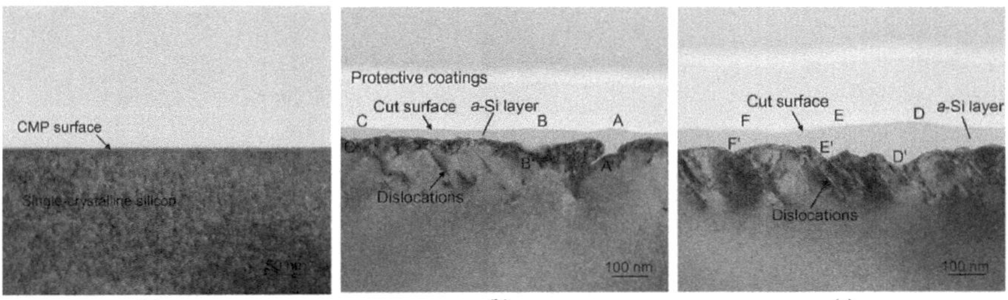


(a) (b) (c)

**FIGURE 6.19** Cross-sectional TEM observations of subsurface of (a) the uncut region, (b) the region cut at a depth of 30 nm with a −30° rake angle tool, and (c) the region cut at a depth of 50 nm with a −45° rake angle tool [18].

is approximately 31 nm. In the crystalline region below the amorphous layer, dislocations can be clearly seen. The dislocation density near the amorphous-crystalline interface is higher than that in deeper regions. The thickness of the dislocated layer, namely, the distance from the amorphous-crystalline boundary to the deepest dislocation, is approximately 165 nm. The dislocations are oriented along the [110] directions. Figure 6.19(c) shows a TEM micrograph of the region cut at a depth of 50 nm with a −45° rake angle tool. Compared to that in Figure 6.19(b) the amorphous Si (*a*-Si) layer was thicker, and the thickness variation was more significant. Besides, despite the significantly higher dislocation density, the thickness of the dislocation layer is almost the same.

### 6.3.2 Germanium

Germanium (Ge) has a strong directional covalent bond of the diamond structure, like silicon. In the past few decades, germanium has been widely used in the electronic fields as a semiconductor material. It is also a versatile infrared material commonly used in imaging systems and instruments in the wavelength range of 2–12 μm as lenses and windows. It is also used as the output coupler for low-power continuous wave (CW) and transversely excited atmosphere carbon dioxide (TEA-$CO_2$) lasers. Germanium is non-hygroscopic and non-toxic. It has good thermal conductivity and strength. The material property of germanium is shown in Table 6.2.

Figure 6.20(a) –(c) show the plots of critical undeformed chip thicknesses $h_c$ versus the orientation angle of single-crystal Ge (100), (110) and (111) planes, respectively, under dry and wet conditions [34]. The orientation angle is the angular deviation of the measurement point from the orientation flat (OF) that

**TABLE 6.2**
**Properties of Single-Crystal Germanium [36]**

| Parameters | Value |
|---|---|
| Crystal structure | Diamond |
| Density | 5.323 $g/cm^{-3}$ |
| Lattice constant | 5.658 Å |
| Optical phonon energy | 0.037 eV |
| Hardness | 6 on the Mohs scale |
| Surface microhardness (using Knoop's pyramid test) | 780 $kg/mm^2$ |
| Cleavage plane | {001} |
| Bulk modulus (compressibility $^{-1}$) | $Bs = 7.5 \times 10^{11}$ $dyn/cm^2$ |
| Shear modulus | $C' = 4.1 \times 10^{11}$ $dyn/cm^2$ |
| Young's modulus [100] | $Y_{100} = 10.3 \times 10^{11}$ $dyn/cm^2$ |
| Poisson ratio [100] | $\sigma_{100} = 0.26$ |

corresponds to <110>. In all the figures, periodical variations of $h_c$ are observed during one rotation of the workpieces. In Figure 6.20(a), there are 8 peaks, 4 deep valleys, and 4 shallow valleys. The maximum value of $h_c$ ($h_{cmax}$) was ~270 nm, and the minimum value ($h_{cmin}$) was ~60 nm. Orientations corresponding to $h_{cmin}$ are 45°, 135°, 225°, and 315°, which are the most difficult to be ductile-machined for the Ge (100) plane. In Figure 6.20(b), $h_c$ fluctuates more significantly than that in Figure 6.20(a). The $h_{cmax}$ was 450 nm, while the $h_{cmin}$ was only 60 nm. Orientations corresponding to $h_{cmin}$ are 60°, 120°, 240°, and 300°. In Figure 6.20(c), the variation amplitude of $h_c$ is smaller than that in Figures 6.20(a) and (b). Three valleys appeared at 0°, 120°, and 240°, corresponding to the $h_{cmin}$ of 60 nm. These results illustrate that the brittle–ductile transition boundary varies significantly with the crystal orientation angle for germanium crystals.

On the other hand, despite that the variation patterns of $h_c$ versus orientation angles are significantly different among the (100), (110), and (111) planes, the $h_{cmin}$ for all these planes was approximately the same, about 60 nm. This value is slightly smaller than those of the single-crystal silicon, indicating higher brittleness of germanium. Moreover, a comparison between the results of dry cuts and wet cuts in Figure 6.20 shows that the dry cuts can produce a larger $h_{cmax}$ at a few

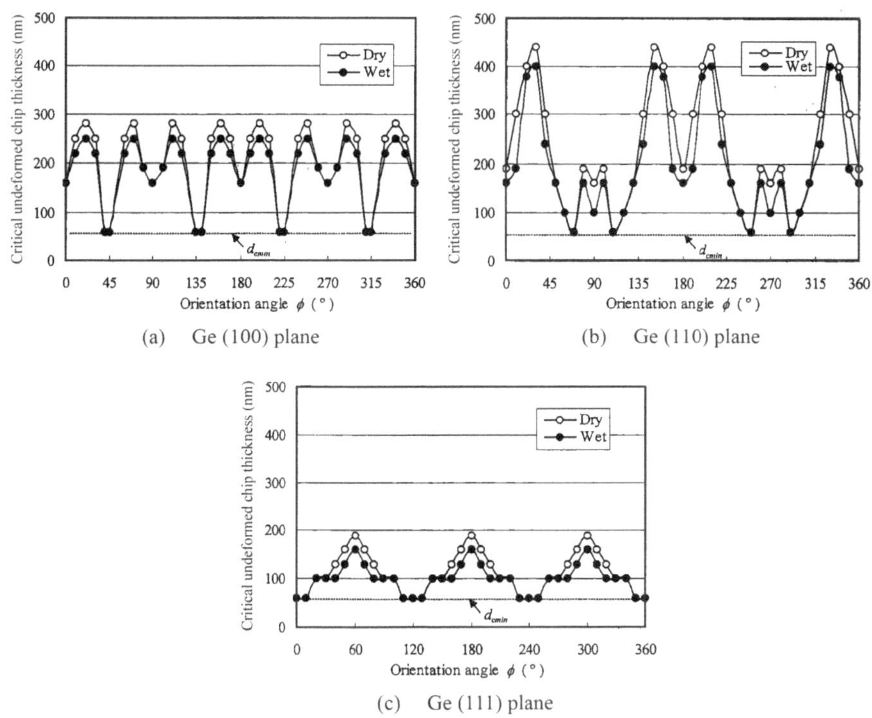


**FIGURE 6.20** Variation of critical undeformed chip thickness of germanium with crystal orientation angle [34].

specific orientations, while the value of $h_{cmin}$ shows no obvious difference between the dry cut and the wet cuts.

Figures 6.21(a)–(c) is micrographs of the workpiece centers of Ge (100), (110) and (111), respectively, which were machined at an undeformed chip thickness of 160 nm. Because this undeformed chip thickness is beyond the $h_{cmin}$ (60 nm) and below the $h_{cmax}$, microfractures occurred on some orientations, while on other orientations ductile-cut surface were obtained. As a result, micro-fractured regions distributing in a radial pattern from the workpiece center were produced. The distribution patterns of the micro-fractured regions agree exactly with the positions of the valleys on the $h_c$ curves in Figure 6.20.

Figure 6.22(a) is a micrograph of the workpiece center after machining at an undeformed chip thickness of 55 nm. This undeformed chip thickness is close to the $h_{cmin}$; in other words, it approaches the highest material removal rate that enables the entire surface to be generated in a ductile mode. Under these conditions, the machined surface was uniform in texture, with no microfracture across the entire surface. The surface roughness is 58 nm Ry, and 7 nm Ra, respectively. Similar to silicon, the ductile response of germanium is thought to originate from a high-pressure-induced phase transformation, which has been observed in nanoindentation test [41]. During hardness indentation with a sharp diamond indenter, the germanium single crystal (Ge-I) transforms into the simple

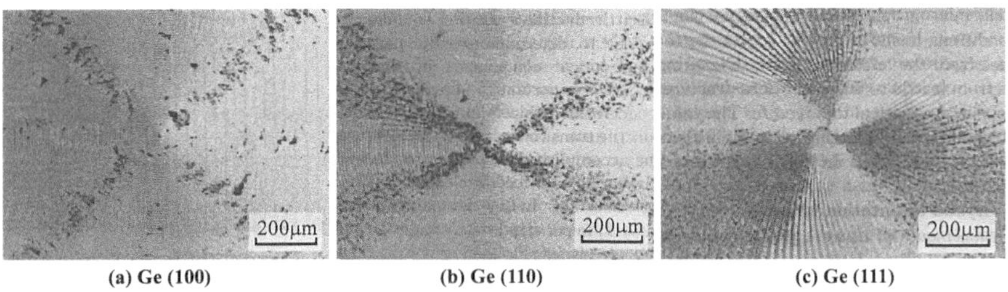


**FIGURE 6.21** Micrographs of the center surfaces of germanium machined at an undeformed chip thickness of 160 nm [34].

**FIGURE 6.22** (a) Micrograph of the germanium (111) workpiece center machined at an undeformed chip thickness of 55 nm [34]; (b) a ductile-cut germanium aspherical lens.

tetragonal ST12-phase (Ge-III) and the amorphous germanium (a-Ge) phase due to the high local pressure [43]. Figure 6.22(b) shows a single-crystal germanium aspherical lens ductile-cut using a straight-nosed tool, the surface of which is very smooth with nanometric surface roughness.

### 6.3.3 Calcium Fluoride

Calcium fluoride ($CaF_2$) is a transparent colorless crystal with a fluorite-type crystalline structure. It has extremely high permeability and refractive index from the infrared range to the ultraviolet range and excellent color-aberration compensation ability. Therefore, $CaF_2$ is not only an optical substrate material for infrared imaging systems, but also an indispensable lens substrate material for large-scale semiconductor photolithography systems. Especially when 193 and 157 nm lithography systems are developed where $CaF_2$ components are utilized to replace fused silica ($SiO_2$), $CaF_2$ is becoming the most commercially important material within the optics community. $CaF_2$ is the only robust and highly developed material with sufficient transmissibility at 157 nm. Applications for $CaF_2$ include 193 nm excimer laser components, chromatic aberration correction of 193 nm stepper lens systems, and stepper lenses for 157 nm systems. Some of these $CaF_2$ elements have already been fabricated for prototype and preproduction lithographic lenses [44]. The optical performance of $CaF_2$, such as laser damage threshold and light transmission, is strongly influenced by its surface quality. However, $CaF_2$ is very difficult to machine for high-quality optical surfaces, because it is a nominally brittle material, having very low fracture toughness, relatively low hardness, and very high thermal expansion coefficient.

Figures 6.23(a)–(c) are SEM photographs of the machined surfaces on $CaF_2$ (111) plane along the [11−2], [−1−12], and [−101] orientations, respectively [45]. In Figure 6.22(a), microfractures of up to a few tens of micrometers in size are observed, and most of the microfractures have triangular appearances. Cross-sectional surface profiling results showed that the depth of the microfractures averages in the range of 5–10 μm, far larger than the undeformed chip thickness (1 μm). In Figure 6.23(b), microfractures occurred more densely, but the size and the depth of the microfractures are distinctly smaller than that in Figure 6.23(a). In addition, the contours of the microfractures are twisted, exhibiting a plastic flow appearance. In contrast, in Figure 6.23(c), the entire surface is smooth, with periodical tool feed marks but no microfractures.

Since $CaF_2$ has a fluorite-type crystal structure, its cleavage planes is {111} and the preferred slip system is {001}<1−10>. The relationships among the cutting directions, the cleavage planes and the slip systems are schematically shown in Figure 6.24. In the figure, the surface to be cut is parallel to the (111) plane (ABC). $V_1$, $V_2$, and $V_3$ indicate the cutting directions along [11−2], [−101] and [−1−12], respectively. As known from the cutting principles, the cutting force in cutting direction, dominates the material removal mechanism. When cutting along [11−2] ($V_1$), the resolved tensile stress acting on the cleavage plane (11−1) (ADB) becomes predominant; hence, brittle fracture will immediately occur along these

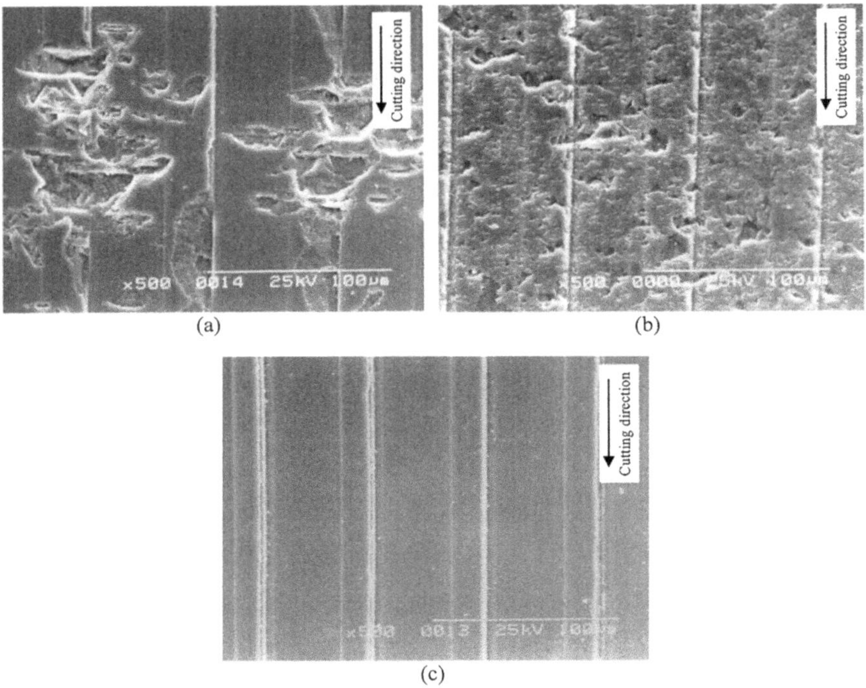


**FIGURE 6.23** SEM photographs of machined surface of $CaF_2$ (111) plane showing three types of surface morphologies at different regions: (a) pitted region cut in [11−2] direction, (b) cloudy region cut in [−1−12] direction, (c) clear region cut in [10−1] direction [45].

planes. The triangular microfractures shown in Figure 6.23(a) are thought to be formed along the cleavage planes (11−1), (−111), and (1−11) (ABD, ACD, and BCD). While when cutting along [10−1] ($V_2$), the shear stress acting along the slip system (010)[10−1] becomes predominant. According to Schmid's law, this slip system is liable to be activated; thus, slip deformation occurs, leading to a ductile mode surface as shown in Figure 6.23(c). However, when cutting along [−1−12] ($V_3$), two slip systems, (010)[10−1] and (100)[0−11], become symmetrical to the cutting condition. In such a case, slip may occur, but the slip deformations along the two symmetrical slip systems will be restricted by each other due to their intersection. Similar phenomenon has been confirmed in the transmission electron microscopic (TEM) investigation of the subsurface damage layer of silicon wafers generated by diamond turning [46]. The restriction between two symmetrical slip systems results in a decrease in the microscopic ductility of work material. Therefore, the material removal mechanism along the [−1−12] orientation may involve both the restricted slip along the two symmetrical slip systems and the cleavage fracture along the (11−1) (ADB) plane, leading to the surface texture shown in Figure 6.23(b).

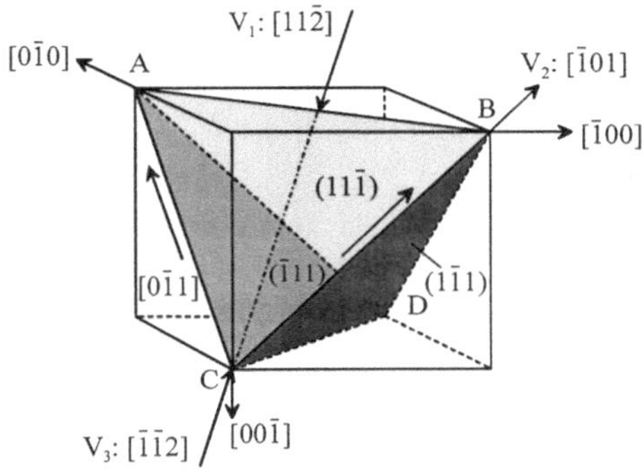


**FIGURE 6.24** Schematic of the relationships among the cutting directions, the cleavage planes, and the slip systems during machining of $CaF_2$ (111) plane.

Apart from cleavage and slip deformation, whether phase transformation occurs for $CaF_2$ or not is a question. It was found in the diamond anvil cell experiment of $CaF_2$ that under a pressure of 8.8 GPa, $CaF_2$ underwent a phase transformation to its high-pressure (orthorhombic) phase, with a corresponding volume decrease of 10.4%. The high-pressure phase is found to withstand a significantly larger differential stress than the low-pressure fluorite phase, with a large degree of lattice anisotropy. In addition, a time-dependent evolution of the lattice parameters was observed during the phase transformation [47]. However, according to the TEM observations of machined surfaces of $CaF_2$, there is no amorphous phase in the subsurface region like that of silicon [48,49] This might be because the hardness of $CaF_2$ is relatively low, and thus the contact pressure during processing is not high enough to cause phase transformation.

The low thermal conductivity and high thermal expansion coefficient also make machining characteristics of $CaF_2$ different from those of silicon and germanium. Figure 6.25 shows SEM photographs of the surface of $CaF_2$ (111) plane machined under wet and dry conditions [50]. Under the wet condition, as shown in Figure 6.25(a), from left to right in the figure, the tool feed was decreased from 40 μm to 1 μm and then increased to 40 μm again. The corresponding undeformed chip thickness varies from 26 nm to 1 μm. The surface can be divided into two kinds of regions: high-feed regions on the left and right (region A), and low-feed regions in the middle (region B). It can be seen that in region A, microfractures of 1 to 10 μm size occurred densely (named as A-type microfractures). In contrast, in region B, fractures as large as a few hundred micrometers occurred (named as B-type microfractures). The value of the critical tool feed ($f_{c1}$), below which A-type microfractures occur, and the value of the critical tool feed ($f_{c2}$), above which B-type microfractures occur, are changed with tool rake angle, as shown in Figure 6.26. These two types of microfractures limit the ductile mode machining, and the widest tool feed range of 3.5–5.5 μm is achieved when tool take angle is −20°.

Figure 6.25(b) shows the SEM photograph of the surface machined under dry conditions, where all other conditions are the same as in Figure 6.25(a). In the

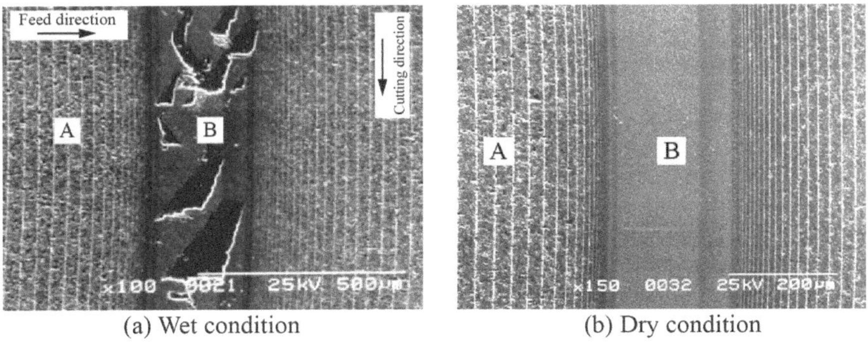


**FIGURE 6.25** SEM photograph of the CaF2 surface machined at continuously varied tool feed under (a) wet condition of kerosene coolant, (b) dry condition. (Regions A and B correspond to high tool feed region and low tool feed region.) [50]

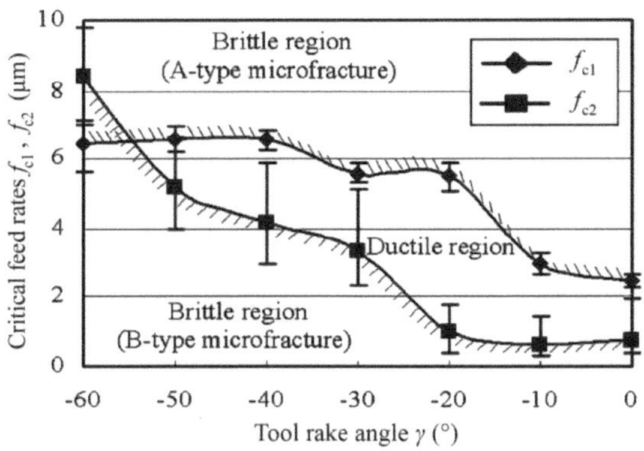


**FIGURE 6.26** Effect of tool rake angle on the critical tool feeds for generating two types of microfractures [50].

region A with high feed rate, A-type microfractures occur, the same as in Figure 6.25(a). However, in the region B with low feed, smooth ductile surfaces were obtained without any visible damage. Therefore, it is supposed that one possible reason for generating B-type microfractures is thermal effects because the thermal conductivity of $CaF_2$ is extremely low (0.06 times that of silicon), and the thermal expansion coefficient is extremely high (5.8 times that of silicon). Under an extremely low tool feed, both the specific cutting energy per tool pass and the number of tool pass per unit surface area increase significantly, which results in a high temperature rise. Similarly, a higher negative rake angle also leads to a higher temperature rise, increasing the possibility of the B-type microfracturing.

Under optimized cutting conditions, that is, dry cutting with −20° rake angle tool at the tool feed of 1 μm, depth of cut of 1 μm and cutting edge angle of 0.6°, the whole surface of the workpiece was ductile cut. Figure 6.27(a) shows an AFM

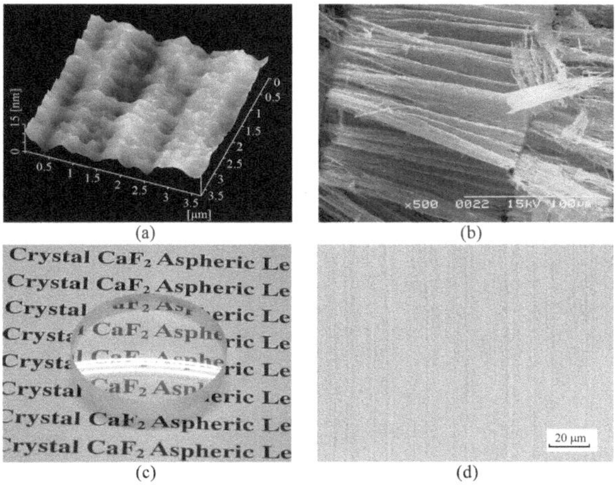


**FIGURE 6.27** (a) AFM image of a $CaF_2$ surface cut at the optimized conditions; (b) SEM image of cutting chips generated in the same process [50]; (c) a ductile-cut aspherical lens of CaF2; (d) micrograph of the surface of the lens in (c).

image of the surface where the surface consists of periodical tool feed marks with a surface roughness of 15 nm $R_y$ and 2.4 nm Ra. Figure 6.27(b) is an SEM photograph of the chips collected under the above-cutting conditions. The chips are formed as continuous ribbons, similar to those produced by metal cutting. These chips indicate that the material has been removed in a completely ductile mode. Figure 6.27(c) shows a ductile-cut aspherical lens of $CaF_2$, and the micrograph of the surface of the lens is shown in Figure 6.27(d).

### 6.3.4 Indium Phosphide

Indium phosphide (InP) is an emerging compound semiconductor material for high-performance optoelectronic communication components, such as light-sensitive sensors, optical integrated circuits, and laser diode elements. The material property of InP is shown in Table 6.3. The fabrication of InP substrates involves a series of material removal processes such as slicing, grinding, lapping, and polishing. Most of the manufacturing processes of InP follow those of silicon, which are well established. However, owing to distinct differences in material properties, the optimum machining conditions for silicon may not be suitable for InP. Thus it is important to clarify the micromechanical properties, deformation and fracture mechanisms, and precision machining characteristics of InP at the micro/nanometer level.

Indentation is an effective method for characterizing the mechanical properties and machinability of a material. Figure 6.28(a) is an SEM image of an indent obtained at a load of 150 mN with the Vickers indenter at an orientation angle of zero (Vickers-0) [51]. Under this condition, the indent surface is very smooth, and

## TABLE 6.3
## Properties of Single-Crystal Indium Phosphide [36]

| Parameters | Value |
|---|---|
| Crystal structure | Zinc blende |
| Density | 4.81 g/cm$^{-3}$ |
| Lattice constant | 5.8687 Å |
| Optical phonon energy | 0.043 eV |
| Surface microhardness (using Knoop's pyramid test) | 460 kg/mm$^2$ |
| Cleavage plane | {100} |
| Bulk modulus (compressibility $^{-1}$) | $Bs = 7.11 \times 10^{11}$ dyn/cm$^2$ |
| Shear modulus | $C' = 2.25 \times 10^{11}$ dyn/cm$^2$ |
| Young's modulus [100] | $Y_{100} = 6.11 \times 10^{11}$ dyn/cm$^2$ |
| Poisson ratio [100] | $\sigma_{100} = 0.36$ |

no fractures can be found. This demonstrates that although InP is a nominally brittle material, a ductile response can be obtained under suitable conditions. Figures 6.27(b) and (c) are SEM images of indents obtained at loads of 300 mN and 500 mN, respectively. Cracks have been generated around the indents. As the indentation load increases, the number and length of the cracks also increase. Most of the cracks are initiated near the corners of the indent and propagate radially. The paths of crack propagation are not straight, and the propagation directions are somewhat random. We can roughly divide the crack propagation directions into three kinds: <100> directions, <110> directions, and directions between these two. In addition, it is noteworthy that the shape of all indents is not square. The four margins of the indents have expanded outward, as schematized in Figure 6.28(d). This phenomenon is due to the anisotropy in elastic recovery of single-crystalline InP after unloading of the indenter. That is, the elastic recovery in the four <100> directions is more significant than that in the <110> directions. In other words, plastic deformation is more significant along <110> than along <100>. The anisotropic phenomenon in elastic recovery of material may cause microscopic undulations on the workpiece surface during machining if the workpiece is rotating about the tool.

Figures 6.29 (a) and (b) are SEM images of indents obtained using the same indenter but at an orientation angle of 45° (Vickers-45). At a load of 300 mN, only one radial crack is generated from the left corner of the indent along the <110> direction. As the indentation load increases to 500 mN, more cracks appear at other corners. Compared to Figure 6.28, the cracks in Figure 6.29 are very straight, and all are oriented along the <110> directions. Another point worth noting is the indent profile. The four margins of the indent have been contracted inward, as schematized in Figure 6.29(c). This phenomenon can also be explained by the anisotropy in elastic recovery.

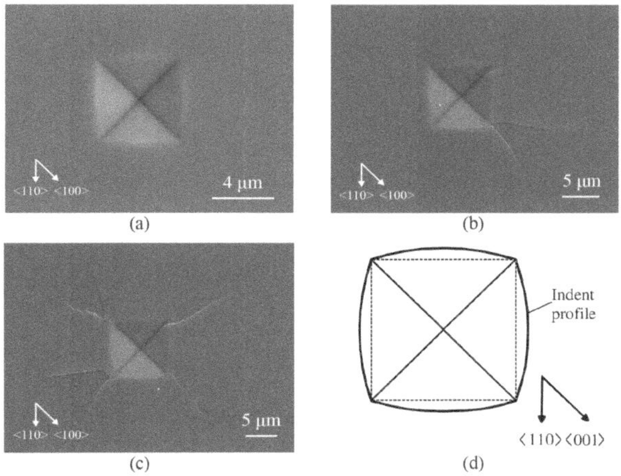


**FIGURE 6.28** SEM images of indents obtained at the load of (a) 150 mN, (b) 300 mN, and (c) 500 mN with a Vickers indenter at an orientation angle of zero. (d) Schematic profile drawing of the indents [51].

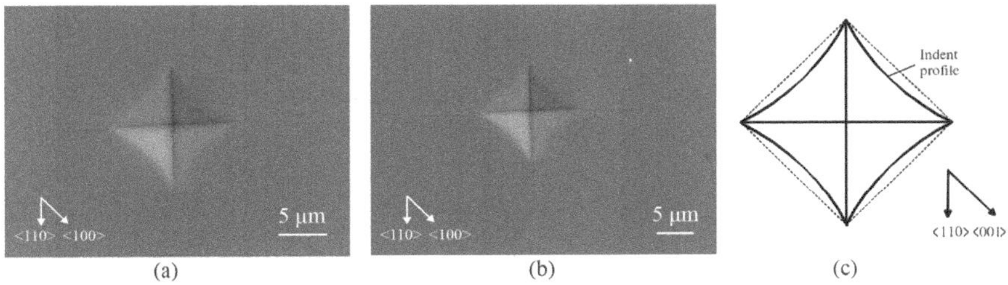


**FIGURE 6.29** SEM images of indents obtained at (a) 300 mN and (b) 500 mN with a Vickers indenter at a 45° orientation angle. (c) Schematic profile drawing of the indents [51].

On the basis of the above results, it is clear that the deformation and microfracturing behaviors of InP are strongly dependent on the geometry and orientation of the indenter. Compared to other directions, <110> is more sensitive to the indenter ridge. If an indenter ridge is orientated along a <110> direction, a crack readily forms along this direction under a small load before cracks along other directions are generated. If the indenter ridge does not coincide with a <110> direction, cracks form under a higher load in a random manner, with <100> as the primary cracking direction. The <110> directions behave very uniquely. They are easy to deform plastically under shear stress conditions, while under tensile stress conditions, cracks are easily generated along this direction. This is because InP has a zinc-blende structure whose preferential cleavage plane is {110} [52], unlike the diamond-structure crystals silicon and germanium, where the preferential slip systems are <110> directions on {111} planes.

Figure 6.30(a) is a cross-sectional TEM image of two indents obtained at loads of 10 mN and 50 mN. No cracks are formed in the subsurface, and the long stripes seen across the whole image are the contrast caused by elastic bending of the TEM sample in FIB processing. In addition, unlike the silicon indentation where amorphous regions are generated below the indenters, there are no amorphization regions in the subsurface of InP crystals. This can be confirmed by the selected area analysis results of the material below the indents, as shown in Figures 6.30(b) and (c). The diffraction pattern of both indents shows a zinc-blende lattice structure with an incident direction of <110>. However, the diffraction spots at higher load are no longer rounded, but distorted and stretched, indicating that the crystalline periodicity has been severely disordered.

For both indents, high-density dislocation regions (enclosed by the dotted lines) are formed beneath the indenters, extending downward. In these regions, dislocations generated on different slip planes interact intensively with each other to form sessile dislocations, which is a primary reason for work hardening. There is an obvious increase in the depth and area of the high-density dislocation region as the indentation load changes from 10 mN to 50 mN. Below the high-density dislocation region of the indent made at 50 mN, long slip bands are clearly visible. The slip bands are oriented at an angle of 54° to the sample surface (100), demonstrating that slip deformation of InP occurs within the {111} planes, which are vertical to the TEM photograph plane, namely, (110).

Figure 6.31(a) shows an overview of the machined InP surface, which was diamond turned with six different undeformed chip thicknesses ($h$): 300, 200, 150, 100, 50, and 20 nm. The surface morphology varied depending on the crystallographic orientation and the $h$, resulting in a mixture of brittle regions (appearing white in the image) and ductile regions (appearing black). When the $h$ was 150 nm or greater, the surface exhibited a fully brittle region. In contrast, at $h$ of 50 and

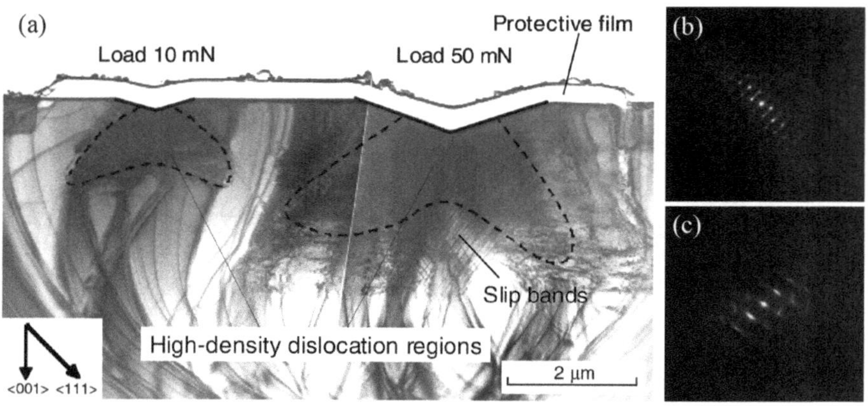


**FIGURE 6.30** (a) Cross-sectional TEM image of two indents obtained at 10 mN (left) and 50 mN (right). (b) and (c) are diffraction patterns below the indents obtained at 10 mN and 50 mN [51].

100 nm, the surface morphology varied along the circumferential direction, revealing a fan-shaped pattern. The surface morphologies were classified into three distinct regions: (A) fractured region, where severe brittle fracture occurred; (B) cloudy region, characterized by minimal brittle fracture; and (C) clear region, where no brittle fracture was observed. Figure 6.31(b) presents schematic illustrations of the surface textures machined at $h$ of 100, 50, and 20 nm. At the $h$ of 100 nm the [−100] [100] direction exhibited a brittle region, while the [0−10] [010] direction corresponded to a cloudy region. When the $h$ was reduced to 50 nm, the [−100] [100] direction became a cloudy region, and the remaining areas were classified as clear regions. Further reducing the $h$ to 20 nm resulted in clear regions across all crystallographic directions.

Figure 6.32 shows Nomarski optical micrographs of the [010] orientation surfaces at the $h$ of 300, 100, and 50 nm. In Figure 6.32(a), the machined surface exhibits significant brittle fracture, resulting in a rough, matte texture. In Figure 6.32(b), brittle fractures are still present but are less prominent compared to Figure 6.32(a). In contrast, the surface in Figure 6.32(c) shows no signs of brittle fracture, displaying a smooth surface formed by the replication of the

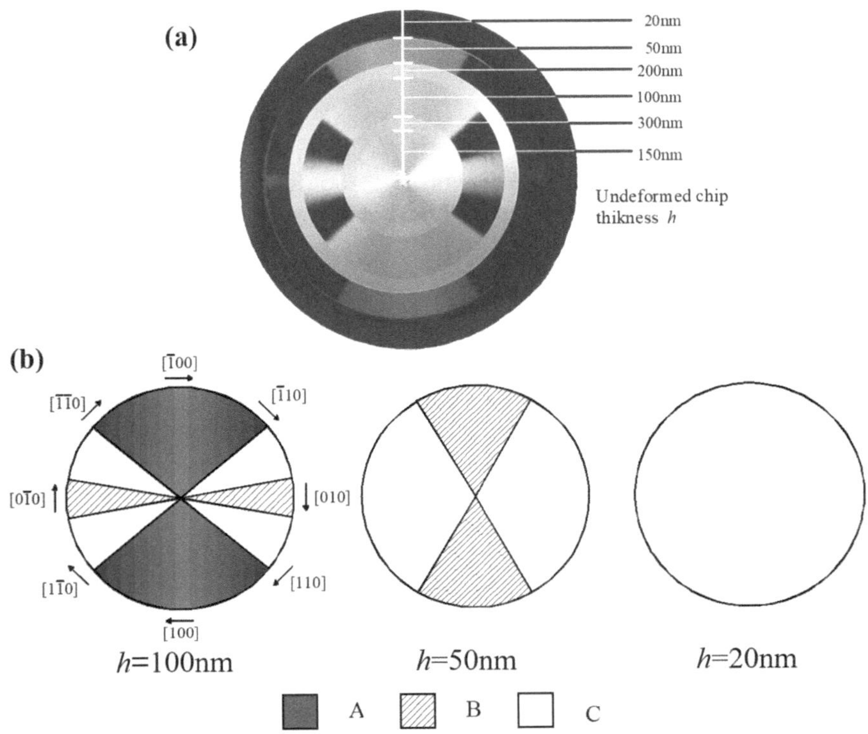


**FIGURE 6.31** (a) Photograph of machined surface of InP crystals and (b) Schematic presentations of surface textures.

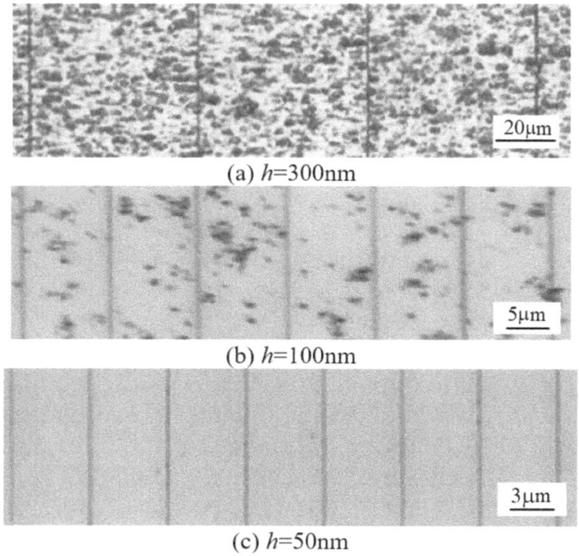


**FIGURE 6.32** Nomarski optical micrographs of the InP surfaces cut along the [010] orientation at the undeformed chip thickness ($h$) of (a) 300, (b) 100, and (c) 50 nm.

cutting edge. The chips formed under $h$ = 50 nm conditions are shown in Figure 6.33, appearing as nearly continuous flow-type chips. This morphology indicates that ductile-mode cutting was achieved. Additionally, when the $h$ was further reduced to 20 nm, the surface roughness of the machined area was measured to be 26 nm Ry.

### 6.3.5 Gallium Arsenide

Gallium arsenide (GaAs) is a semi-insulator, which can be used in large-power continuous $CO_2$ laser systems in lens or mirror. The hardness and strength of GaAs make it a good choice for use in environments where dust or abrasive particles tend to build up or bombard the optical surface. GaAs optics are limited by crystal growth technology to diameters typically less than 10 cm. The material is non-hygroscopic, safe to use in laboratory and field applications, and chemically stable except when contacted with strong acids.

Similar to other crystalline materials, GaAs also show strong anisotropy in machinability, as shown in Figure 6.34 [53]. The machined surface of GaAs (001) plane exhibited four obvious fan-shaped sections. Cutting along the <110> direction cannot gain a complete ductile response even with the extremely small feed rate of 0.5μm/rev, while it is relatively easy to achieve ductile machining when cutting along the <100> direction. The understanding of the anisotropic machinability of GaAs materials and optimal process parameters for obtaining the ductile-regime diamond turning of GaAs will be introduced in this section.

**FIGURE 6.33** SEM micrograph of cutting chips generated under the undeformed chip thickness of 50 nm.

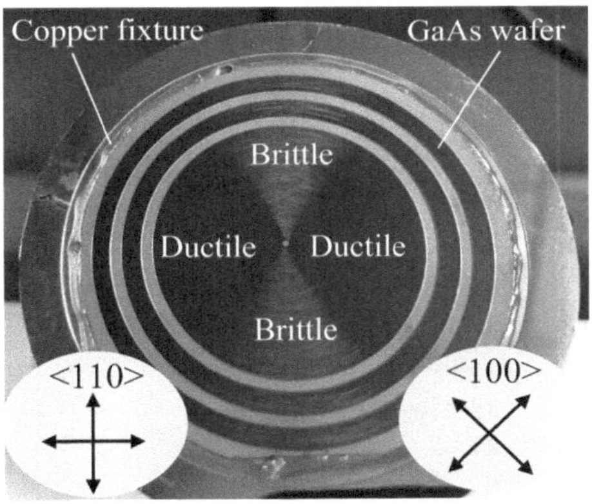


**FIGURE 6.34** Anisotropic machinability of GaAs (001) plane [53].

Figure 6.35 shows SEM images of the indents obtained at 500 mN with a Vickers indenter at 45° (Vickers-45) and 0° (Vickers-0) orientation angles. Obvious elastic recovery for the edge of the impression was observed along the <100> orientation for Vickers-45 indent. In addition, since the external applied energy cannot be completely absorbed by elastic and plastic deformation, cracks formed at middle point of each edge along the <100> orientation, which is the intersection of the {110} plane and the (001) plane, as shown in Figure 6.35(a). In contrast, owing to the fact that {110} plane is the preferential cleavage plane of GaAs crystal, for Vickers-0 indent, the cracks mainly propagated along the <100> orientation, and no elastic recovery was found in the edge of impression as shown in Figure 6.35(b). It can be explained that <110> orientation in Figure 6.35(b) is the intersection of the {111} planes and the (001) plane, where the {111} slip planes are easier to produce dislocation and plastic deformation than elastic deformation.

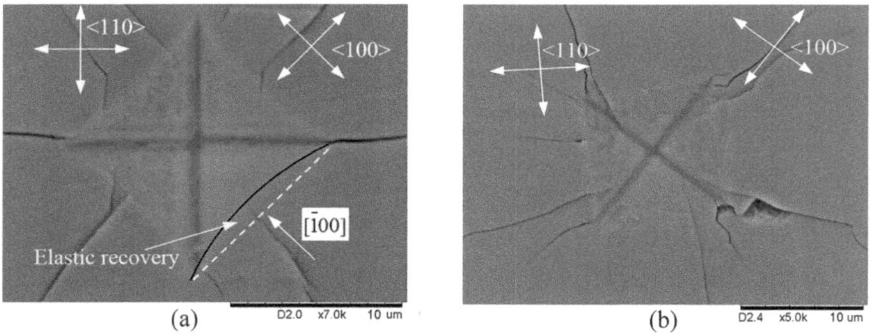


**FIGURE 6.35** SEM images of the indents obtained at 500 mN with a Vickers indenter at (a) 45° and (b) 0° orientation angles [53].

Figure 6.36 shows the surface roughness distribution in the circumferential direction of the diamond-turned surface with a –30° rake angle tool at 2.5 μm/rev tool feed rate. The direction of $\theta = 80°$, which is close to the [110] orientation, has the highest surface roughness. This indicates that the surface is most likely to produce cracks when cutting is performed in the [110] orientation. In other words, once this direction can achieve ductile cutting, the entire surface of GaAs (001) plane can be ductile cut. According to the plunge-cut grooving test, the critical chip thickness in [110] orientation is ~10 nm for a tool with –30° rake angle and ~26 nm for a tool with –40° rake angle [53]. This result again demonstrates that the use of a negative tool rake angle facilitates ductile mode cutting by enhancing the hydrostatic pressure. Figure 6.37 presents the crack-free surface machined by a –40° rake angle tool at the feed rate of 0.5 μm/rev.

The microstructure below the crack-free machined surface of GaAs is shown in Figure 6.38. Three layers can be clearly observed beneath the cut surface,

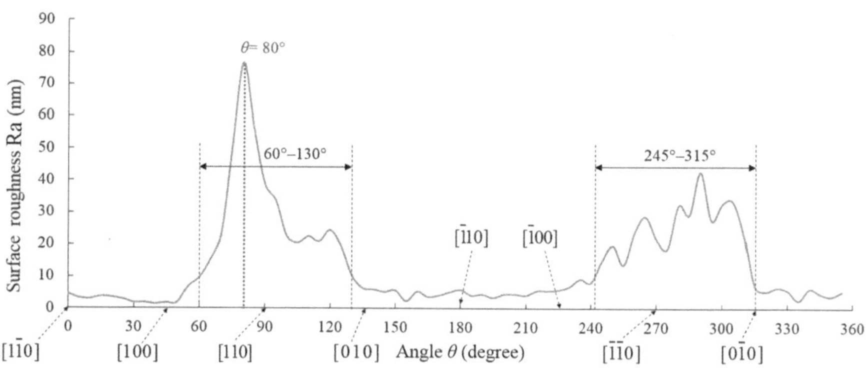


**FIGURE 6.36** Distribution of the machined surface roughness in the circumferential direction after diamond turning [53].

**FIGURE 6.37** A ductile-regime turned single-crystal GaAs wafer and its surface morphology [53].

including the heavily deformed layer, the dislocation pileup layer, and the dislocation extension layer. In the heavily deformed layer just beneath the machined surface, clear crystal structure can be observed, indicating that the material did not transform into an amorphous or nano-crystal structure under the machining pressure. However, the deformed layer with an average thickness of ~43 nm contains a large number of dislocations, meanwhile, lattice deformation can be observed along the cutting direction due to high shear stress and high pressure. As the depth of subsurface layer increases, both the shear stress and pressure decrease, thus the crystal defect is dominated by high-density dislocations and formed dislocation pileup layer. Finally, the density of dislocations gradually decreases as the depth increases in the dislocation extension layer. Additionally, almost all the dislocation extended along the (1−11) plane and a few line dislocations extended to a deeper region.

### 6.3.6 Zinc Selenide

Zinc selenide (ZnSe) single crystal has a cubic (zincblende) structure with high brittleness and low hardness (~1.47 GPa [54]). It is a preferred material for lenses, windows, output couplers, and beam expanders for its low absorptivity at infrared wavelengths and its visible transmission. The most important application of ZnSe crystal is to fabricate optical components for high-power $CO_2$ lasers. Also, it is

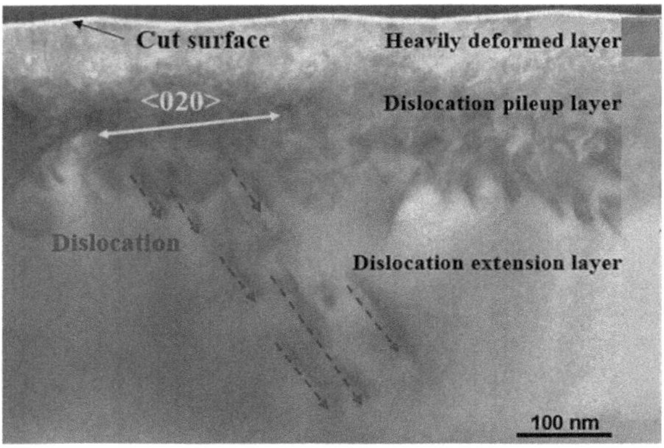


**FIGURE 6.38** Cross-sectional TEM image of diamond-turned GaAs surface, showing a clear three-layer structure: a heavily deformed layer, a dislocation pileup layer, and a dislocation extension layer [53].

used in designing optical elements for various spectral devices and IR lenses, protective windows of devices detecting thermal signals in a broad spectral range, and so on. Among those applications, most of the optical components are made of polycrystalline ZnSe (p-ZnSe) for reducing the material costs. Chemical vapor deposition (CVD) and physical vapor deposition (PVD) are the two main methods for growing bulk p-ZnSe. The size of the grains in the bulk p-ZnSe grown by CVD method is about 50–70 μm, while the size of the grains in the bulk p-ZnSe grown by PVD method is about 300–700 μm [55].

In polycrystalline brittle materials, the grains with varied orientations exhibit different machinability and the effects of the grain boundaries cannot be neglected. These factors make it difficult to produce smooth and uniform surfaces on them. In this section, the machinability of polycrystalline brittle materials will be discussed by presenting examples of p-ZnSe diamond turning conducted by Huang and Yan [15].

Figure 6.39(a) shows the inverse pole figure (IPF) map of a p-ZnSe workpiece grown by CVD method, which was measured by electron backscatter diffraction (EBSD) detector. Many twins are observed in the microstructure of the p-ZnSe, as indicated by the black circles in the figure. The corresponding grain size distribution and misorientation angle distribution are determined using analysis software, as shown in Figures 6.39(b) and (c). The range of the grain size is estimated to be approximately 3–80 μm and the peak grain size to be around 30–50 μm. The microstructure contains a large fraction of high-angle grain boundaries.

Figure 6.40(a) shows an SEM image of the machined surface of p-ZnSe. The crystal grains of different orientations can be differentiated from each other by showing different contrasts. Also, some grains have smooth surface, while others have different degrees of surface cracks. This implies that under the same cutting conditions the machinability of each grain is different. In the cutting of p-ZnSe,

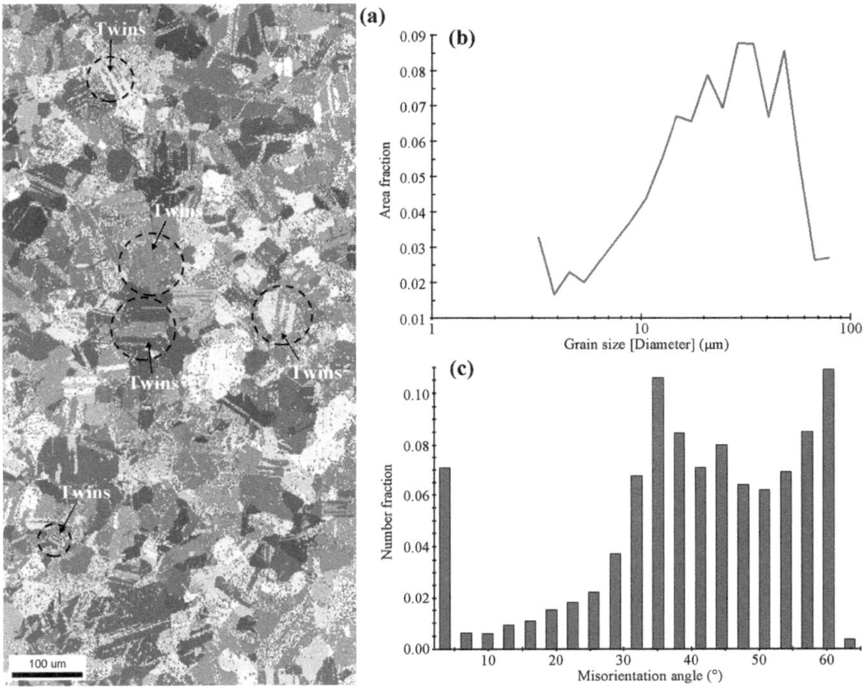


**FIGURE 6.39** (a) EBSD IPF map; (b) grain size distribution; and (c) misorientation angle distribution of p-ZnSe [15].

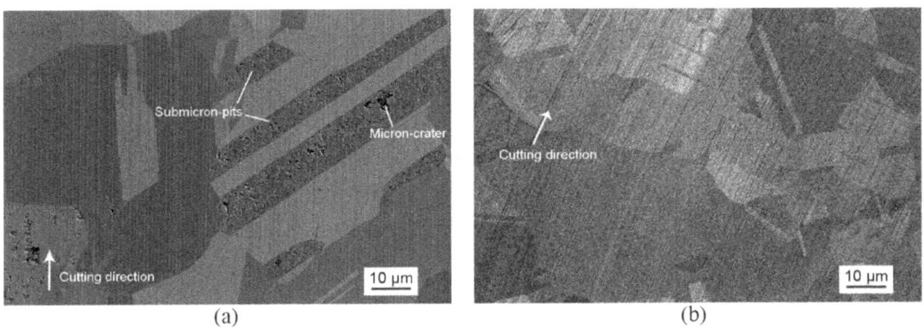


**FIGURE 6.40** SEM images of machined surfaces (a) before and (b) after optimizing cutting parameters [15].

each grain can be considered as a single crystal with a specific crystal orientation. Therefore, to ensure that all grains with various crystal orientations can be removed in a ductile mode, the undeformed chip thickness should be smaller than the critical undeformed chip thickness of the most difficult-to-cut grain. By optimizing cutting parameters to reduce undeformed chip thickness, the whole machined surface becomes free of cracks, as shown in Figure 6.40(b).

Although the machined surface is crack-free, its topography is uneven due to the inconsistent orientation of each grain. Figures 6.41(a) and (b) show the 3D topography of the machined surface, which were measured in a field of 160 × 160 $\mu m^2$ by a white light interferometer. The mean roughness (Sa), root mean square of roughness (Sq), and maximum height of the profile (Sz) of the surface are 5.3 nm, 6.4 nm, and 48.5 nm, respectively. Figures 6.41(c) and (d) give the extracted 2D surface profiles along Lines 1 and 2 that are marked in Figure 6.41(b). In both profiles, grain boundary steps, i.e., a height difference between two adjoining crystal grains appearing at their common border, can be identified clearly. In Figure 6.41(c) the height difference between Grain i and ii is about 14 nm, and that between Grain i and iii is about 9 nm. In Figure 6.41(d) the height difference between Grain iv and v is about 18 nm, and that between Grain iv and vi is about 6 nm. Figure 6.41(e) compares the surface roughness of three different grains that are marked in Figure 6.41(b). The result shows that the surface roughness within each grain is similar, and is smaller than the roughness of the whole measured surface. These indicate that the surface roughness is dominated by the grain boundary steps. In ultraprecision cutting of poly-crystalline germanium, the formation of grain boundary step was also observed [56]. The grain boundary step may result from the different elastic recovery of grains following tool pass or anisotropic plasticity dominated by dislocation slip and grain boundary accommodation. The height of grain boundary step can be reduced as the undeformed chip thickness decreases [15, 56].

EBSD analysis has been made in combination with surface topography measurement to further explore the formation mechanism of grain boundary step. The top view of 3D surface topography is illustrated in Figure 6.42(a), and the IPF map in Figure 6.42(b). The extracted 2D surface profiles along Lines A - E marked in surface topography are given in Figure 6.42(c). The misorientation angle profile along Lines A–E marked in IPF map is given in Figure 6.42(d), as represented by the point-to-origin and point-to-point curves.

As can be seen from Figure 6.42(c), the height difference between grains I and II in profile A, grains IV and V in profile B, and grains IV and X in profile C are obvious. On the other hand, the corresponding misorientation angle profiles (see Figure 6.42(d)) show that the misorientation angles between the two grains are 60°, indicating that these boundaries are twin boundaries around <111>. In addition, it is noted that in surface profile B there is a small height difference between grains III and IV. The corresponding misorientation angle is 39°, which indicates that the boundary is twin boundary around <110>. In misorientation angle profile C, though there is a high angle grain boundary between grains IX and IV of 43.6° which is greater than 39°, the height difference between the two grains is unobvious. That is to say, the grain boundary step is easily formed at the twin boundary, and the higher the twin boundary misorientation angle is, the higher the grain boundary step is.

Besides, it is noted that the boundary between grains VIII and III and that between grains VII and VI are also twin boundaries as indicated by misorientation

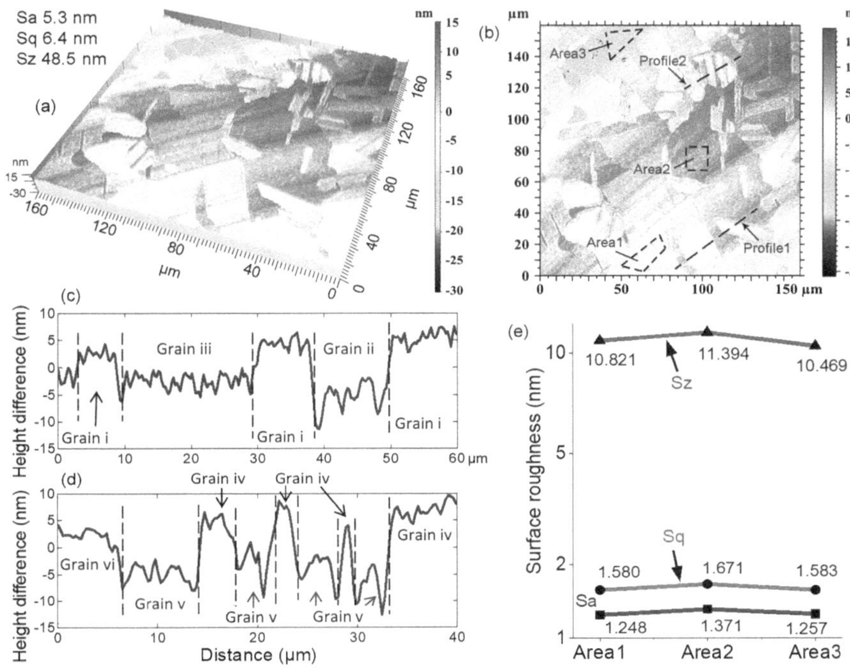


**FIGURE 6.41** (a)(b) isometric view and top view of 3D topography of the surface machined at a feed rate of 1 μm/rev; (c)(d) 2D surface profile along Lines 1 and 2 marked in (b); (e) surface roughness of Areas1 to 3 marked in (b) [15].

angle profiles D and E, but the corresponding grain boundary steps shown in surface profiles D and E are very small. This may be because the cutting direction is parallel to the twin boundary. Therefore, it can be deduced that the grain boundary step tends to form at the place where the cutting tool path crosses the twin boundary with a large angle.

### 6.3.7 KDP

Single-crystal potassium dihydrogen phosphate (KDP) is an important material for laser nuclear fusion. The wavelength of laser used in nuclear fusion is becoming shorter and shorter in order to improve the absorption coefficient of laser into the target. It is important to perform wavelength conversion from long-wavelength (for example, $\lambda = 1.053$ μm) high-power lasers to short-wavelength ones (for example, $\lambda = 0.351$ μm), i.e., the third harmonic. KDP is an ideal material for such laser wavelength conversion. KDP crystal has a tetragonal crystal structure. The (001) plane, doubler plane, and tripler plane of KDP crystal shown in Figure 6.43 are the most commonly used crystal planes for electro-optical switches and frequency converters. KDP is brittle and soft (hardness 1.6–2.0 GPa depending

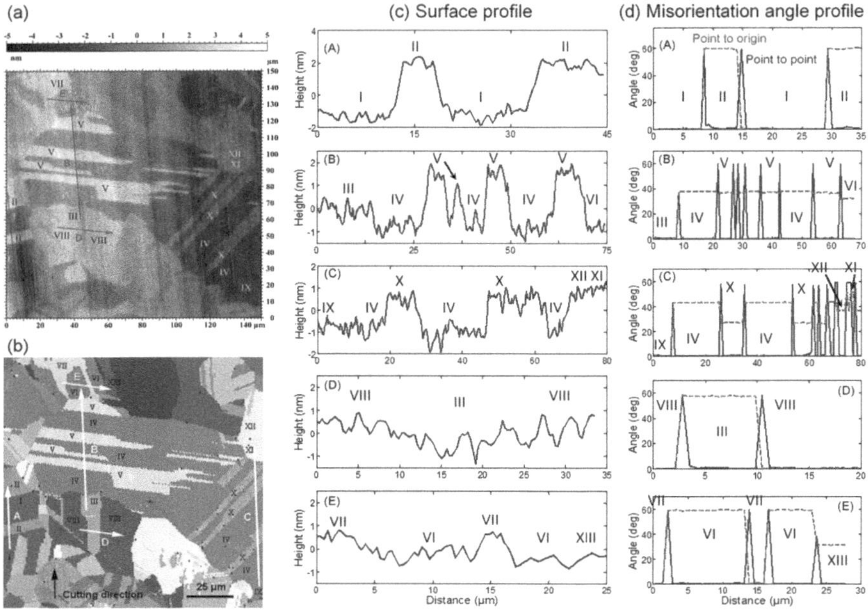

**FIGURE 6.42** (a) Top view of 3D topography of the surface; (b) IPF map of the same surface shown in (a); (c) 2D surface profile along Lines A – E marked in (a); (d) misorientation angle profile along Lines A–E marked in (b) [15].

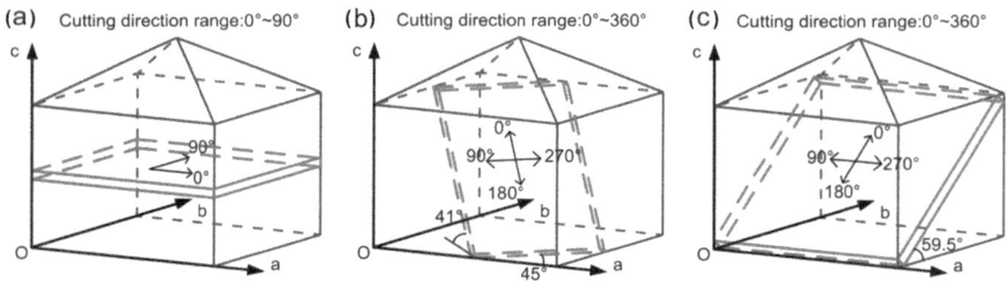


**FIGURE 6.43** Crystal orientation and cutting direction definition on (a) (001) plane, (b) Doubler plane, and (c) Tripler plane of KDP crystal [58].

on the crystal plane [57]) and thus very difficult to machine. It is a deliquescent material, hence machining processes involving coolants and lubricants cannot be used. Diamond turning under dry condition is the most common way to the ultra-precision machining of KDP.

Like machining other brittle materials, a small uncut chip thickness and a negative tool rake angle are beneficial for KDP cutting. A tool with −45° rake angle could generate minimal subsurface damage and surface roughness of better than 8-A rms and 36-A P-V was achieved [59].

Figure 6.44 shows the plots of brittle-to-ductile transition (BDT) chip thickness versus the orientation angle for the (001), doubler, and tripler planes of the KDP crystal, respectively when a tool with −45° rake angle was used [58]. On (001) and doubler planes, the BDT chip thickness reaches the minimum value when cut along 0° and 180°; while on tripler plane, BDT chip thickness is minimum only when cut along 0°. Besides, 90° on the (001) plane, 75° and 285° on the Tripler plane are also the cutting directions with smaller BDT chip thickness. These directions should be avoided in the cutting process. In contrast, 45° and 135° on the (001) plane, 90° and 270° on the Doubler plane, 180° and 225° on the Tripler plane are the cutting directions with larger BDT chip thickness. Thus, in fly-cutting of KDP crystal, it is recommended to cut along these directions to improve the processing efficiency as well as the surface quality.

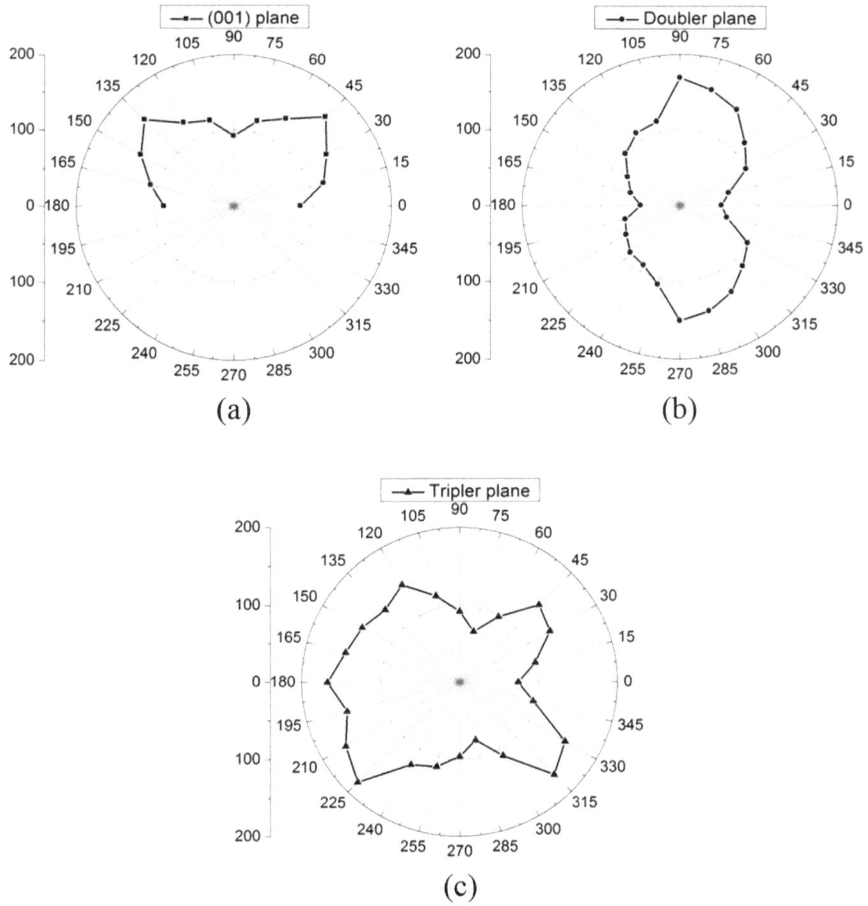


**FIGURE 6.44** Brittle-to-ductile transition depth of (a) (001) plane, (b) Doubler plane, and (c) Tripler plane of KDP crystal [58].

### 6.3.8 Silicon Carbide

Single-crystal silicon carbide (SiC) is a promising wide bandgap semiconductor material for future high-power devices operable at high temperatures (>200 °C), at high frequencies, and in harsh environments (chemical and radiation) due to its unique physical and crystallographic properties. Some potential applications are large-scale quantum computers and vacuum ultraviolet (VUV) telescopes. There is also a growing interest in the use of SiC in the optics industry for space-based laser mirrors as a replacement for beryllium [60].

SiC has equal parts of silicon and carbon, both of which are group IV elements. The distance ($a$) between neighboring silicon or carbon atom is approximately 3.08 Å for all polytypes. The carbon atom is situated at the center of mass of the tetragonal structure outlined by the four neighboring Si atoms, as shown in Figure 6.45(a). The distance between the C atom and each of the Si atoms is approximately 2.52 Å. The height of the unit cell, called $c$, varies between the different polytypes. Therefore, the ratio of $c/a$ differs from polytype to polytype. The main properties of typical types of SiC, including 3C-SiC, 4H-SiC, and 6H-SiC, are listed in Table 6.4 [60]. Figure 6.45(b) shows the stacking sequence of silicon and carbon atoms in typical polytypes of SiC.

In the early 2000s, Patten et al. [61] conducted a study to demonstrate that the ductile machining of single-crystal SiC can be achieved by diamond turning. After that, several studies attempted to machine flat surfaces of SiC by diamond turning, but it was found that chipping and flank wear were significant on the tool edge within a short cutting distance [62]. The rapid tool wear due to high hardness impeded the ductile-mode cutting of a larger-size SiC workpiece. Efforts have been made to suppress the tool wear in the diamond cutting of SiC, such as

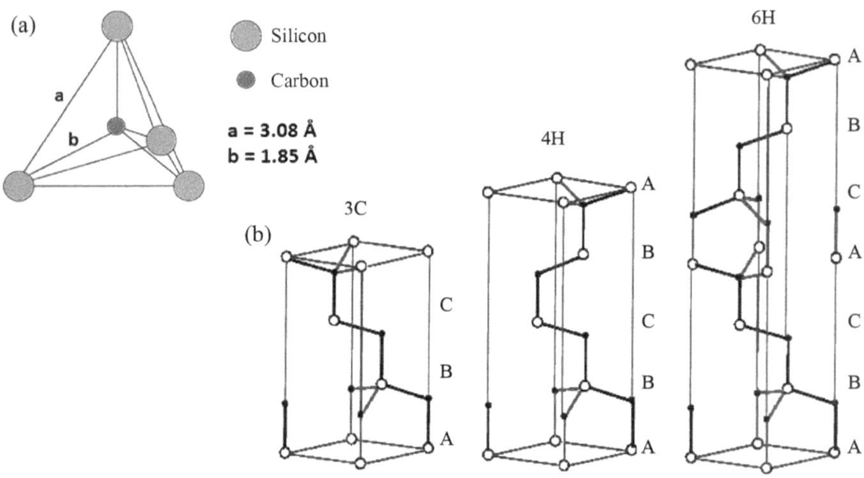


**FIGURE 6.45** (a) The tetragonal bonding of a carbon atom with the four nearest silicon. (b) The staking sequence of common 3C-, 4H-, and 6H-silicon carbide [60].

**TABLE 6.4**
**Properties of Single-crystal Silicon Carbide Polytypes [60]**

| Parameter | 3C-SiC | 4H-SiC | 6H-SiC |
|---|---|---|---|
| Experimental lattice parameter (Å) | $a_0 = 4.359$ | $a_0 = 3.079$<br>$c_0 = 10.254$ | $a_0 = 3.0817$<br>$c_0 = 15.1183$ |
| Mechanical properties | | | |
| Bulk modulus (GPa) | 225 | 215 | 215 |
| Shear modulus (GPa) | 124 | 131.4 | 131.4 |
| Hardness on (100) plane (GPa) | 25–30 | 26 | 20–26 |
| Poisson's ratio | 0.267 | 0.231 | 0.231 |
| Young's modulus (GPa) | 314.55 | 347.01 | 347.01 |
| Fracture toughness (MPa $m^{1/2}$) | 2.02 | 1.9 | 1.9 |
| Electronic properties | | | |
| Band gap (eV) | 2.3 | 3.2 | 3 |
| Thermal conductivity (W/cm-K) | 4.9 | 3–5 | 3–5 |

tool-swinging cutting [63], vibration-assisted cutting [64], laser-assisted cutting [65], and applying nanoparticle lubrication [66]. However, until now, ductile-mode machining of a large area on SiC using diamond turning has not yet been achieved due to severe tool wear.

However, it is possible to use diamond turning to fabricate microlens arrays, where the diameter of the lenses is few hundred micrometers. Figure 6.46 shows the machined dimples located in different crystal directions on 4H-SiC (0001) basal plane [67]. As illustrated in Figure 6.46(a), 4H-SiC has a hexagonal close-packed lattice structure, and the basal plane shows 60° symmetry. The <−12−10> and <10−10> directions are normal to the primary planes {−12−10} and secondary planes {10−10}, respectively. For simplicity, the <−12−10> and <10−10> directions are referred to as the "P" (primary) and "S" (secondary) directions, respectively, and the in-between directions are referred to as the "I" (intermediate) directions.

The SEM images of P-, I-, and S-dimples are presented in Figures 6.46(b)–(d), respectively. In P-dimple, long parallel cracks are generated perpendicularly to the cutting direction and follow the same crystal direction <−12−10>, indicating that the cracks propagate parallel to the S planes {10−10}. These cracks are deep and wide, causing partial material to spill off from the substrate. In I-dimple, the cracks also formed perpendicularly to the cutting direction. However, numerous short and narrow cracks were generated, which are distinctly different from the cracks formed in P-dimples. Unlike P- and I-dimples, there are no evident damages on the surfaces of S-dimple.

Figure 6.47 shows the subsurface damage of the machined P-, I-, and S-dimples. In P-dimple, as shown in Figure 6.47(a), a clear subsurface damage layer with a depth of ~200 nm can be observed in the near machined surface. The main forms

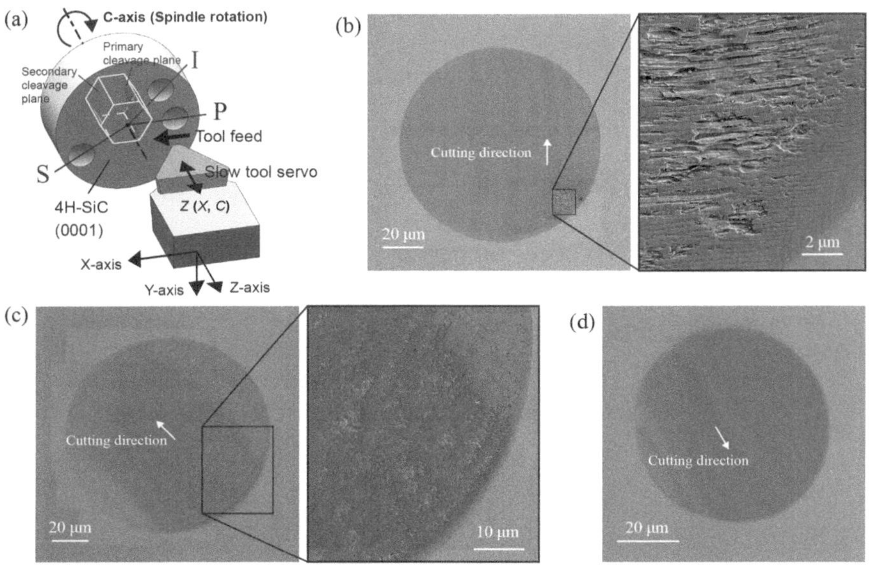


**FIGURE 6.46** (a) Schematic of diamond turning for spherical dimples on 4H-SiC. SEM images of the machined dimples located in (b) P-, (c) I-, and (d) S-directions [67].

of the subsurface damage are cracks, disordered layer, and dislocations. In I-dimple, as shown in Figure 6.47(b), the subsurface damage layer is much thinner than in P-dimple, with a depth of ~60 nm. The main forms of the subsurface damage are the same as those in P-dimple, including cracks, disordered layer, and dislocations. However, in I-dimple the disordered layer is thicker, and the depth of cracks is smaller, compared to P-dimple. In S-dimple, as shown in Figure 6.47(c), a subsurface damage layer with a depth of ~220 nm can be observed in the near-machined surface, which has the thickest damage among the three dimples. However, no cracks were generated in the material. Instead, stacking faults appear

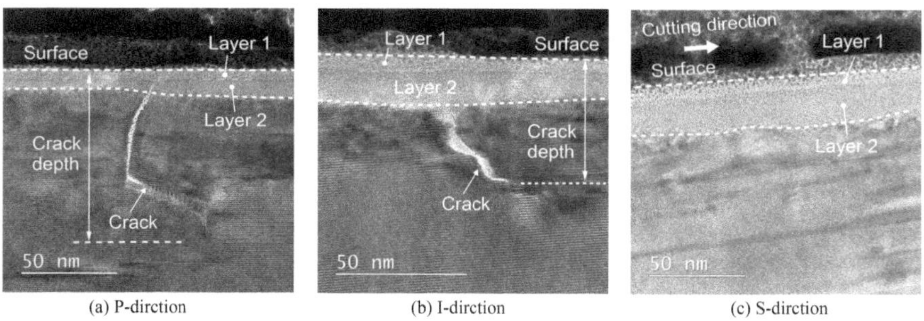


**FIGURE 6.47** Subsurface damage of the spherical dimples on 4H-SiC located in (a) P-, (b) I-, and (c) S-directions [67].

as a form of subsurface damage, together with disordered layers and dislocations. The stacking faults extend deeply into the machined subsurface, even deeper than the cracks and dislocations formed in P- and I-dimples.

These results demonstrate the anisotropic machinability of 4H-SiC (0001) plane. The optimal cutting direction for preventing crack formation is along the <−12−10> crystal orientation. However, a thick subsurface damage layer is formed when cutting along this direction. Cutting along <10−10> crystal orientation is likely to cause both severe surface cracks and thick subsurface damage layers. Cutting along the direction between <−12−10> and <10−10> allows for the generation of a shallower subsurface damage layer, but surface cracks are more likely to form. The choice of preferred cutting direction should be determined by the trade-off between machined surface quality and subsurface quality.

## 6.4 MACHINABILITY OF GLASS

In optical area, glass lenses have many predominant advantages over the plastic counterparts on aspects of hardness, refractive index, light permeability, stability to environmental changes in terms of temperature and humidity, and so on. For this reason, glass lenses have been used increasingly in the field of high-resolution digital cameras, mobile phone cameras, and other optical systems. Conventionally, glass lenses have been fabricated by a series of material removal processes such as grinding, lapping, and polishing, which requires a long production cycle and results in high production cost [68]. In recent years, high-temperature glass molding has been accepted as an effective way to efficiently produce precision optical elements with complex shapes, such as aspherical lenses, Fresnel lenses, diffractive optical elements (DOEs), micro lens arrays, and so on [69,70]. Although glass molding is beneficial for mass production, the fabrication and maintenance of high-quality molds, usually made of tungsten carbide or silicon carbide, is time-consuming and expensive. Direct machining of glass by cutting or grinding is necessary especially when fabricating test pieces or prototype optical elements. Thus, the understanding of the machining mechanism and machinability of glass materials is important.

Unlike crystalline materials, which have a difference in machinability according to the crystal orientation, glass is an amorphous and isotropic material. The machining mechanisms and machinability of glass depends significantly on the chemical composition and thermal/mechanical properties of the glass material. In this chapter, the machinability of three typical glasses, including boron silicate glass, soda lime glass, and BK7 glass, will be introduced.

### 6.4.1 Borosilicate Glass

Borosilicate glass is a type of glass that contains $SiO_2$, $Na_2O$, $Al_2O_3$, and $B_2O_3$ (boron trioxide) which allows for a very low thermal expansion, high chemical resistance, and low electrical conductivity. Unlike regular glasses, borosilicate glass will not crack even under extreme temperature changes. Due to its excellent properties, it is widely used in not only optical products, but also biomedical fields, such as micro-electromechanical systems, and microfluidic devices for chemical and biological micro total analysis systems [71].

Figure 6.48(a) shows AFM measurement results for borosilicate glass surface after constant-loading scratching with various normal loads 0.4, 0.6, 0.8, 1, 3, 5, and 10 mN [72]. A crack-free surface with small pile-ups along the groove edges can be observed. For an amorphous material with a low ratio of elastic modulus to plastic strength, the material does not significantly increase in strength (or hardness) as it is plastically deformed; thus the plastic zone is locally contained within the circle of contact and the elastic deformation that accommodates the volume of the indent is spread to a greater distance from the indenter; as a result, the pile-up formed in the scratching of borosilicate glass is smaller than the crystalline materials, such as silicon [72]. For more precise observation, the scratched groove with a normal load of 5 mN was measured, as shown in Figure 6.48(b). No surface protuberance was observed and the scratched surface is very smooth, which proves that it is possible to machine the borosilicate glass in a ductile mode.

To achieve ductile mode cutting of borosilicate glass, Kuriyama et al [73,74] developed a constant pressure cutting device that has air static pressure bearings mounted in three directions so that constant pressure cutting can be applied to the workpiece. Therefore, the depth of cut and the position of the cutting edge of the tool can be controlled stably and will not be affected by the kinematic performance of the machine tool. Figure 6.49(a) shows the micrograph of the finished surfaces as cut in V-grooves. There are no chippings or other defects observed and

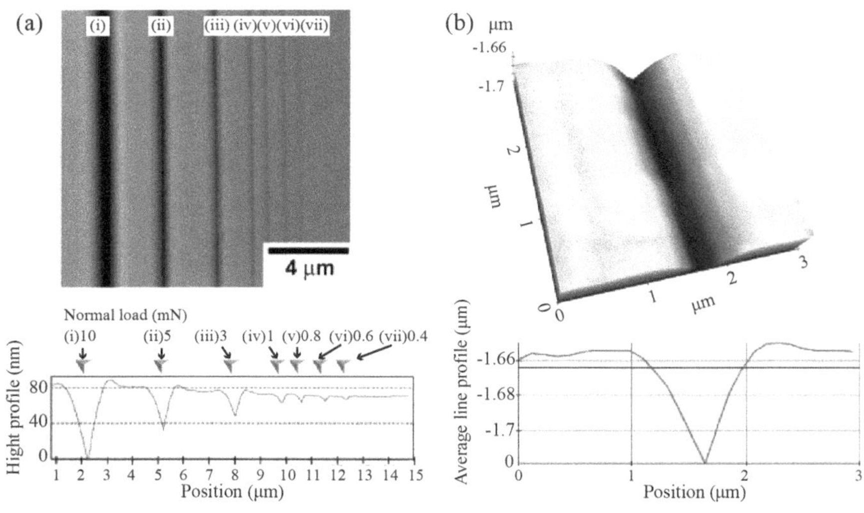


**FIGURE 6.48** AFM analysis results of borosilicate surface after constant-loading scratching: (a) general view under various normal loads 0.4, 0.6, 0.8, 1, 3, 5, and 10 mN; (b) typical groove morphology under the normal load of 5 mN [72].

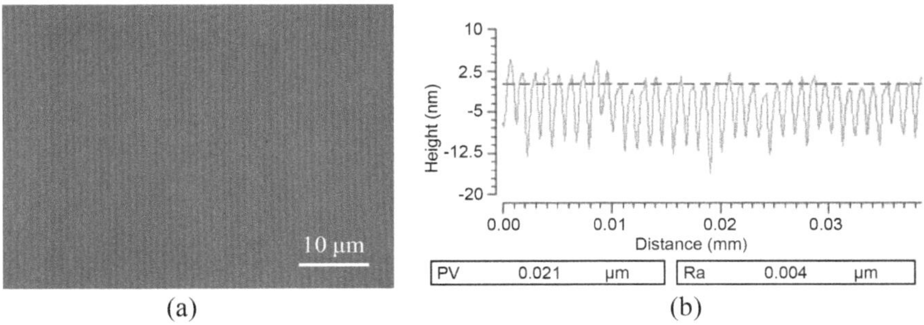


**FIGURE 6.49** (a) Micrograph of V grooves machined on borosilicate glass by diamond turning under constant pressure; (b) cross-sectional profile of V-grooves [73].

uniform grooves have been created at a pitch of 1 μm. Figure 6.49(b) shows the cross-sectional profiles obtained from the surface texture measurements. Actual grooves created on the finished surfaces are less than 20 nm deep, which is smaller than the critical chip thickness for brittle–ductile transition (BDT). This proves that constant pressure cutting can be applied to optical glass to realize stable ductile mode cutting.

An elevated temperature can increase critical chip thickness for BDT and contribute to the ductile removal of glass. Laser heating is an effective method for assisting ductile mode cutting of glass materials. Borosilicate glass's properties are strongly affected by thermal effects. The elastic modulus and the fracture toughness increase with temperature by its strain point (525 °C), and the hardness shows a decreasing tendency with temperature although remains constant in the range up to about 300 °C due to a densifying behavior [75]. Figure 6.50 shows SEM images of a channel machined using a flat-end mill without and with $CO_2$ laser irradiation. As shown in Figure 6.50(a), the harsh milling condition caused large fractures at the sides of the channel and many cracks were observed at the bottom surface. However, no side fractures or bottom cracks were generated when a $CO_2$ laser was irradiated during the milling process, as shown in Figure 6.50(b). Furthermore, it is hard to find any resolidified layer near the side of the channel. The cross-section profiles of the machined channels in Figure 6.50(c) show that precise square-shaped channels were fabricated using laser heating. A surface roughness of about 0.1 μm was achieved in laser-assisted machining, which is a 96% reduction from that without laser irradiation. Cutting forces under laser-assisted machining were also found to be smaller and more stable compared with those without laser heating.

### 6.4.2 Soda Lime Glass

Soda lime glass is made from a mixture of compounds like lime (calcium dioxide $CaCO_3$), silica (silicon dioxide, $SiO_2$), soda (sodium carbonate, $Na_2CO_3$), and other elements that can imbue the glass with different properties. It is characterized

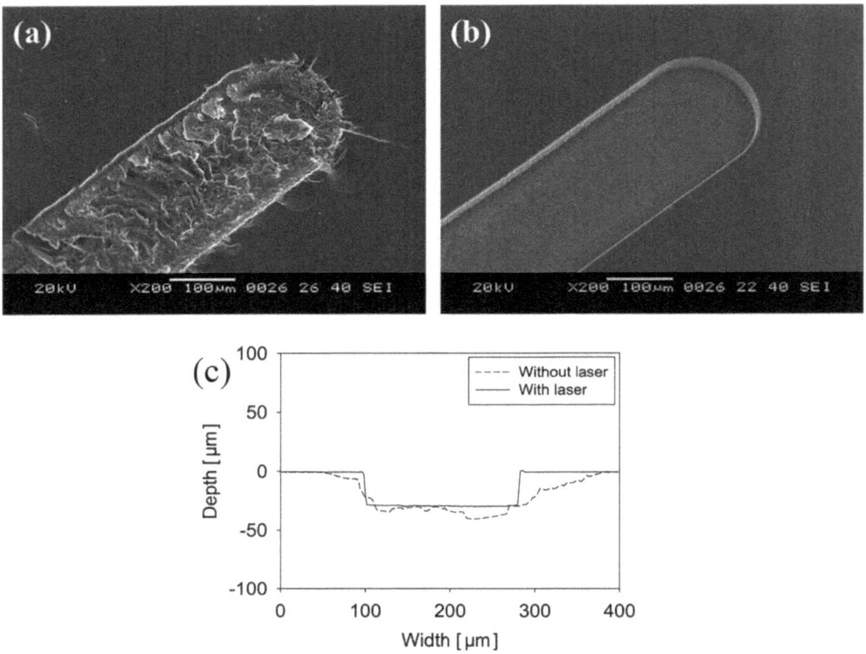


**FIGURE 6.50** SEM images of machined channel (a) without and (b) with assistance of laser irradiation. (c) Comparison of the cross-section profiles of machined channel without and with laser irradiation assistance [75].

by a greenish tinge. Soda lime glass is relatively inexpensive, chemically stable, reasonably hard, and can be recycled by softening, remelting, and reforming it multiple times without the risk of diminishing its physical properties. Some of the most common applications of soda lime glass include food containers, windows, and high-voltage insulator.

Like other brittle materials, soda lime glass can be cut in ductile mode cutting when the critical chip thickness is sufficiently small. In grooving tests, the critical chip thickness for the ductile-brittle transition was about 560 nm [76]. Soda lime glass has a Mohs hardness of 6–7, which is similar to the hardness of silicon. Thus, the tool wear is significant as the cutting distance increases. Figure 6.51 shows the SEM photographs of the tool wear after cutting soda lime glass [77]. After cutting for 50.3 m, a crater wear near the cutting edge was formed on the rake face, as shown in Figure 6.51(a), and the maximum width was approximated as 600 nm. Meanwhile, a wear land of width exceeding 5 µm was formed on the flank face. Both the crater wear and the flank wear land are smooth and uniform, indicating that the tool wear is a stable and gradual process. When the cutting distance was increased to 150.8 m, as shown in Figure 6.51(b), the crater itself had no significant increase compared to that in Figure 6.51(a); however, the width of crater narrowed due to edge recession. In addition, the flank wear land is covered

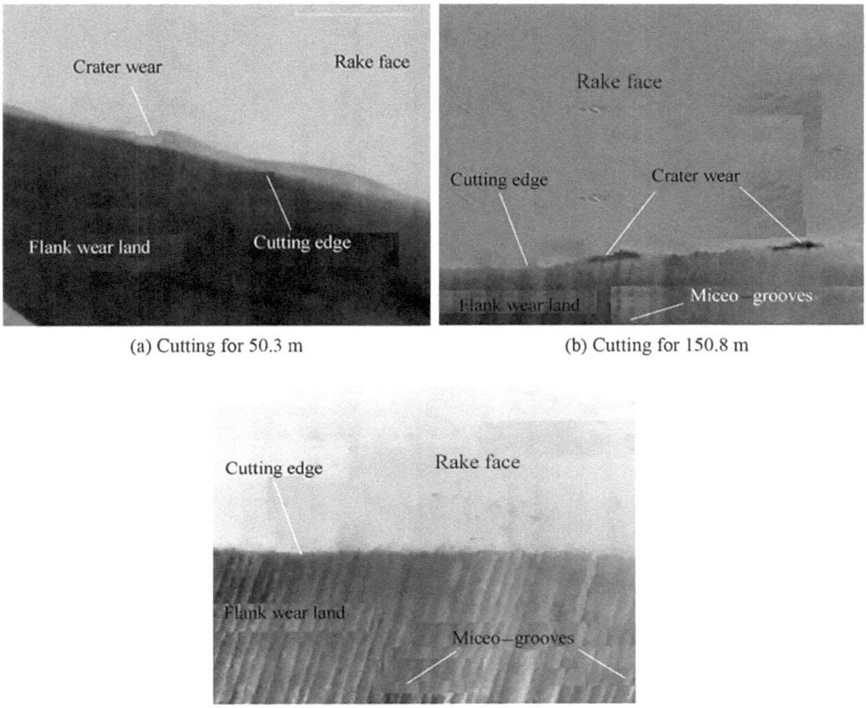


(a) Cutting for 50.3 m

(b) Cutting for 150.8 m

(c) Cutting for 251.3 m

**FIGURE 6.51** SEM photographs of the tool wear after cutting soda lime glass [77].

with numerous micro-grooves oriented along the cutting direction, and the interspaces of micro-grooves were 250–400 nm. The integrity and sharpness of the cutting edge have also descended. After cutting for a distance of 251.3 m, as shown in Figure 6.51(c), no obvious rake face wear can be observed anymore, because the significant edge recession has made the crater wear disappear. The flank wear became dominant and the wear zone increased in length and width. Furthermore, micro-grooves on the wear land increased in both amount and depth, nevertheless the interspaces of micro-grooves changed little compared with that in Figure 6.51(b).

The tool wear greatly affected the machined surface quality. When the cutting distance was 50 m, the surface was very smooth with no apparent fractures, as shown in Figure 6.52(a), which demonstrated material removal was performed in ductile mode. The surface roughness Rq, Ra, and Rmax were 19 nm, 15 nm, and 170 nm, respectively. When the cutting distance was over 250 m, the quality of machined surface became obviously poor, as shown in Figure 6.52(b). Although tool marks remain visible on some parts of the surface, most of the surface has

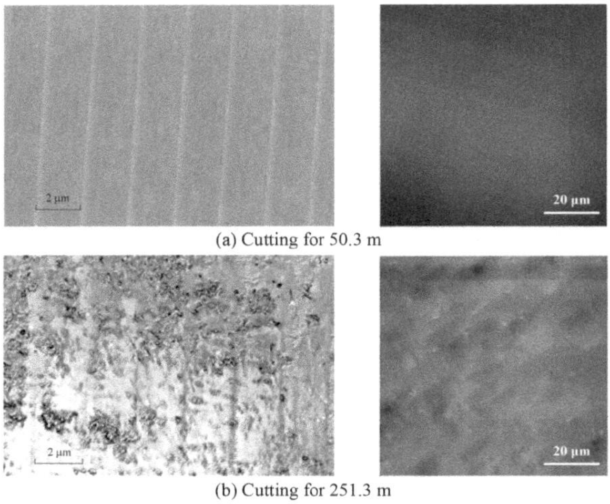


**FIGURE 6.52** Surface morphologies of the machined soda lime glass surface [77].

been damaged by microfractures. The surface roughness Rq, Ra, and Rmax increased substantially to 303 nm, 234 nm, and 3 μm, respectively.

### 6.4.3 BK7 Glass

BK7 glass is mainly composed of $SiO_2$ and $B_2O_3$. It is characterized by extremely low bubble and inclusion content, thus it is a high-quality optical glass providing excellent transmittance through the visible and near-infrared spectra and down to 350 nm in the ultraviolet. On the other hand, BK7 glass is a typical hard-brittle material, which is difficult to cut.

A nanoindentation test on BK7 glass was attempted to understand its plastic deformation behavior [78]. When the indentation load was below 0.1 N and the indentation depth smaller than 1 μm, the material did not crack, as shown in Figure 6.53(a), indicating that ductile responses can occur in sufficiently small scales. It is also obvious from the photograph that plastic flow is involved in

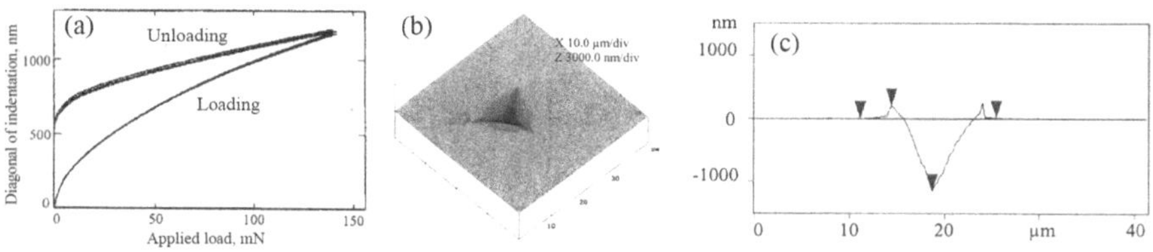


**FIGURE 6.53** Nanoindentation analysis of BK7 glass: (a) diagonal of indentation versus load, (b) topography of a residual indent, and (c) sectional profile of the indent [78].

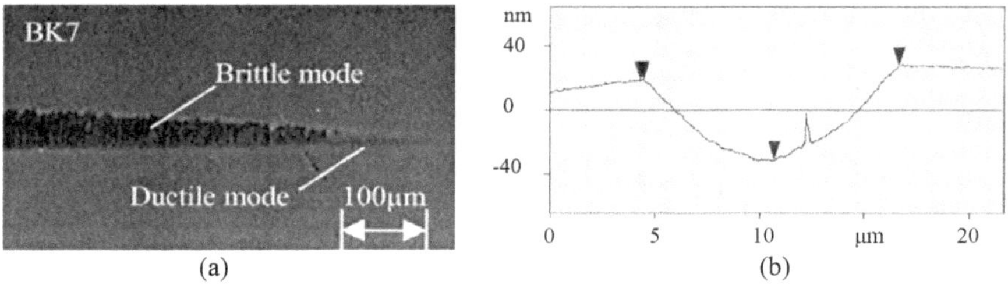


**FIGURE 6.54** (a) Micrograph of plunge-cut groove for analysis of the brittle–ductile transition of BK7, (b) cross-sectional profile of the groove at the transition point [78].

indentation as permanent "pile-up" along the residual indent as shown in Figures 6.53(b) and (c).

Figure 6.54(a) shows a result of plunge-cutting test on BK7 glass using a 0° rake angle tool. Ductile cutting can be obtained when the depth of cut is below a threshold value. According to the cross-sectional profile analysis, if the undeformed chip thickness $h$ is smaller than the threshold $h_c$ = ~62 nm, the groove exhibits a very smooth topography indicating ductile mode cutting, as shown in Figure 6.54(b).

Different from crystalline materials, a unique characteristic of glass materials is viscoplasticity. In other words, the mechanical properties of glass are strongly influenced by temperature, which in turn, affects the processing mechanism and machinability of the material. In the machining of glass with high cutting speed, more cutting heat will be generated at the cutting region, and the thermal effects cannot be neglected.

Feng et al. [79] performed single and consecutive scratch tests on BK7 under the scratching speeds of 1, 5, and 20 m/s to investigate the thermal effect on scratched surface morphology. Figure 6.55 shows the scratch profiles at the transition point where the depth rapidly increased with scratch speed. In the 1 m/s scratch, the residual depth for the transition point was nearly 150 nm, while the depths for the 5 m/s and the 20 m/s scratches were 100 nm and 70 nm, respectively. The morphology of the scratch groove at the transition stage also varied with the scratch speeds. In Figures 6.55(a) and (b), the scratch groove was smooth before and after the transition point at the scratch speed of 1 m/s. When the scratch speed was 5 m/s and 20 m/s, the scratch depth first increased rapidly then decreased while the profile became rough, as shown in Figures 6.55(d) and (f). In the scratch with speed of 20 m/s, a large pileup of material was observed at the center of the scratch. In Figures 6.55(c) and (e), many small slits expanding laterally across the groove can be observed after the transition, which was totally different from the smooth surface before the transition. Such behavior was supposed to be a ductile deformation because the opening of these long "slits" did not have crack tips propagating at the edge of the scratch-like radial cracks. In Schinker's study [80], these small slits were called "micro-shear" and were considered as the result of viscous flow when glass got softened during high-speed machining.

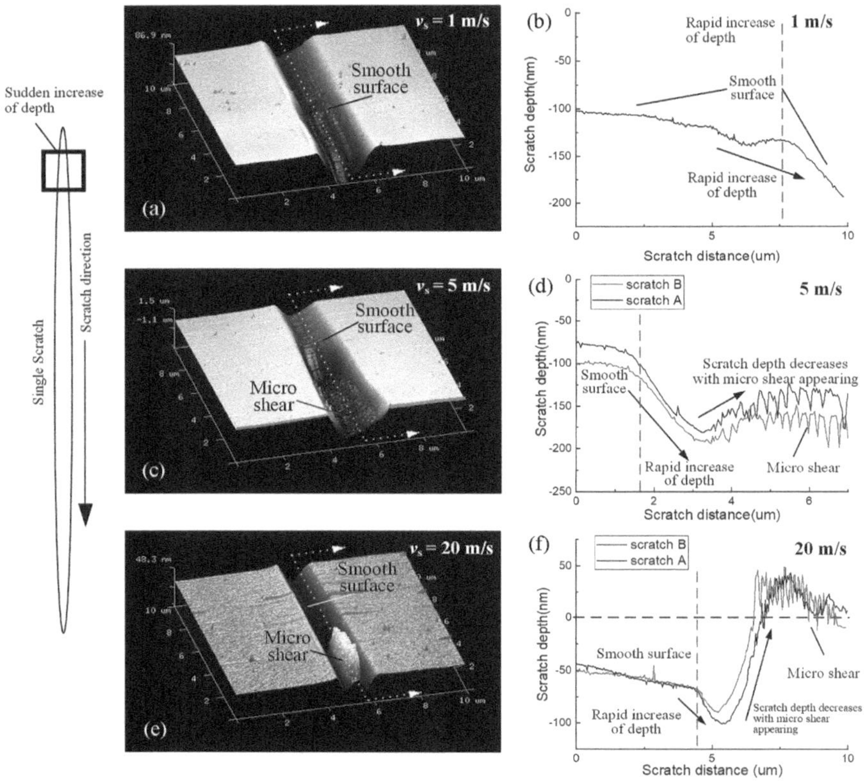


**FIGURE 6.55** AFM images for the single scratch tests under scratch speeds of (a, b) 1 m/s, (c, d) 5 m/s, and (e, f) 20 m/s at the transition point where scratch depth suddenly increased. The curve in (b) is the depth at the center cross-section in (a); the depth curve of "scratch A" in (d, f) is the center cross-section in (c) and (e) respectively; while the depth curve of "scratch B" in (d, f) was derived from another scratch at the same transition point for comparison [79].

In addition, the critical depth $h_s$ for brittle–ductile transition changes with scratch speeds. When the scratch speed is 1 m/s, the $h_s$ is about 493 nm, and the brittle chippings formed as deep lateral crack appeared from the start to the end of a scratch, as shown in Figure 6.56(a). When the scratch speed increases to 5 m/s, smaller brittle chippings do not occur until the $h_s$ up to 1.2 μm. The size of the chipping also changed, as shown in Figure 6.56(b). The lateral cracks that caused large chippings found in 1 m/s scratches reached the nearby scratches, forming a combined 30 μm wide chipping zone, while the chippings in the 5 m/s scratches are much smaller surrounding the scratched grooves they originate. Meanwhile, the quantity of the chippings is also lower in the 5 m/s scratches. When the scratch speed increases to 20 m/s, the chippings start to appear earlier again, where the $h_s$ is deceased to ~624 nm, and the chippings are larger than those in the 5 m/s

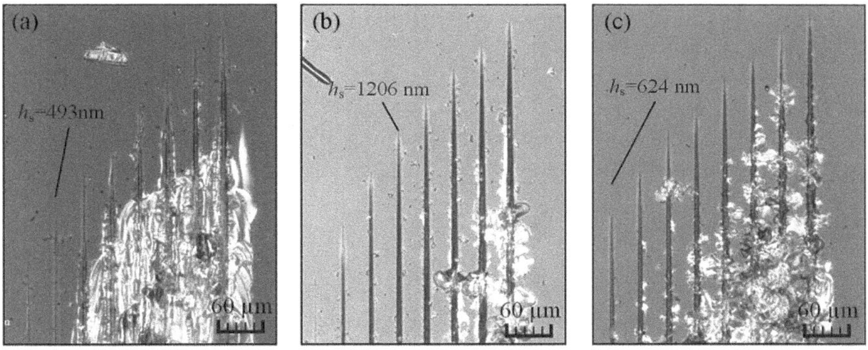


**FIGURE 6.56** The surface morphology of single scratch test on BK7 glass under speeds of (a) 1 m/s, (b) 5 m/s, and (c) 20 m/s [79].

scratches, as shown in Figure 6.56(c). The coupling effects of temperature and strain rate in the material, which is induced by the high-speed cutting, makes the crack formation mechanisms complex.

### 6.4.4 Fluoride Glass

By now, diamond turning of optical glasses has been attempted by many researchers, but in most of those studies severe diamond tool wear has been reported. Tool wear in diamond turning of silica glass is especially significant so there is not yet successful report of ductile machining a whole lens surface. Most glass materials, such as silica glass, soda lime glass, and BK7, are based on the $SiO_2$ constituent, thus the chemical reaction between $SiO_2$ in glass and C in diamond might be one of the reasons for the severe tool wear.

Ishizuka et al. [81] attempted diamond turning of fluoride glass, which has no $SiO_2$ constituent, and found that the tool wear characterized by edge retreat was reduced to 1/22 of that in the cutting of silica glass. Ductile mode cutting of

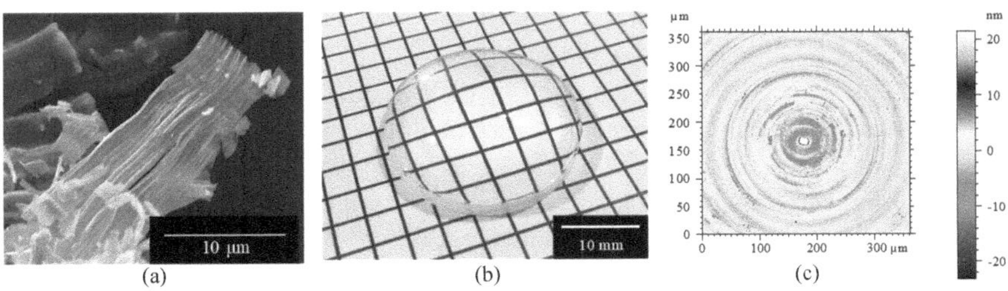


**FIGURE 6.57** Results for ductile mode diamond turning of fluoride glass: (a) SEM image of cutting chips; (b) photograph of a diamond-turned aspherical lens; (c) surface topography of lens center [81].

fluoride glass was realized for a considerably long cutting distance. As a demonstration, an aspherical glass lens of fluoride glass, which has a diameter of 50 mm, was successfully produced by ductile mode diamond turning using a straight-nosed diamond tool. Figure 6.57(a) shows SEM image of cutting chips and Figure 6.57(b) is a photograph of the machined aspherical lens. The surface roughness of the lens is 2 nm Sa, as can be seen from Figure 6.57(c). More details can be found in Ref. [82].

From this example, it is presumable that silicon dioxide in glass materials causes severe tool wear due to the thermal-chemical reaction between the diamond tool and the glass material. By using a silicon dioxide-free optical glass, the diamond tool wear can be significantly suppressed. From this point of view, it is highly expected that by optimizing the chemical composition, new kinds of glasses will be developed, which have good optical properties and high diamond machinability.

## 6.5 MACHINABILITY OF CERAMICS

Ceramics is a group of composite materials artificially made from a mixture of multiple different materials via high-temperature and/or high-pressure methods. The material properties of ceramics can be distinctly different depending on the material composition and manufacturing conditions. Most ceramics materials have high hardness, high strength, and stiffness, thus having many applications as structural components, mold substrates, and dental/biomedical parts. For some mold and biomedical applications, the ultraprecision machining of the ceramic surfaces is required. However, due to that ceramics are multiphase composite materials, the machinability varies among different phases, which makes it very difficult to obtain a uniformly smooth surface. In addition, the high hardness of ceramics causes severe tool wear, leading to a high production cost. In this chapter, the machining mechanisms and machinability of typical ceramics materials will be introduced.

### 6.5.1 Reaction-Bonded Silicon Carbide

#### 6.5.1.1 Material Property

There are many kinds of silicon carbide ceramics, the mechanical, thermal, and chemical properties of which depend on the material manufacturing methods. Conventional methods for the manufacture of SiC ceramics include pressureless sintering, gas pressure sintering, and hot pressing, where SiC grains are bonded together by a variety of other materials, such as boron (B), carbon (C), and aluminum (Al), etc. As a relatively new method, reaction bonding of SiC has been popularly used in industry. Reaction-bonded silicon carbide (RB-SiC) is fabricated by infiltrating silicon (Si) melt into a green compact consisting of C and SiC powders. The liquid Si reacts with the C particles, forming new SiC particles. In this way, a very dense SiC body is produced without the formation of micro pores.

However, the infiltrated Si melt cannot react with C completely; therefore, excess Si remains in the body of material as a bonding component [82].

RB-SiC has been now considered to be a most promising material of the SiC ceramics family. It can be manufactured at relatively low temperatures and with low production costs. It has high purity, dense structure, and excellent thermo-mechanical characteristics, such as high thermal stability and conductivity, high specific heat capacity, and low thermal expansion. In particular, RB-SiC is characterized by its low density (slightly higher than aluminum) and high stiffness (bending strength over 1 GPa and Young's modulus higher than 400 MPa). The unique properties of RB-SiC also enable it to be used under cryogenic temperatures, for example, as large-diameter light mirrors of infrared universe imaging systems used in space [83].

Like single-crystal SiC, the RB-SiC is very hard and brittle, which is a difficult-to-machine material. The material properties of a typical type of RB-SiC, produced by Japan Fine Ceramics Co., Ltd are listed in Table 6.5. Figure 6.58(a) shows an SEM micrograph of the fast atom beam etched sample surface. The smooth regions correspond to SiC grains, and the micropits correspond to residual silicon. Most of the SiC grains are directly bonded to each other without the presence of silicon at grain boundaries. A schematic model of the RB-SiC material microstructure is depicted in Figure 6.58(b).

### 6.5.1.2 Machined Surface Morphology

The machinability of RB-SiC was investigated by precision cutting tests using a single-crystal diamond tool [84]. Figures 6.59(a) and (b) show two SEM micrograph pairs (at different magnifications) of the surfaces machined at the low (3 μm/rev) and high (72 μm/rev) tool feed rates, respectively, using a 0° rake angle tool. In Figure 6.59(a), periodical tool feed marks are observed on the surface. Although the majority of the surface is smooth and exhibits a ductile-cut appearance, a few small pits are clearly observed, the sizes of which are in the order of 1 μm, similar to that of the SiC grains. Therefore, these micropits may have been caused by the dislodgement of SiC grains from the surface. In contrast to

**TABLE 6.5**
**Properties of Reaction-bonded Silicon Carbide [63]**

| Parameters | Value |
|---|---|
| Si/SiC volume ratio (% : %) | 12 : 88 |
| Density | 3.12 $g/cm^{-3}$ |
| Young's modulus | 407 GPa |
| Bending strength | 780 MPa |
| Thermal expansion coefficient | $3.23 \times 10^{-6}$/K |
| Thermal conductivity | 143 W/m K |
| Porosity | <0.1% |

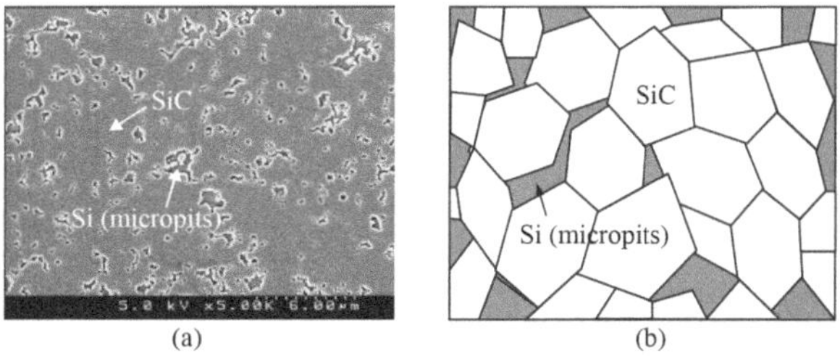


**FIGURE 6.58** (a) SEM micrograph of a fast atom beam etched surface of RB-SiC sample. (b) Schematic model of the RB-SiC material microstructure [84].

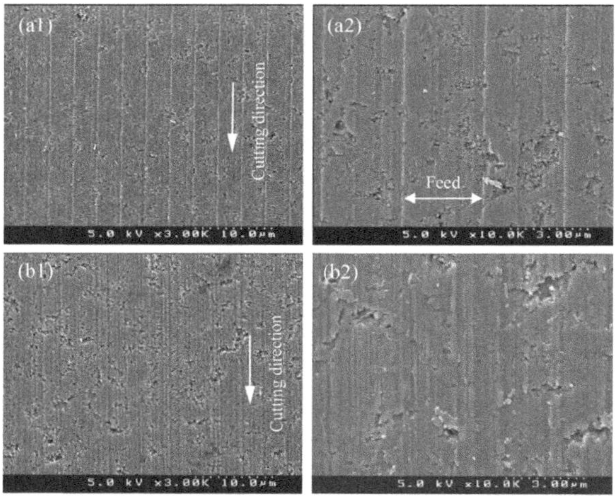


**FIGURE 6.59** SEM micrographs of surfaces machined at (a) $f = 3$ μm/rev and (b) $f = 72$ μm/rev, using 0° rake angle tool. The micrograph pairs were obtained at different magnifications [84].

Figure 6.59(a), no periodical tool feed marks are observed in Figure 6.59(b) because the tool feed rate is very high in this case and the field of view is between two neighboring feed marks. Both the size and the number of micropits have increased compared with those in Figure 6.59(a). However, no brittle fractures as large as a few tens of microns, which are often seen in the machining of single-crystal silicon and germanium in brittle modes, are observed.

### 6.5.1.3 Machining-Induced Microstructural Change

Figure 6.60(a) shows a Raman spectrum of the uncut sample surface (left-side intensity scale) and a replot of the same spectrum at a magnified intensity scale (right-side intensity scale). There is a strong sharp peak at 521 $cm^{-1}$ that corresponds to crystalline Si (c-Si), and four peaks at 154, 767, 788, and 964

$cm^{-1}$ that correspond to the four characteristic peaks of the 6H-SiC polycrystal, which confirms that the RB-SiC sample is composed of crystalline Si and crystalline 6H-SiC. Figures 6.60(b) and (c) show Raman spectra of an area of Si bond on the surface after machining with tools at tool rake angles of 0° and –40°. Although a sharp peak still appears at 521 $cm^{-1}$ in Figure 6.60(b), two broadband peaks are evident at 470 and 150 $cm^{-1}$, indicating the presence of amorphous Si. This Raman spectrum is very similar to that of ductile machined Si wafers [18], which demonstrates that a thin layer of crystalline Si has transformed into the amorphous state. In comparison with Figure 6.60(b), the broadband peaks centered at 470 $cm^{-1}$ and 150 $cm^{-1}$ in Figure 6.60(c) have grown significantly, whereas the peak at 521 $cm^{-1}$ has become extremely small. This indicates that a thick layer of Si bond in the RB-SiC sample has changed from crystalline to the amorphous phase. That is, a higher negative tool rake angle has a more significant effect on Si amorphization. However, comparison of as-received sample surface, the machined surfaces do not reveal obvious changes in the positions of the 6H-SiC Raman peaks, although there is a distinct decrease in the intensity of these peaks after machining. These results suggest that 6H-SiC grains do not undergo phase transformation during machining, even if plastic flow occurs.

### 6.5.1.4 Chip Formation Mechanism

Figures 6.61(a) and (b) are two pairs of different magnification micrographs of cutting chips produced by diamond tools of rake angles of 0° and –40°, respectively. In the low-magnification micrographs, all the chips appear to be continuous and ribbon-like. However, in the high-magnification micrographs, the chips appear to have more separated structures, and there are distinct differences between the structures in Figures 6.61(a2) and (b2). The chip in Figure 6.61(a2) consists of numerous micro particles and small flat pieces that are different in size and shape, while the chip in Figure 6.61(b2) is composed of extremely thin curled slices. These micrographs suggest that dislodgement of SiC grains dominates the chip formation for a 0° tool rake angle, while ductile cutting of the SiC grains dominates the chip formation for a –40° tool rake angle, as with metal cutting.

The material removal mechanism of RB-SiC can be considered from the deformation behavior of each of the two components: the Si bond and 6H-SiC grains. The Raman investigation confirmed that the Si bond undergoes amorphization, which assists the restraint of microfracture formation and facilitates continuous chip formation, due to the ductility of high-pressure phases of Si [18]. However, the amount of Si is too small (12%) to dominate the general machining mechanism of RB-SiC. Therefore, the dominant factor in machining RB-SiC is the mechanical behavior of SiC gains. High-pressure tests have shown that 6H-SiC crystals do not undergo phase transformation even at hydrostatic pressures up to 95 GPa [85]; also in Figure 6.60, no phase transformation was detected in the 6H-SiC grains. Therefore, the deformation of the material is most likely governed by dislocation mobility. It is generally believed that at low temperature, SiC

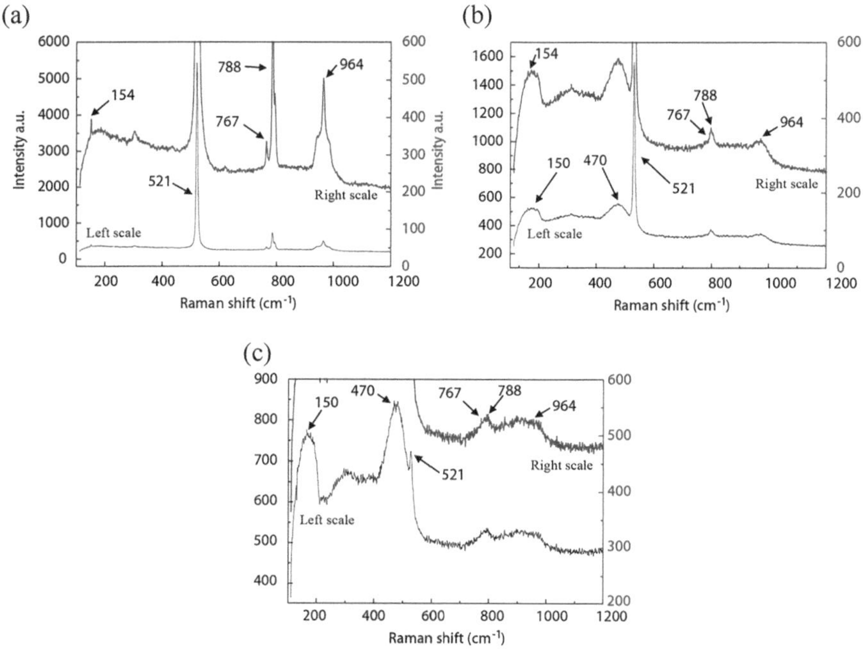


**FIGURE 6.60** Raman spectrum of the as-received RB-SiC sample, plotted in different scales of intensity. (b) and (c) Raman spectra of sample surfaces machined using diamond tools with rake angles of 0° and –40°, respectively [84].

crystals are extremely brittle and cannot be plastically deformed; however, many studies on nanoindentation and abrasive machining of 6H-SiC crystals have shown that even at ambient temperature, plastic deformation can be achieved within a small control volume before fracture occurs [86].

When using a round-nosed tool for cutting, the undeformed chip thickness $h$ varies along the cutting edge from zero to $h_{max}$, thus plastic deformation of 6H-SiC may occur along the tool tip apex where the undeformed chip thickness is smaller than the critical value for ductile-brittle transition ($h_c$). As schematically shown in Figure 6.62(a), when $h < h_c$, 6H-SiC grains are cut in ductile mode, generating very thin plastically deformed chips, such as those in Figure 6.61(b). In other words, some SiC grains will be cut into two parts: one part will be removed as a chip while the other part will remain embedded in the workpiece so that a very smooth surface can be obtained. When $h > h_c$, as shown in the upper region in Figure 6.62(a), the 6H-SiC grains begin to fracture via cleavage cracks. However, the fracture energy can be released at the grain boundaries, thus the microcracks propagate within the grains are then stopped at the grain boundaries. As a result, the microcracks depicted in the upper region of Figure 6.62(a) do not propagate deeply. This may be the reason why large-scale fractures, like those observed in cutting single-crystal materials such as Si and Ge, are not observed in the cutting

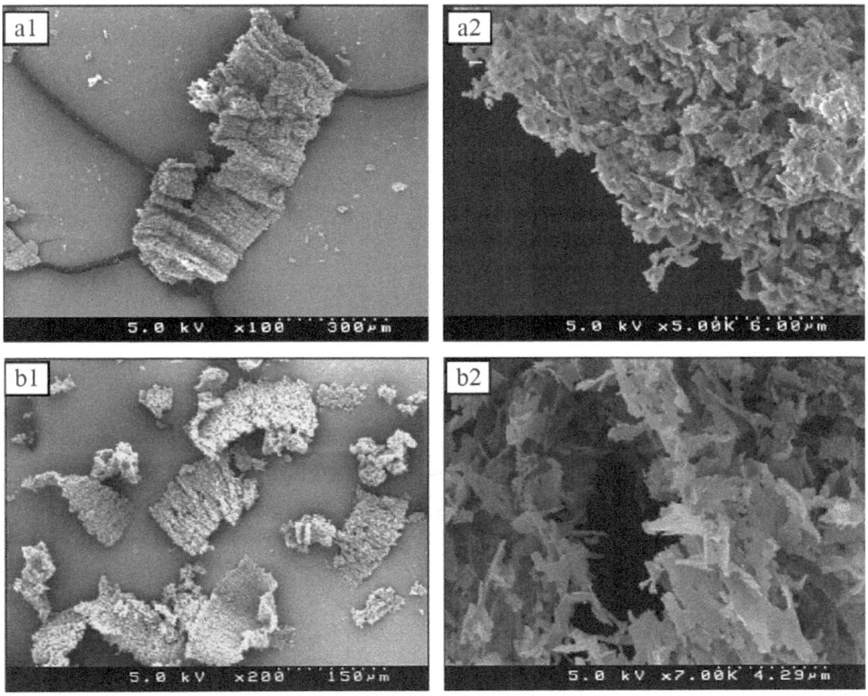


**FIGURE 6.61** SEM micrographs of cutting chips generated by diamond tools of different rake angles of (a) 0° and (b) –40°. Micrographs (a1) and (b1) are low magnification and (a2) and (b2) are high magnification [84].

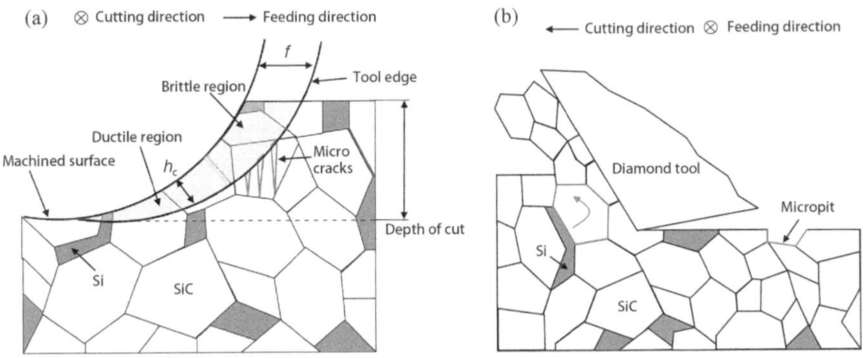


**FIGURE 6.62** Schematic models for material removal during cutting of RB-SiC: (a) ductile cutting when $h < h_c$ and brittle cutting when $h > h_c$; and (b) dislodgement of SiC grains [84].

of RB-SiC. It is also noteworthy that not all the 6H-SiC grains can be cut into two parts. A few shallow grains will be dislodged from the surface by the tool tip, leaving micropits on the machined surface, as shown in Figure 6.62(b). The dislodgement phenomenon is mainly determined by the bonding strength at the grain boundaries. The mechanism for material removal may also depend on pre-existing defects, such as micropores generated during the material fabrication. These results suggest that the mechanism for machining RB-SiC is far more sophisticated than that of single-crystal materials or amorphous materials.

### 6.5.2 Tungsten Carbide

Tungsten carbide (WC), also termed cemented carbide, is another important ceramic material that exhibits excellent mechanical and physical properties such as high hardness (HRA 80–90), high strength (deflecting strength 2–4 GPa), high toughness 6–8 MPa·m$^{1/2}$), and strong resistance to wear and heat. As mentioned in Chapter 4, it is an important material for cutting tools. In addition, it is the most widely used material for precision glass molding dies in optical manufacturing. Usually WC is prepared by sintering with cobalt or nickel as binder materials, thus the properties of WC depend on the binder material and its concentration.

In recent years, binderless WC has been developed by spark plasma sintering process without the addition of any binder phases. Another kind of WC is TiC-WC, which is prepared by high-frequency induction heating sintering using titanium carbide (TiC) as a binder. Figure 6.63 shows an SEM micrograph of a TiC-WC surface after polishing. Grey areas correspond to WC grains and dark ones correspond to TiC [87].

Compared to RB-SiC, the hardness of WC is relatively lower, and the ductile machinability is higher. For the WC using cobalt or nickel as binder materials, the binder material may cause severe tool wear due to chemical reaction between the binder and diamond. To reduce the chemical wear, it is effective to reduce the temperature of the tool tip. An effective method to reduce tool temperature and wear is tool-swing cutting, as illustrated in Figure 6.64, where a swinging movement is

**FIGURE 6.63** SEM micrograph of the polished TiC-WC surface [87].

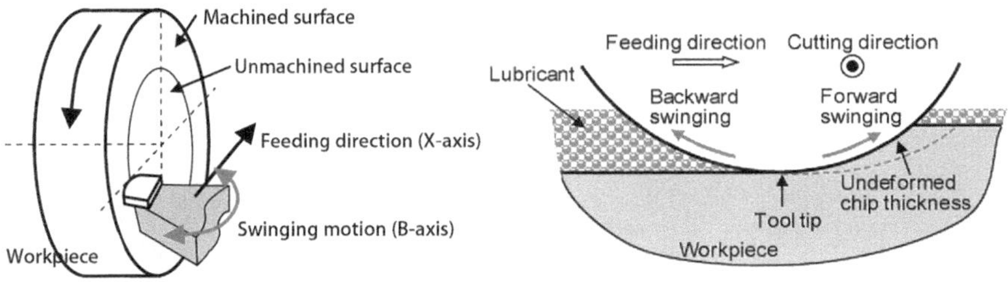


**FIGURE 6.64** Schematic models for tool-swinging cutting method [88].

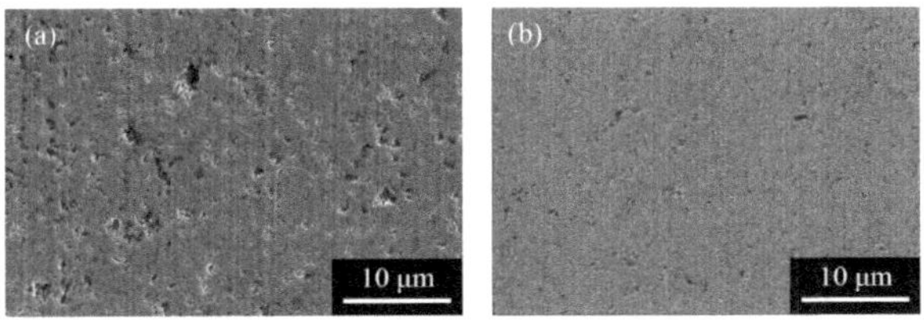


**FIGURE 6.65** Machined surface by applying (a) conventional cutting method, and (b) tool-swinging cutting method [87].

applied to a round-nosed diamond tool in diamond turning [87]. The center of curvature of the cutting tool is adjusted to be in agreement with the center of rotation of the B-axis rotary table of the machine. By swinging the rotary table, the cutting point will constantly vary with respect to the tool edge during machining. As a result, the cutting time for a certain cutting point becomes very short and the temperature increase at this point will be greatly reduced. In addition, the tool can be well lubricated as it swings, because lubricants, especially the solid lubricants such as nanoparticles, can easily penetrate the tool-workpiece interface. The effectiveness of this method has also been demonstrated in machining RB-SiC [88].

Figure 6.65(a) shows an SEM micrograph of a TiC-WC surface by conventional diamond turning method [87]. Some micro pits can be found on the surface, the size of which is about 1 μm, and their shapes are irregular. Figure 6.65(b) is a TiC-WC surface by diamond turning with a tool-swinging movement. The machined surface is very smooth, indicating a stable ductile-cut appearance. The micro pits are distinctly smaller than those in Figure 6.65(a).

Figure 6.66 shows cutting chips removed from the WC sample. The chips generated in the conventional cutting consist of micro particles different in size and shape. However, cutting chips generated in the tool-swinging cutting are composed of extremely thin slices, indicating that the individual grains were cut in a ductile mode. Under this condition, a few micro pits were observed on the machined surface, which might be because some shallow grains having very low bonding strength were inevitably dragged out from the machined surface (grain dislodgement) by the cutting force. This is very similar to the machining behavior

**FIGURE 6.66** Cutting chips in (a) conventional cutting and (b) tool-swinging cutting [87].

of RB-SiC as mentioned in Section 6.5.1. The grain dislodgements will increase significantly after tool edge becomes blunt, because the friction force acting on WC grains increases significantly. However, in the tool-swinging cutting method, the tool wear was suppressed, and thus micro pits on the machined surface were decreased significantly. This demonstrated again the possibility of directly diamond turning of WC to fabricate optical molds with less need for subsequent polishing processes.

Ultrasonic elliptical vibration cutting was also used for enhancing the machineability of WC [89]. As illustrated in Figure 6.67, synchronized two-directional vibration is applied to the cutting edge, and the cutting edge forms an elliptical locus in each cycle of the vibration. In a cycle of the elliptical vibration, the cutting starts from point A, and then the workpiece material is removed in the form of a chip. After the tangent of the tool trajectory becomes parallel to the rake face, the cutting tool separates from the chip at point B. Because of this intermittent process, reduction in the chip thickness and cutting forces can be attained. Due to this separation in each vibration cycle, the cutting tool and the workpiece can be cooled by the surrounding air and/or the cutting fluid. This also allows suppression of adhesion between the tool and the workpiece. As a result, the thermochemical wear is suppressed.

Figure 6.68 shows the machined surface and the collected chips in the conventional cutting and elliptical vibration cutting at the nominal cutting distance of 4.3 m [89].

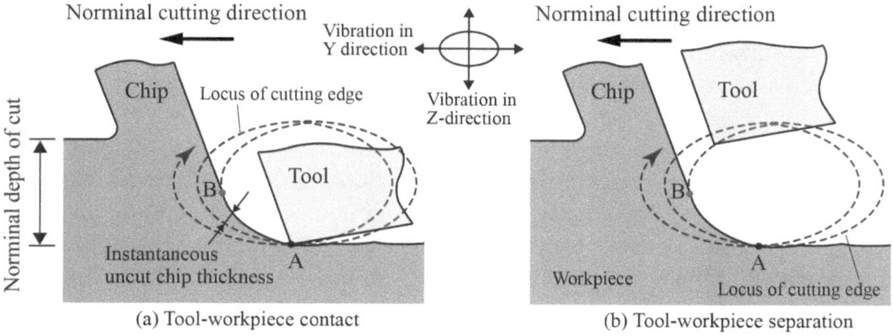


**FIGURE 6.67** Schematic models for elliptical vibration cutting process.

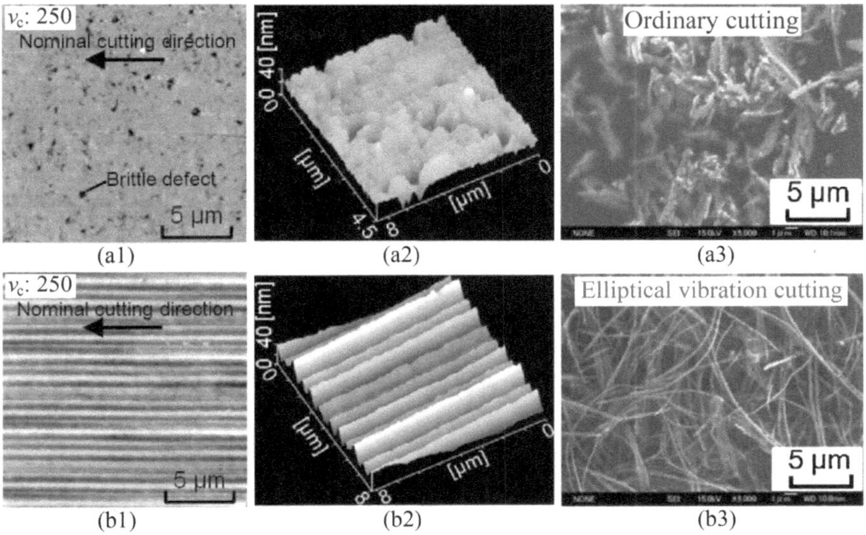


**FIGURE 6.68** Photographs of the machined surface and chips formed in (a) conventional cutting, and (b) elliptical vibration cutting ($v_c$: nominal cutting speed mm/min) [89].

In the conventional cutting, micro pits due to brittle fracture or grain dislodgement are observed on the surface. The chip segment surface is asperous, as shown in Figure 6.68(a3). In contrast, ductile mode machining is attained in the elliptical vibration cutting, resulting in a smooth surface without micro pits, although tool feed marks along the cutting direction are seen. The cutting chips are continuous, as shown in Figure 6.68(b3), like the chips formed in metal cutting. These chips are also thinner than those in conventional cutting, due to the reduction of the chip thickness by the tool vibration.

### 6.5.3 Zirconia

Zirconia is a crystalline dioxide of zirconium ($ZrO_2$). Zirconia stabilized with yttria ($Y_2O_3$), which is known as Y-TZP, is a fine ceramic having the excellent mechanical properties, such as high hardness, stiffness, and chemical stability, as well as low thermal conductivity, among zirconia-based ceramics [90]. Thus, Y-TZP has been used as a biomaterial component material, and in recent years has been used in many other applications such as dental implants, optical fiber connectors, and solid oxide fuel cells [91].

Grinding has been used for machining Y-TZP [92], but it is difficult for grinding to create small 3D features. In recent years, thermally assisted cutting of Y-TZP has been attempted by using tungsten carbide and cubic boron nitride tools [93]. However, the surface quality is low due to the unevenness of tool edge and possible thermal damage to sample surface. As an alternative, ultraprecision diamond cutting is a promising method to generate complex 3D surface shapes with nanometric roughness on Y-TZP.

Figure 6.69 shows SEM images of diamond-cut surfaces of Y-TZP [94]. When feed rate $f > 2\,\mu m/rev$, microfractures and micro pits due to grain dislodgements were found on the surface. In addition, modulated waviness perpendicular to the cutting direction was observed. The modulated waviness was especially significant at $f = 5\,\mu m/rev$, which might be a result of the stick-slip phenomenon occurring at the interface of the tool and workpiece. At $f = 0.5\,\mu m/rev$, however, uniform tool feed marks appeared while no microfractures, pits, or modulated waviness were observed on the surface.

Figure 6.70 shows SEM images of cutting chips obtained at various tool feed rates. The chips are generally continuous under all conditions, indicating that plastic deformation has dominated the material removal of Y-TZP at the micro-nanometer scale. However, cracks were observed at the chip surfaces at high tool feed rates ($f$ = 10 and 5 μm/rev), indicating a partially brittle response of the material. In addition, as the tool feed rate decreased, the period of the lamellar structure and the chip thickness also decreased. For the chip generated at $f$ = 0.5 μm/rev, the entire chip seems to be composed of fine grains, the sizes of which are ~50 nm. Such granular chip morphology was rarely reported in metal cutting. Moreover, the grain size is distinctly smaller than that of the raw material before cutting (~300 nm), indicating that grain refinement is very significant in the cutting process of Y-TZP.

To further examine the subsurface layer of the machined workpiece, cross-sectional TEM was performed. Figure 6.71(a) shows the TEM image of the subsurface machined at $f = 5\,\mu m/rev$. A crack is observed at a depth of ~500 nm from

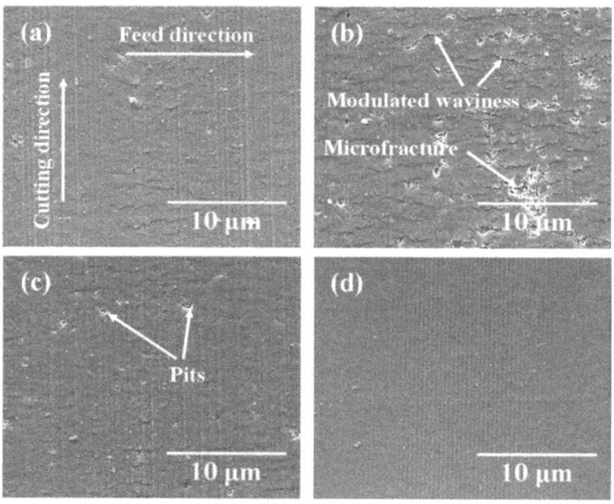


**FIGURE 6.69** SEM images of the machined surface obtained at the feed rates of (a) $f$ = 10, (b) $f$ = 5, (c) $f$ = 2, and (d) $f$ = 0.5 μm/rev [94].

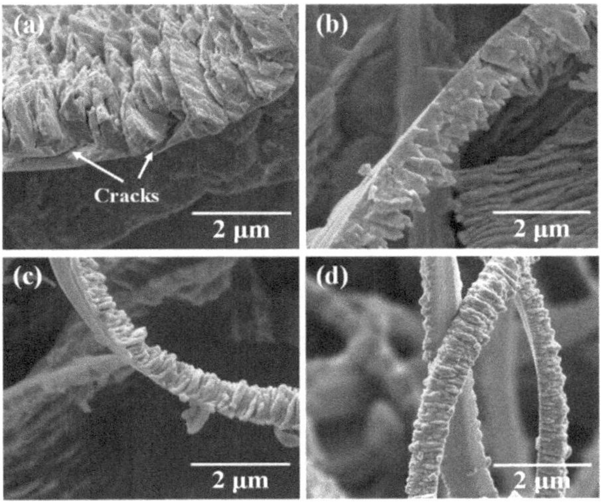


**FIGURE 6.70** SEM images of the cutting chips obtained at the feed rates of (a) $f = 10$, (b) $f = 5$, (c) $f = 2$, and (d) $f = 0.5$ μm/rev [94].

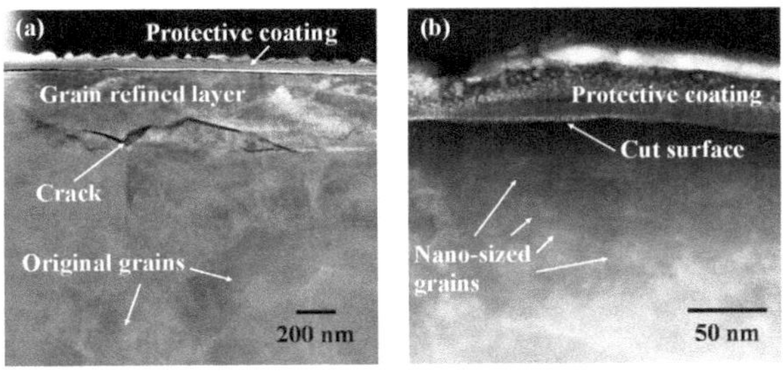


**FIGURE 6.71** Cross-sectional TEM images of the subsurface structure formed at feed rates of (a) $f = 5$ and (b) $f = 0.5$ μm/rev [94].

the surface, above which is a region of refined grains. Figure 6.71(b) is a TEM image of the subsurface machined at $f = 0.5$ μm/rev. In the subsurface region, no crack is observed. A layer of ultra-fine grains is seen below the surface. The grain size is tens of nanometers, similar to that in cutting chips shown in Figure 6.70(d).

The mechanism of chip and surface formation in Y-TZP is schematically shown in Figure 6.72. At large undeformed chip thicknesses, as shown in Figure 6.72(a), cracks occur due to the tensile stress near the tool tip. These cracks propagate along the grain boundaries, causing microfractures and grain dislodgements. However, further crack propagation is prevented by a dilatational stress-induced tetragonal-to-monoclinic phase transformation around the crack tip. This stress-induced toughening mechanism leads to Y-TZP's high fracture toughness. As a result, the microfractures in Y-TZP cutting are distinctly smaller than those in

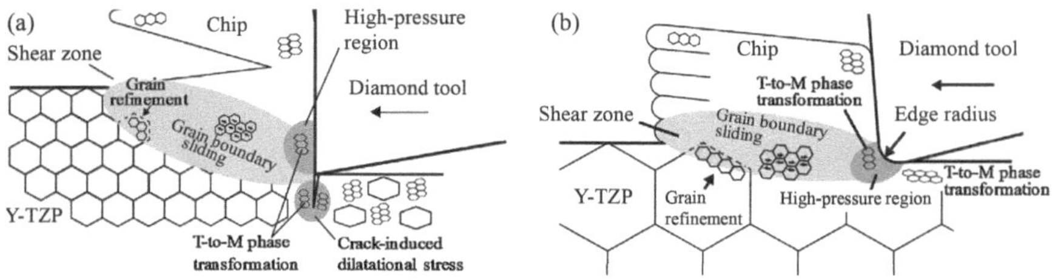


**FIGURE 6.72** Schematic model for chip/surface formation in Y-TZP at (a) large and (b) small undeformed chip thicknesses [94].

cutting single crystals such as Si and Ge. At the same time, a high-pressure region forms in front of the tool tip, which also causes tetragonal-to-monoclinic phase transformation. At small undeformed chip thicknesses, however, the effect of tool edge radius becomes significant, and the effective rake angle becomes negative, creating a high pressure in front of the tool tip, as shown in Figure 6.72(b). This local high pressure not only prevents crack generation but also causes tetragonal-to-monoclinic phase transformation. At the same time, the high shear stress caused by the friction between the tool flank face and workpiece material causes grain refinement in the near-surface layer. Such grain refinement improves the mechanical properties of the machined surface.

## REFERENCES


[1] Astakhov, V.P., Shvets, S.V. (1998) A system concept in metal cutting. *Journal of Materials Processing Technology* 79 (1): 189–199.

[2] Nakasuji, T., Kodera, S., Hara, S., Matsunaga, H., Ikawa, N., Shimada, S. (1990) Diamond turning of brittle materials for optical components. *CIRP Annals* 39 (1): 89–92.

[3] Puttick, K.E., Franks, A. (1990) The physics of ductile-brittle machining transitions: Single-point theory and experiment. *Journal of the Japan Society for Precision Engineering* 56 (5): 788–793.

[4] Blackley, W.S., Scattergood, R.O. (1991) Ductile-regime machining model for diamond turning of brittle materials. *Precision Engineering* 13 (2): 95–103.

[5] Bifano, T.G., Dow, T.A., Scattergood, R.O. (1991) Ductile-regime grinding: A new technology for machining brittle materials. *Journal of Engineering for Industry* 113 (2): 184–189.

[6] Moriwaki, T., Shamoto, E., Inoue, K. (1992) Ultraprecision ductile cutting of glass by applying ultrasonic vibration. *CIRP Annals* 41 (1): 141–144.

[7] Puttick, K.E. (1979) Energy scaling, size effects and ductile-brittle transitions in fracture. *Journal of Physics D: Applied Physics* 12 (3): L19.

[8] Morris, J.C., Callahan, D.L., Kulik, J., Patten, J.A., Scattergood, R.O. (1995) Origins of the ductile regime in single-point diamond turning of semiconductors. *Journal of the American Ceramic Society* 78 (8): 2015–2020.

[9] Yan, J., Syoji, K., Kuriyagawa, T., Suzuki, H. (2002) Ductile regime turning at large tool feed. *Journal of Materials Processing Technology* 121 (2): 363–372.

[10] Yan, J., Zhao, H., Kuriyagawa, T. (2009) Effects of tool edge radius on ductile machining of silicon: an investigation by FEM. *Semiconductor Science and Technology* 24 (7): 075018.
[11] Yan, J., Yoshino, M., Kuriagawa, T., Shirakashi, T., Syoji, K., Komanduri, R. (2001) On the ductile machining of silicon for micro electro-mechanical systems (MEMS), opto-electronic and optical applications. *Materials Science and Engineering A* 297 (1–2): 230–234.
[12] Komanduri, R., Chandrasekaran, N., Raff, L.M. (1998) Effect of tool geometry in nanometric cutting: a molecular dynamics simulation approach. *Wear* 219 (1): 84–97.
[13] Blake, P.N., Scattergood, R.O. (1990) Ductile-regime machining of germanium and silicon. *Journal of the American Ceramic Society* 73 (4): 949–957.
[14] Heidari, M., Akbari, J., Yan, J. (2019) Effects of tool rake angle and tool nose radius on surface quality of ultraprecision diamond-turned porous silicon. *Journal of Manufacturing Processes* 37: 321–331.
[15] Huang, W., Yan, J. (2020) Surface formation mechanism in ultraprecision diamond turning of coarse-grained polycrystalline ZnSe. *International Journal of Machine Tools and Manufacture* 153: 103554.
[16] Ohta, T., Yan, J., Yajima, S., Takahashi, Y., Horikawa, N., Kuriyagawa, T. (2007) High-efficiency machining of single-crystal germanium using large-radius diamond tools. *International Journal of Surface Science and Engineering* 1 (4): 374–392.
[17] Huang, W., Yan, J. (2021) Fundamental investigation of diamond cutting of micro V-shaped grooves on a polycrystalline soft-brittle material. *Journal of Manufacturing and Materials Processing* 5 (1): 17.
[18] Yan, J., Asami, T., Harada, H., Kuriyagawa, T. (2009) Fundamental investigation of subsurface damage in single crystalline silicon caused by diamond machining. *Precision Engineering* 33 (4): 378–386.
[19] Fuchs, B.A., Hed, P.P., Baker, P.C. (1986) Fine diamond turning of KDP crystals. *Applied Optics* 25 (11): 1733–1735.
[20] Yan, J., Syoji, K., Kuriyagawa, T. (2000) Ductile-brittle transition at large negative tool rake angles. *Journal of the Japan Society for Precision Engineering* 66 (7): 1130–1134.
[21] Kita, Y., Ido, M., Kawasaki, N. (1982) A study of metal flow ahead of tool face with large negative rake angle. *Journal of Engineering for Industry* 104 (3): 319–325.
[22] Chiang, S.S., Marshall, D.B., Evans, A.G. (1982) The response of solids to elastic/plastic indentation. I. Stresses and residual stresses. *Journal of Applied Physics* 53 (1): 298–311.
[23] Asai, S., Taguchi, Y., Horio, K., Kasai, T., Kobayashi, A. (1990) Measuring and analysis on cutting edge radius of single point diamond tools using newly developed Scanning Electron Microscope (SEM). *Journal of the Japan Society for Precision Engineering* 56 (7): 1311–1316.
[24] Fang, F., Xu, F. (2018) Recent advances in micro/nano-cutting: Effect of tool edge and material properties. *Nanomanufacturing and Metrology* 1 (1): 4–31.
[25] Rahman, M.A., Amrun, M.R., Rahman, M., Kumar, A.S. (2017) Variation of surface generation mechanisms in ultra-precision machining due to relative tool sharpness (RTS) and material properties. *International Journal of Machine Tools and Manufacture* 115: 15–28.
[26] Cai, M.B., Li, X.P., Rahman, M. (2007) Study of the mechanism of nanoscale ductile mode cutting of silicon using molecular dynamics simulation. *International Journal of Machine Tools and Manufacture* 47 (1): 75–80.

[27] Fang, F.Z., Wu, H., Liu, Y.C. (2005) Modelling and experimental investigation on nanometric cutting of monocrystalline silicon. *International Journal of Machine Tools and Manufacture* 45 (15): 1681–1686.

[28] Liu, B., Xu, Z., Chen, C., Pang, K., Wang, Y., Ruan, Q. (2019) Effect of tool edge radius on material removal mechanism of single-crystal silicon: Numerical and experimental study. *Computational Materials Science* 163: 127–133.

[29] Ameli Kalkhoran, S.N., Vahdati, M., Yan, J. (2020) Effect of relative tool sharpness on subsurface damage and material recovery in nanometric cutting of mono-crystalline silicon: A molecular dynamics approach. *Materials Science in Semiconductor Processing* 108: 104868.

[30] Dai, H., Chen, G., Fang, Q., Yin, J. (2016) The effect of tool geometry on subsurface damage and material removal in nanometric cutting single-crystal silicon by a molecular dynamics simulation. *Applied Physics A* 122 (9): 804.

[31] Zong, W.J., Li, Z.Q., Zhang, L., Liang, Y.C., Sun, T., An, C.H., Zhang, J.F., Zhou, L., Wang, J. (2013) Finite element simulation of diamond tool geometries affecting the 3D surface topography in fly cutting of KDP crystals. *The International Journal of Advanced Manufacturing Technology* 68 (9): 1927–1936.

[32] Wang, H., Dong, Z., Yuan, S., Guo, X., Kang, R., Bao, Y. (2022) Effects of tool geometry on tungsten removal behavior during nano-cutting. *International Journal of Mechanical Sciences* 225: 107384.

[33] Yan, J., Asami, T., Harada, H., Kuriyagawa, T. (2012) Crystallographic effect on subsurface damage formation in silicon microcutting. *CIRP Annals* 61 (1): 131–134.

[34] Yan, J., Maekawa, K., Tamaki, J., Kubo, A. (2004) Experimental study on the ultraprecision ductile machinability of single-crystal germanium. *JSME International Journal Series C Mechanical Systems, Machine Elements and Manufacturing* 47 (1): 29–36.

[35] Mizumoto, Y., Kakinuma, Y. (2018) Revisit of the anisotropic deformation behavior of single-crystal CaF2 in orthogonal cutting. *Precision Engineering* 53: 9–16.

[36] Ioffe Physical Technical Institute NSM Archive-Physical Properties of Semiconductors. http://www.ioffe.ru/SVA/NSM/Semicond/index.html (accessed May 2023).

[37] Suzuki, T., Ohmura, T. (1996) Ultra-microindentation of silicon at elevated temperatures. *Philosophical Magazine A* 74 (5): 1073–1084.

[38] Minor, A.M., Lilleodden, E.T., Jin, M., Stach, E.A., Chrzan, D.C., Morris, J.W. (2005) Room temperature dislocation plasticity in silicon. *Philosophical Magazine* 85 (2–3): 323–330.

[39] Juliano, T., Gogotsi, Y., Domnich, V. (2003) Effect of indentation unloading conditions on phase transformation induced events in silicon. *Journal of Materials Research* 18 (5): 1192–1201.

[40] Kailer, A., Gogotsi, Y.G., Nickel, K.G. (1997) Phase transformations of silicon caused by contact loading. *Journal of Applied Physics* 81 (7): 3057–3063.

[41] Kroll, M.C., Kirchner, P.D., Cook, R.F., Hockey, B.J., Clarke, D.R. (1988) Amorphization and conductivity of silicon and germanium induced by indentation. *Physical Review Letters* 60 (21): 2156–2159.

[42] Ruffell, S., Bradby, J.E., Williams, J.S., Warren, O.L. (2007) An in situ electrical measurement technique via a conducting diamond tip for nanoindentation in silicon. *Journal of Materials Research* 22 (3): 578–586.

[43] Khayyat, M.M.O., Chaudhri, M.M., Patriarche, G., Le Bourhis, E. (2009) Nanoindentation-induced structural phase transformations in crystalline and amorphous germanium. *International Journal of Nano and Biomaterials* 2 (1–5): 91–99.

[44] Rothschild, M. (2005) Projection optical lithography. *Materials Today* 8 (2): 18–24.

[45] Yan, J., Syoji, K., Tamaki, J.I. (2003) Crystallographic effects in micro/nanomachining of single-crystal calcium fluoride. *Journal of Vacuum Science & Technology B: Microelectronics and Nanometer Structures Processing, Measurement, and Phenomena* 22 (1): 46–51.

[46] Shibata, T., Fujii, S., Makino, E., Ikeda, M. (1996) Ductile-regime turning mechanism of single-crystal silicon. *Precision Engineering* 18 (2): 129–137.

[47] Kavner, A. (2008) Radial diffraction strength and elastic behavior of CaF2 in low- and high-pressure phases. *Physical Review B* 77 (22): 224102.

[48] Mizumoto, Y., Kangawa, H., Itobe, H., Tanabe, T., Kakinuma, Y. (2017) Influence of crystal anisotropy on subsurface damage in ultra-precision cylindrical turning of CaF2. *Precision Engineering* 49: 104–114.

[49] Lee, Y.J., Chong, J.Y., Chaudhari, A., Wang, H. (2019) Enhancing ductile-mode cutting of calcium fluoride single crystals with solidified coating. *International Journal of Precision Engineering and Manufacturing-Green Technology* 7: 1019–1029.

[50] Yan, J., Tamaki, J.I., Syoji, K., Kuriyagawa, T. (2004) Single-point diamond turning of CaF2 for nanometric surface. *The International Journal of Advanced Manufacturing Technology* 24 (9): 640–646.

[51] Yan, J., Tamaki, J., Zhao, H., Kuriyagawa, T. (2008) Surface and subsurface damages in nanoindentation tests of compound semiconductor InP. *Journal of Micromechanics and Microengineering* 18 (10): 105018.

[52] Brasen, D. (1976) Hardness anisotropy of InP. *Journal of Materials Science* 11 (5): 791–793.

[53] Chen, J., Ding, F., Luo, X., Rao, X., Sun, J. (2020) Fundamental study of ductile-regime diamond turning of single crystal gallium arsenide. *Precision Engineering* 62: 71–82.

[54] Grillo, S.E., Ducarroir, M., Nadal, M., Tourni, E., Faurie, J. (2002) Nanoindentation of Si, GaP, GaAs and ZnSe single crystals. *Journal of Physics D: Applied Physics* 36 (1): L5–L9.

[55] Li, W., Yang, K., Wang, P., Zhang, G., Liu, D. (2016) Experimental investigation of the ultra-precision turning capability of PVD ZnSe. Proceedings of the 8th International Symposium on Advanced Optical Manufacturing and Testing Technologies: Advanced Optical Manufacturing Technologies, SPIE, 9683: 968319.

[56] Yan, J., Takahashi, Y., Tamaki, J., Kubo, A., Kuriyagawa, T., Sato, Y. (2006) Ultraprecision machining characteristics of poly-crystalline germanium. *JSME International Journal Series C Mechanical Systems, Machine Elements and Manufacturing* 49 (1): 63–69.

[57] Kucheyev, S.O., Siekhaus, W.J., Land, T.A., Demos, S.G. (2004) Mechanical response of KD2xH2(1−x)PO4 crystals during nanoindentation. *Applied Physics Letters* 84 (13): 2274–2276.

[58] Wang, S., An, C., Zhang, F., Wang, J., Lei, X., Zhang, J. (2016) An experimental and theoretical investigation on the brittle ductile transition and cutting force anisotropy in cutting KDP crystal. *International Journal of Machine Tools and Manufacture* 106: 98–108.

[59] Fuchs, B.A., Hed, P.P., Baker, P.C. (1986) Fine diamond turning of KDP crystals. *Applied Optics* 25 (11): 1733–1735.

[60] Luo, X., Goel, S., Reuben, R.L. (2012) A quantitative assessment of nanometric machinability of major polytypes of single crystal silicon carbide. *Journal of the European Ceramic Society* 32 (12): 3423–3434.

[61] Patten, J., Gao, W., Yasuto, K. (2004) Ductile Regime Nanomachining of Single-Crystal Silicon Carbide. *Journal of Manufacturing Science and Engineering* 127 (3): 522–532.

[62] Goel, S., Luo, X., Comley, P., Reuben, R.L., Cox, A. (2013) Brittle–ductile transition during diamond turning of single crystal silicon carbide. *International Journal of Machine Tools and Manufacture* 65: 15–21.

[63] Zhang, Z., Yan, J., Kuriyagawa, T. (2011) Study on tool wear characteristics in diamond turning of reaction-bonded silicon carbide. *The International Journal of Advanced Manufacturing Technology* 57 (1): 117–125.
[64] Zhang, J., Han, L., Zhang, J., Liu, H., Yan, Y., Sun, T. (2019) Brittle-to-ductile transition in elliptical vibration-assisted diamond cutting of reaction-bonded silicon carbide. *Journal of Manufacturing Processes* 45: 670–681.
[65] You, K., Yan, G., Luo, X., Gilchrist, M.D., Fang, F. (2020) Advances in laser assisted machining of hard and brittle materials. *Journal of Manufacturing Processes* 58: 677–692.
[66] Yan, J., Zhang, Z., Kuriyagawa, T. (2011) Effect of nanoparticle lubrication in diamond turning of reaction-bonded SiC. *International Journal of Automation Technology* 5 (3): 307–312.
[67] Huang, W., Yan, J. (2023) Mechanisms of tool-workpiece interaction in ultraprecision diamond turning of single-crystal SiC for curved microstructures. *International Journal of Machine Tools and Manufacture* 191: 104063.
[68] Zhong, Z. (1997) Partial-ductile grinding, lapping, and polishing of aspheric and spherical surfaces on glass. *Materials and Manufacturing Processes* 12 (6): 1063–1073.
[69] Chen, Y., Yi, A.Y., Yao, D., Klocke, F., Pongs, G. (2008) A reflow process for glass microlens array fabrication by use of precision compression molding. *Journal of Micromechanics and Microengineering* 18 (5): 055022.
[70] Zhou, T., Liu, X., Liang, Z., Liu, Y., Xie, J., Wang, X. (2017) Recent advancements in optical microstructure fabrication through glass molding process. *Frontiers of Mechanical Engineering* 12 (1): 46–65.
[71] Pawar, P., Ballav, R., Kumar, A. (2017) Micromachining of borosilicate glass: A state of art review. *Materials Today: Proceedings* 4 (2, Part A): 2813–2821.
[72] Youn, S.W., Kang, C.G. (2004) A study of nanoscratch experiments of the silicon and borosilicate in air. *Materials Science and Engineering: A* 384 (1): 275–283.
[73] Kuriyama, K., Fukuta, M., Sekiya, K., Yamada, K., Yamane, Y. (2013) Applying constant pressure unit to ductile mode cutting of hard and brittle materials. *International Journal of Automation Technology* 7 (3): 278–284.
[74] Kuriyama, K., Fukuta, M., Sekiya, K., Yamada, K., Yamane, Y. (2010) Forming of microstructure on hard brittle materials in ductile mode cutting. *Key Engineering Materials* 447–448: 91–95.
[75] Kim, T., Kwon, K., Chu, C.N., Song, K.Y. (2020) Experimental investigation on CO2 laser-assisted micro-slot milling characteristics of borosilicate glass. *Precision Engineering* 63: 137–147.
[76] Liu, K., Li, X., Liang, S.Y., Dong Liu, X. (2005) Nanometer-scale, ductile-mode cutting of soda-lime glass. *Journal of Manufacturing Processes* 7 (2): 95–101.
[77] Jia, P., Zhou, M. (2012) Tool wear and its effect on surface roughness in diamond cutting of glass soda-lime. *Chinese Journal of Mechanical Engineering* 25 (6): 1224–1230.
[78] Fang, F., Zhang, G. (2004) An experimental study of optical glass machining. *The International Journal of Advanced Manufacturing Technology* 23 (3): 155–160.
[79] Feng, J., Huang, X., Yang, S., Qing, J., Tang, P., Wan, Z. (2021) Speed effect on the material behavior in high-speed scratching of BK7 glass. *Ceramics International* 47 (14): 19978–19988.
[80] Schinker, M.G. (1991) Subsurface damage mechanisms at high-speed ductile machining of optical glasses. *Precision Engineering* 13 (3): 208–218.
[81] Ishizuka, J., Nagasawa, K., Mikami, M., Yan, J. (2018) Ultraprecision cutting of fluoride glass. - Fundamental investigation on surface formation and tool damage. *Journal of Japan Society of Abrasive Technology* 62 (2): 88–93.

[82] Suyama, S., Kameda, T., Itoh, Y. (2003) Development of high-strength reaction-sintered silicon carbide. *Diamond and Related Materials* 12 (3): 1201–1204.
[83] Shoko, S., Yoshiyasu, I., Katsuhiko, T., Kazuhiko, O. (2005) NT-SiC (new-technology silicon carbide) : *Φ* 650mm optical space mirror substrate of high-strength reaction-sintered silicon carbide. Proceedings of the Optical Materials and Structures Technologies II, SPIE, 5868: 58680E.
[84] Yan, J., Zhang, Z., Kuriyagawa, T. (2009) Mechanism for material removal in diamond turning of reaction-bonded silicon carbide. *International Journal of Machine Tools and Manufacture* 49 (5): 366–374.
[85] Onodera, A., Ueno, M., Takemura, K., Shimomura, O., Yoshida, M. (1993) Pressure-induced phase transition in SiC. *Physical Review B* 48 (14): 10587–10590.
[86] Yin, L., Vancoille, E.Y.J., Ramesh, K., Huang, H. (2004) Surface characterization of 6H-SiC (0001) substrates in indentation and abrasive machining. *International Journal of Machine Tools and Manufacture* 44 (6): 607–615.
[87] Zhang, Z., Yan, J., Kuriyagawa, T. (2010) Tool-swinging cutting of binderless tungsten carbide. *Advanced Materials Research* 139–141: 727–730.
[88] Yan, J., Zhang, Z., Kuriyagawa, T. (2010) Tool wear control in diamond turning of high-strength mold materials by means of tool swinging. *CIRP Annals - Manufacturing Technology* 59 (1): 109–112.
[89] Zhang, J., Suzuki, N., Wang, Y., Shamoto, E. (2014) Fundamental investigation of ultra-precision ductile machining of tungsten carbide by applying elliptical vibration cutting with single crystal diamond. *Journal of Materials Processing Technology* 214 (11): 2644–2659.
[90] Manicone, P.F., Rossi Iommetti, P., Raffaelli, L. (2007) An overview of zirconia ceramics: Basic properties and clinical applications. *Journal of Dentistry* 35 (11): 819–826.
[91] Yamamuro, Y., Shimoyama, T., Yan, J. (2023) Generation of nanopore structures in yttria-stabilized zirconia by femtosecond pulsed laser irradiation. *Journal of Materials Research and Technology* 23: 1155–1176.
[92] Krishnamurthy, R., Arunachalam, L.M., Gokularathnam, C.V., Venkatesh, V.C. (1991) Grinding of Transformation-Toughened Y-TZP Ceramics. *CIRP Annals* 40 (1): 331–333.
[93] Kizaki, T., Harada, K., Mitsuishi, M. (2014) Efficient and precise cutting of zirconia ceramics using heated cutting tool. *CIRP Annals* 63 (1): 105–108.
[94] Yan, J., Okuuchi, T. (2019) Chip morphology and surface integrity in ultraprecision cutting of yttria-stabilized tetragonal zirconia polycrystal. *CIRP Annals* 68 (1): 53–56.

# 7 Surface Metrology

## 7.1 INTRODUCTION

Metrology, the discipline dedicated to measurement science, has great significance in ensuring the quality control of ultraprecision machining. The importance of metrology rests on two key aspects. Firstly, metrology plays a crucial role in assessing whether a product meets the quality standards. Secondly, analyzing metrology results provides valuable information that is subsequently fed back into the manufacturing process. This iterative feedback loop serves to enhance and refine the manufacturing processes, ultimately enabling the production of higher-quality outcomes.

Typically, surface metrology in machining involves the measurement of surface geometrical parameters, including but not limited to size, shape, form errors, and surface roughness. It can be performed either after machining through offline measurement or concurrently during machining, known as in-process measurement, or on-machine measurement. The distinction between these approaches lies in the timing of measurement, with in-process measurement providing real-time data during the machining operation itself. Each approach has its advantages, and in-process measurement is particularly valuable for immediate feedback to optimize and control the machining process. However, achieving high-resolution results with in-process measurement is challenging, primarily due to the disruptions introduced by the machining environment and the limitation of equipping high-specification measurement devices to a machine tool. The expense and complexity associated with an in-process measurement facility are considerable. Therefore, offline measurement remains the predominant and crucial method for achieving ultra-precision metrology. This chapter introduces typical offline measurement technologies employed in ultraprecision machining, some of which have potential to be used, or are already being attempted for use, in in-process measurement.

In most of the machining methods, the tool's quality greatly determines the quality of the machined parts. To ensure the ultraprecision machining of a surface, not only the metrology of the machined surface is important, but sometimes we also need to confirm the accuracy and quality of the tools used for machining. In other words, tool metrology is as important as surface metrology. This chapter encompasses the measurement of both machined workpieces and the tools utilized in the machining processes.


DOI: 10.1201/9781003292388-7

## 7.2 CONTACT MEASUREMENT

### 7.2.1 Stylus Profilometry

The most popular and also the standard method for measuring workpiece geometrical features, such as form error and surface roughness, is stylus profiling. Stylus profilometers utilize a probe to detect the surface, physically moving it along the surface to acquire height information. This process is carried out mechanically through a feedback loop that monitors the force exerted by the sample against the probe as it scans along the surface. A typical stylus profilometer is the Form Talysurf® and Talyrond® series made by Taylor Hobson Co. Ltd. For examples, Figure 7.1 shows a 3D freeform optics profilometer for surface finish and form measurement, as well as an ultra-high precision measurement system for roundness, cylindricity, and straightness.

In stylus profiling measurements, the geometry of the stylus tip significantly affects the measurement results. Different tip shapes and tip radii lead to greatly different measurement results. For example, Figure 7.2 illustrates how the measurement results change with variations in stylus geometry. The sharper stylus can enter into the micro craters much deeper, thus can measure the depth of craters, i.e., the real surface roughness, more accurately. Especially for the brittle-regime

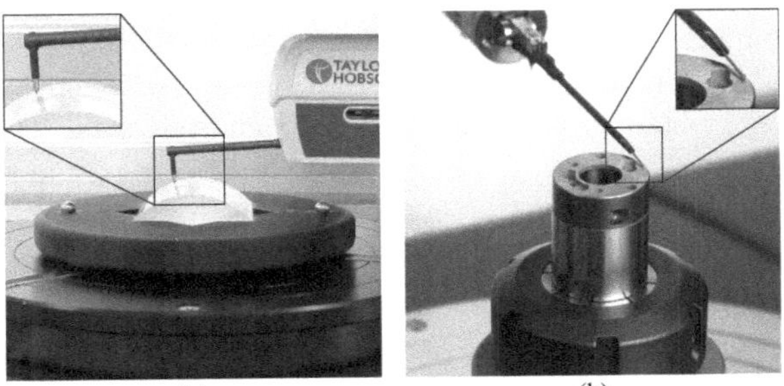

(a) (b)

**FIGURE 7.1** Typical stylus profiling devices: (a) Form Talysurf® PGI Freeform [1]; (b) Talyrond® 595H PRO [2]. (Courtesy of AMETEK.)

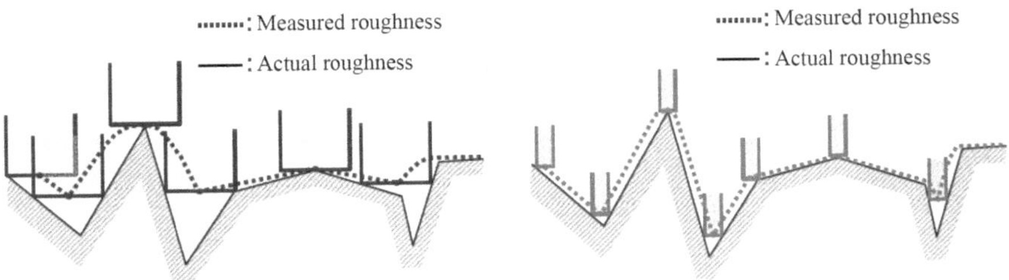


**FIGURE 7.2** Schematic of surface profiling with different stylus sizes.

machined surface, where very narrow and deep cracks occur in the surface, or surface with scratches by abrasive grains, the difference in surface roughness measured by different styli becomes very significant.

Theoretically, a sharper stylus can always yield more accurate measurement results. However, a sharper stylus typically has less strength, which may lead to failure when measuring very rough surfaces. For example, in the measurement of a surface machined by grinding, a 0.1 × 2.5 μm rectangular stylus and a 0.2 × 0.2 μm square stylus were employed and compared [3]. For the rectangular stylus, its long side (2.5 μm) traversed perpendicularly to the grinding marks, while with the square stylus, one side (0.2 μm) moved perpendicular to the grinding marks. It was found that the small square stylus can be only used to evaluate the roughness of surfaces ground by grinding wheels with 8 and 10 μm abrasive grains because using the stylus on rougher surfaces ground by larger abrasives damaged the stylus [3].

Using a different stylus, different surface profiles, and in turn, different values of surface roughness will be obtained. Figure 7.3(a) and (b) show micrographs and measurement results of germanium and silicon surfaces ground by 20 μm and 10 μm diameter abrasive wheels, respectively. Both germanium and silicon surfaces ground by the 20 μm diameter abrasive show brittle regime appearances. The surface roughness was 2.44–3.22 μm Ry. On the other hand, the brittle fractures decreased on the surfaces ground by the 10 μm abrasives and smooth grinding traces become clearly seen. The surface roughness was 0.67 to 0.87 μm Ry. It is worth noting that for the same surface ground by the 10 μm abrasive, the surface roughness profile obtained with the 0.2 × 0.2 μm stylus indicates some deep scratches and cracks. In contrast, the result from the 0.1 × 2.5 μm stylus shows a relatively smooth surface without indicating any deep surface features.

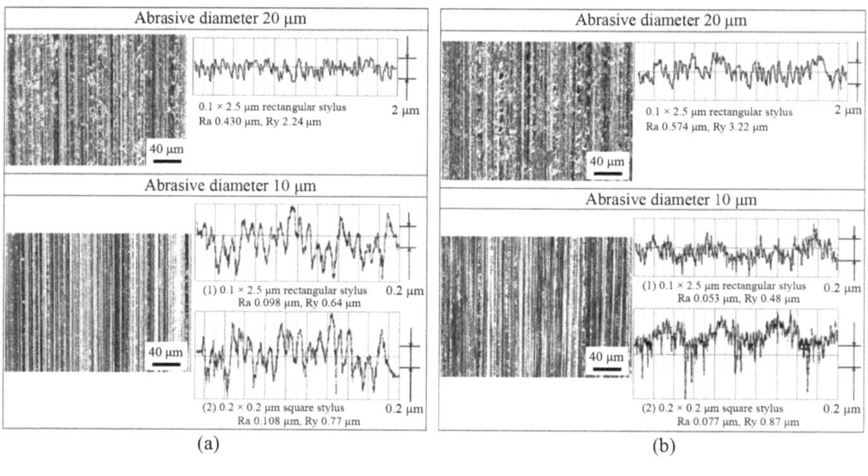


**FIGURE 7.3** Micrographs and cross-sectional profiles of ground surfaces of (a) germanium and (b) silicon measured by different stylus shapes.

### 7.2.2 Scanning Probe Microscopy

Scanning probe microscopy (SPM) is a relatively new technique used for imaging surfaces at the nanoscale level. There are two major types of SPM, scanning tunneling microscope (STM) and atomic force microscope (AFM). These techniques employ a sharp physical probe that scans the surface of a sample to gather information about its topography, properties, and even its atomic-level structure. The primary principles behind SPM involve the physical interaction (tunneling current for STM, atomic force for AFM) between the probe and the sample, which is then translated into images or other measurable data. As STM is only useable for measuring electrically conductive materials, its applications are relatively limited compared with AFM.

In AFM, a sharp probe is brought very close to the sample surface without making direct contact (noncontact type AFM) or making slight contact (contact-type AFM). The interaction forces between the probe and the atoms or molecules on the sample surface are detected, providing detailed information about the sample's topography. In this way, AFM can detect three-dimensional surfaces with very high precision (< 1 nm) in the height direction. It can be used for measuring any kind of solids including conductive, nonconductive, and semiconductor materials. However, it is essential to note that while AFM are excellent for high-resolution imaging and surface profiling, their primary strength lies in measuring features on the order of nanometers. The measurement range in the height direction is limited (several μm), and the measurement area is usually very small, less than 100 × 100 μm.

For example, AFM is a powerful tool in measurement of coatings, allowing for the characterization of the nanostructure and morphology of coated surfaces. This provides quantitative insights into the quality and uniformity of the applied coating layer. Figure 7.4(a) illustrates the SEM image of a silicon wafer coated with a nano-graphitic carbon layer. Although homogeneous nanoscale island-like domains covering the Si wafer without noticeable gaps can be observed, there is a lack of height information. Figure 7.4(b) displays the AFM image of the same surface, providing nano-morphological details of the carbon coating, where the height information is indicated by the color contrast in the scale bar.

Figure 7.4(c) shows an AFM image of a ductile-machined silicon surface (the same surface as that shown in Figure 6.18(c)). The surface has periodical tool feed marks generated in diamond turning. Through the 3D presentation of the surface topography, the height of the tool feed marks could be clearly separated from the surface roughness components in the nanometer level. This height information cannot be obtained in optical imaging and SEM observation.

Another application example of AFM is in nanoindentation studies. Researchers employ AFM to precisely measure the topography of the indents and the deformation responses at the nanoscale. For instance, AFM can be used to create a topographic map of a material's surface before and after indentation, allowing for detailed analysis of features such as indentation depth, surface deformation, and recovery. This information is crucial for understanding a material's plasticity,

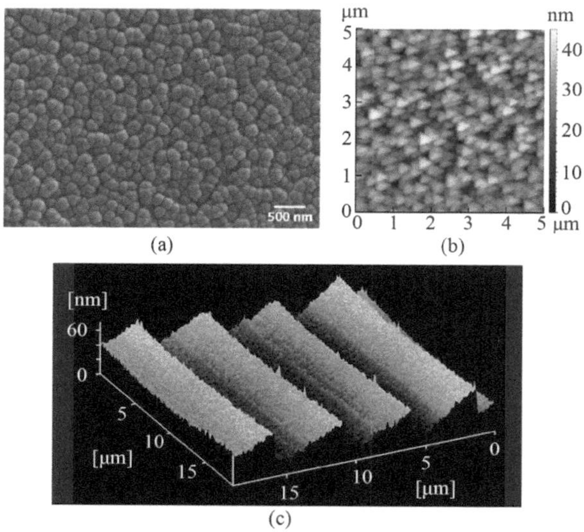


**FIGURE 7.4** Examples of AFM measurements. (a) SEM image and (b) AFM image of a nano-graphitic carbon coating on a silicon wafer [4]. (c) AFM image of a diamond-turned silicon surface [5].

elasticity, and other mechanical characteristics at the nanoscale, providing valuable insights for various applications, including materials science and engineering.

Figure 7.5(a) shows an SEM image of a residual indent after indentation [6]. Radial cracks propagating perpendicular to the surface along three corners are observed. Additionally, the lines parallel to the edges of the indent are clearly observed on two of the three sides of the indent surface, which is attributed to the shear stress-induced plastic flow of the material. The corresponding surface topography measured with AFM is shown in Figure 7.5(b). Significant pile-ups were detected near the two indent surfaces where the plastic flow lines were generated. Moreover, the cross-sectional profile of the residual indent can be extracted to analyze the depth of the residual indent and the height of the pile-ups, as shown in Figure 7.5(c).

### 7.2.3 Ultrahigh Accuracy 3D Profilometer

For conventional stylus profilometry, there are some critical issues including the stylus-induced surface damage in the form of scratches and difficulty of measuring a sloped surface with an angle larger than ~30°. Both of the issues originated from the contact force between the stylus and surface, usually in the level of mN or larger, which cannot be prevented by the conventional cantilever beam unit for holding the stylus. On the other hand, for the AFM method, although the contact force is extremely small, approaching zero, the measurement range is very small, limiting its application to metrology of large workpieces and curved surfaces as well as 3D surface structures.

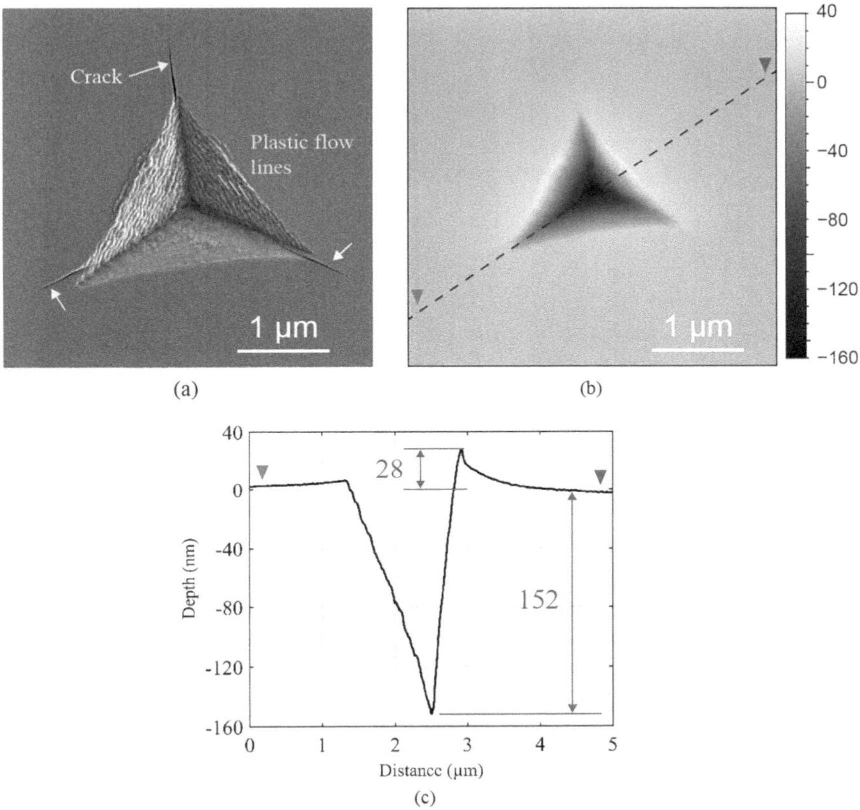


**FIGURE 7.5** A residual indent after nanoindentation: (a) SEM image, (b) AFM image, and (c) a cross-sectional profile of the indent extracted along the line indicated in (b) [6].

To close these gaps, a new type of atomic force probe profilometer called UA3P (Ultrahigh Accuracy 3D Profilometer) was developed by Panasonic in the 1980s [7]. In UA3P, the probe is supported by an air-bearing slider to realize an extremely small contact force between the probe and the workpiece surface based on a similar working principle to an AFM probe, as shown in Figure 7.6. In this way, the contact force could be controlled below 0.3 mN. The measurement is performed under a constant force mode and the probe is scanned in XY horizontal directions by numerical control while the displacement in vertical Z position is measured by a He–Ne laser interferometer system. This profilometer enables the measurements of steep surfaces with an inclination angle of up to 75°. This surface profilometer has been used widely for measuring complicated surfaces such as aspherical lenses and other freeform optics. An example of measuring an optical sphere using UA3P is shown in Figure 7.7 [7].

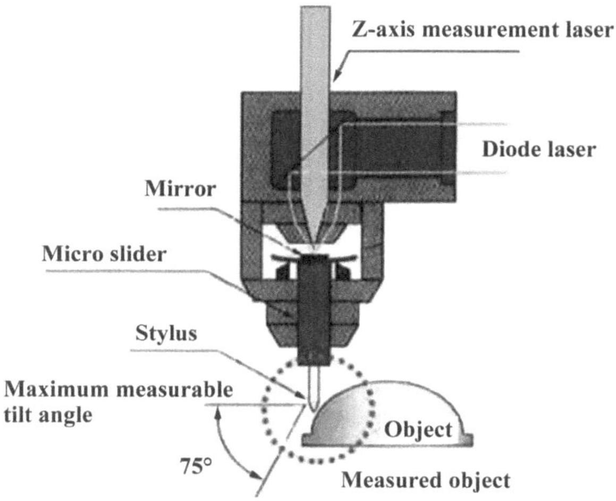


**FIGURE 7.6** Schematic of the structure of the probe unit of UA3P. (Courtesy of Panasonic.)

**FIGURE 7.7** Measurement of an optical sphere using UA3P. (Courtesy of Panasonic.)

## 7.3 NONCONTACT MEASUREMENT

### 7.3.1 Scanning Electron Microscopy

A scanning electron microscope (SEM) is an instrument that produces a highly magnified image of a surface by using accelerated electrons instead of a light beam to detect the surface features. At the top of an SEM, an electron gun generates a beam of electrons in a vacuum chamber. The electron beam passes through electromagnetic fields and lenses and is then focused onto the sample. Upon the electron beam striking the sample, various signals are emitted via secondary electrons (SEs), backscattered electrons (BSEs), photons (X-rays used for elemental analysis), and visible light (cathodoluminescence, CL). Electron collectors (detectors) gather these signals, which are then processed by a computer to create the desired image. However, the electron beam used in an SEM can induce the accumulation of electrical charge on the surface of non-conductive samples. This charge buildup can distort the imaging and may lead to poor image quality. Therefore, non-conductive samples are often coated with a thin conductive layer with the nanometer thickness, using materials such as gold, carbon, or osmium, for preparing samples for SEM observation.

SE and BSE detectors are two major electron detectors commonly employed for creating sample images in an SEM. For example, Figure 7.8(a) shows the groove surface of polycrystalline ZnSe after diamond cutting, the image of which was captured using an SE detector. The images produced by SE detectors are generally more surface-sensitive compared to those produced by BSE detectors, making SE detectors suitable for observing the topmost layers of a sample. In addition, SE images often exhibit high contrast, which is effective for distinguishing different materials and structures based on their surface properties. Thus, SE images are particularly useful for visualizing surface features, such as roughness, texture, and the shape of microstructures. In contrast, BSE detectors have a greater penetration depth, enabling the detection of signals from deeper layers of the sample. They provide valuable information about variations in material density within the sample. The signal intensity is directly related to the density of the material, allowing for the visualization of subsurface features. Figure 7.8(b) displays the same surface as that in Figure 7.8(a), captured using Inlens Duo mode. This mode enables both Inlens SE and BSE detectors, facilitating high-resolution topographical imaging of surfaces to distinguish grain boundaries.

An energy-dispersive X-ray (EDS) detector in an SEM is another analytical tool used to detect and quantify the characteristic X-rays emitted from a sample when bombarded by electrons. EDS detectors allow for the identification and quantification of the elements present in the sample. Each element emits X-rays at

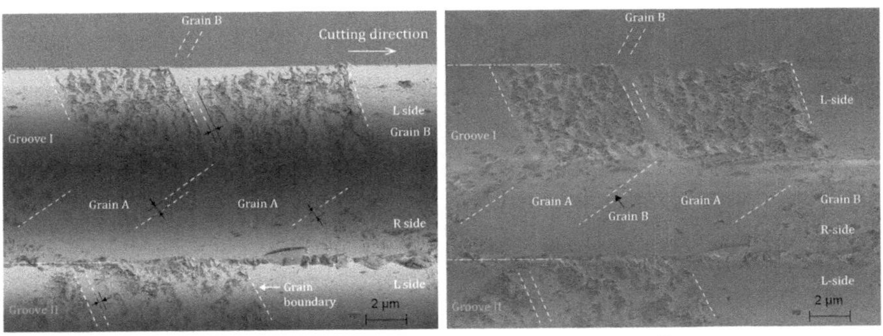


(a) (b)

**FIGURE 7.8** SEM images of a groove surface of polycrystalline ZnSe taken in (a) HE-SE mode and (b) Inlens Duo mode [8].

characteristic energies, and the detector can measure the intensity of these X-rays to determine the elemental composition. For example, Figure 7.9 shows the EDS analysis results of the surface of a high-entropy alloy FeCrCoMnNi after diamond turning. According to the SEM image, deep scratches were generated by the movement of hard particles in the workpiece material. The EDS results indicated that the particles were rich in Cr, which has higher hardness than other elements (Mohs hardness: Cr 8.5, Co 5.0, Mn 6.0, Fe 4.0, Ni 4.0).

A traditional SEM is mainly designed for "observation", through two-dimensional imaging, and not for precise "measurement". However, by equipping multiple SE detectors (up to four), an SEM enables achieving high resolution in both vertical and horizontal directions for the three-dimensional analysis of surface roughness and topography, namely, 3D SEM. A typical 3D SEM providing exceptional spatial resolution is the ERA series made by ELIONIX Inc.

An important application of 3D SEM is the precise evaluation of the geometry of a diamond tool edge which typically has an extremely sharp edge. The quantitative measurement of the tool edge radius has become possible by using a 3D SEM. Figure 7.10 shows the cross-sectional profiles of the main cutting edges of diamond tools after cutting for a long distance [10]. For the tool shown in Figure 7.10(a), the tool edge radius is 0.1 μm, and the depth and width of the crater wear on the rake face are 0.06 and 0.5 μm, respectively. The flank wear, which is parallel to the cutting direction, is about 0.17 μm, considerably smaller than the crater wear.

Figure 7.10(b) and (c) show the cross-sectional profiles of two other tools. These two profiles exhibit similarities. In both cases, the cutting edge has retreated by approximately 0.1 μm from its initial state. The total width of flank wear reached about 1.0 μm. Figure 7.10(d) shows the cross-sectional profiles of another tool. A shallow crater wear has extended to approximately 1 μm away from the cutting edge, indicating a more significant retreat compared to Figures 7.10(b) and (c).

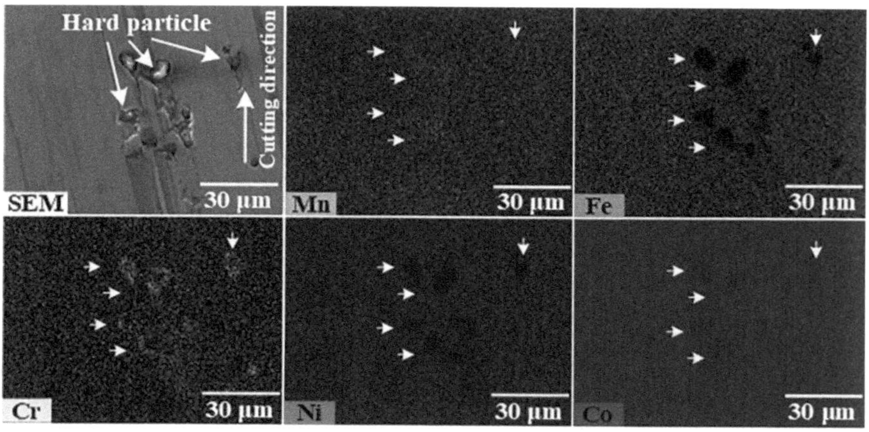


**FIGURE 7.9** EDS analysis of a machined surface of high-entropy alloy FeCrCoMnNi [9].

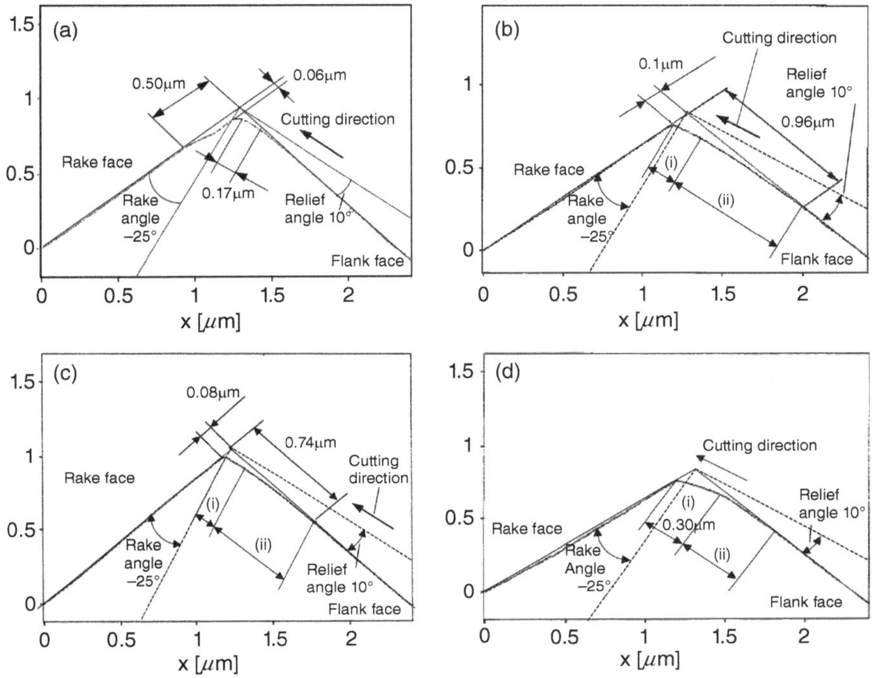


**FIGURE 7.10** Cross-sectional profiles of the main cutting edges of diamond tools obtained using 3D SEM: (a) 1.15 mm nose radius tool after cutting for 1020 km, (b) 10.91 mm nose radius tool after cutting for 8.5 km, (c) 11.61 mm nose radius tool after cutting for 38.4 km, and (d) 11.22 mm nose radius tool after cutting for 160 km.

Simultaneously, notable flank wear has developed, marked by a distinct boundary between primary and secondary flank wear. The primary flank wear width measures approximately 0.3 μm, while the secondary flank wear width is 0.5 μm.

### 7.3.2 Confocal Laser Scanning Microscope

A confocal laser scanning microscope is a specialized type of microscope that uses laser light sources emitting a focused and intense beam of light for illumination. It employs a pinhole aperture in front of the photodetector. This pinhole allows only the light originating from the focal plane to pass through while rejecting out-of-focus light, resulting in improved optical sectioning and resolution. A scanning system, typically composed of mirrors, scans the laser beam across the specimen in a raster pattern. This scanning process enables the collection of image data point by point. The use of a pinhole and point scanning imparts high spatial resolution to confocal microscopes. In other words, it is capable of collecting optical sections at different depths, allowing for the reconstruction of high-resolution three-dimensional imaging of specimens [11].

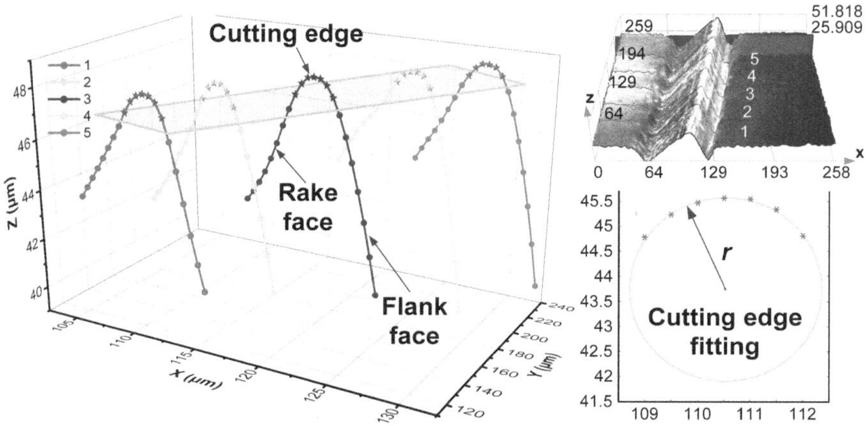


**FIGURE 7.11** Measurement of the tool edge radius using a confocal laser scanning microscope [12].

For example, a confocal laser scanning microscope (LEXT OLS4000, Olympus, Japan) with a lateral resolution of 120 nm and a vertical resolution of 10 nm was employed to measure the tool tip radius. The scanning distance along the cutting edge was 260 μm, and five sectional curves with an interval of 40 μm were analyzed, as depicted in Figure 7.11. The points on the cutting edge between the rake face and the flank face were marked with stars, and their coordinates were extracted. Subsequently, the edge radius was calculated using MATLAB and the least-squares fitting method by assuming that the edge radius is uniform along the cutting edge. After calculation, an average value of 1.84 μm was obtained, which falls within an acceptable range based on a reference value (2 μm) provided by the tool producer. It is worth noting that within a distance of 3 μm, only 7 data points are available, which is insufficient for accurately measuring the edge radius of a diamond tool like that illustrated in Figure 7.10. Although the resolution of confocal laser scanning microscopes in the XY plane is relatively lower than that of AFM and 3D SEM, it compensates with a larger measurement range.

Another key feature of a confocal laser scanning microscope is its ability to produce optical images similar to those obtained with conventional optical microscopes, while providing the capability for sub-micron resolution for 3D observation. It allows researchers to visualize specimens with high resolution in both the XY-plane and the Z-direction. In other words, a confocal laser scanning microscope combines the benefits of optical microscopy with high-resolution 3D imaging, making it a valuable tool in various fields for characterizing biological structures, new materials, and other microscopic surface features. For example, an optical image and a colored surface height topography are presented in Figure 7.12, which were acquired by the same laser beam scan across the specimen. The optical image provides a more intuitive impression than the colored height topography, while the latter gives more quantitative surface topography information.

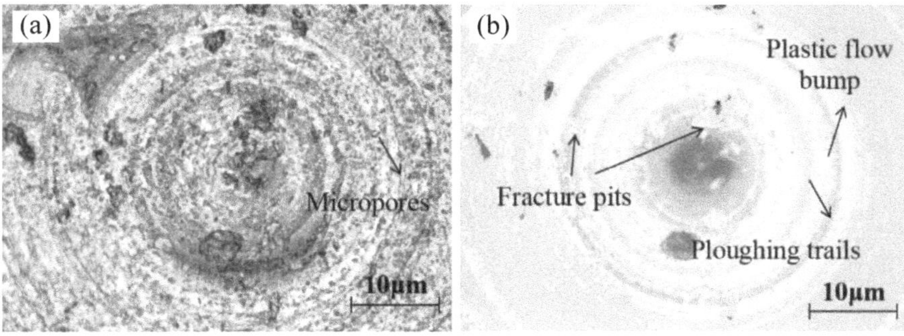


**FIGURE 7.12** Laser confocal scanning microscopy results of a lapped microlens cavity surface: (a) optical image, and (b) colored surface height topography [13].

### 7.3.3 White Light Interferometer

A white light interferometer (WLI) is an optical instrument used for surface profiling and measurements. It is commonly employed in applications such as surface roughness measurement, step height measurement, and characterization of microstructures. The key principle of WLI involves splitting a beam of light into two that travel along different optical paths and subsequently combining them to produce interference. This is achieved using a Michelson interferometer setup, where a beam splitter directs light to both a reference mirror and a sample surface [14]. The light waves traverse these paths and recombine, leading to the formation of interference fringes, as illustrated in Figure 7.13. Importantly, WLI employs a broadband light source, ensuring the utilization of a range of wavelengths. This results in a multitude of interference patterns that, when analyzed, yield comprehensive surface information. The optical path difference (OPD), a crucial parameter in interferometry, represents the disparity in the lengths traveled by the light waves in the reference and test arms. By capturing and analyzing the resulting interference patterns, WLI translates minute OPD variations into three-dimensional surface profiles, thus enabling high-precision measurements of surface roughness, step heights, and other topographical features [14].

A white light interferometer plays a crucial role in ensuring precision and quality control in the fabrication process of microstructured optical components. For example, in the diamond turning of micro-dimples on a single crystal 4H-SiC wafer, tool wear is severe, resulting in increased surface roughness and form errors. Figure 7.14(a) shows an optical microscope image of the machined micro-dimples. To evaluate surface roughness and form errors, WLI measurement was performed, and the results are presented in Figure 7.14(b). Due to the fact that these dimples were fabricated one by one, the surface roughness for the first machined dimple (No. 1) is the smallest (3.36 nm Sa), while for the last machined dimple (No. 4) is the largest (6.30 nm Sa) due to the gradually increased tool wear. Besides, the edge contour of the No. 1 dimple is almost a perfect circle,

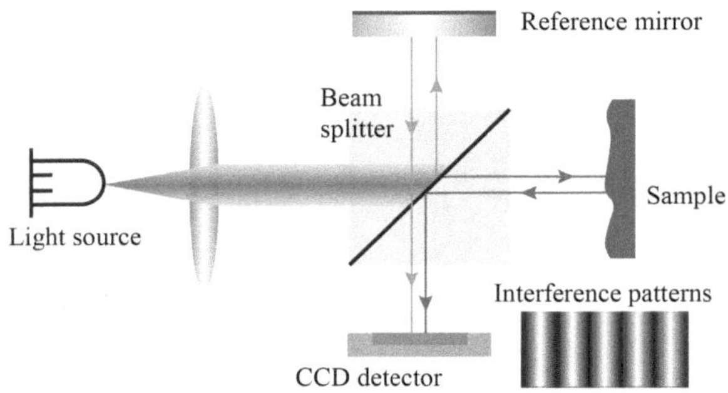


**FIGURE 7.13** Schematic of the principle of the white light interferometer.

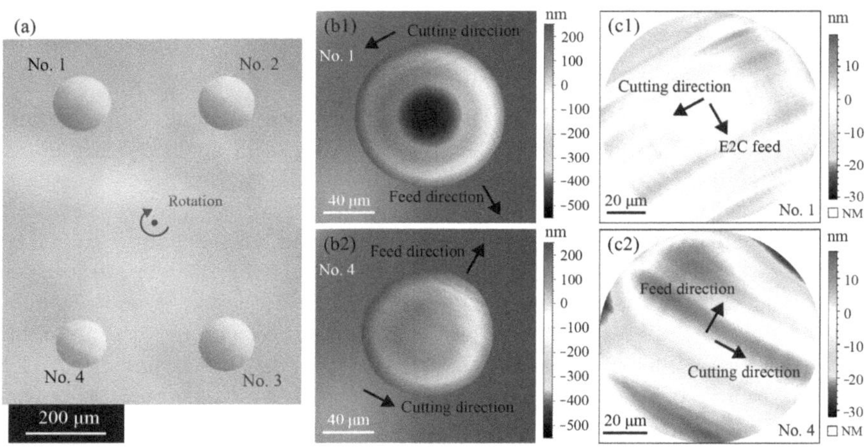


**FIGURE 7.14** (a) Optical image of micro-dimples machined on a single crystal 4H-SiC wafer. (b1) and (b2) are the three-dimensional topographies of the No. 1 and No.4 dimples shown in (a) measured by white light interferometry. (c1)–(c2) are the calculated form errors of the corresponding dimples [15].

while the edge contours of the subsequently machined dimples are gradually distorted. The edges around cut-in and cut-out areas of the dimple have the most severe distortion. This is because the center point of the cutting edge is the most involved in the cutting process, and thus tool wear at the center point is the most severe. The form errors of the No. 1 and No.4 dimples are presented in Figure 7.14(c). The peak-to-valley form error is 26 and 45 nm, respectively. The form error increases with increasing cutting distance because of the accumulated tool wear. It can be inferred from the results that diamond turning can fabricate crack-free microlens array surfaces on SiC even after tool wear has occurred. However, the surface quality and form accuracy of the surface will be reduced as the tool wear accumulates.

### 7.3.4 Laser Probe Profilometer

A laser probe profiling system typically refers to a scanning measurement device that utilizes laser to capture detailed information of the surface profile or geometry of an object. In laser triangulation, a laser beam is projected onto the surface of an object. The reflected beam is then captured by a sensor at an angle to the incident beam. The position of the reflected beam on the sensor is used to calculate the distance between the laser probe and the surface based on the principle of triangulation. Laser probe profiling is a noncontact measurement method and thus is advantageous for delicate or sensitive surfaces. In addition, the system collects data at each point as the laser beam scans across the predefined locations. This means that the size of the measurement area in a laser probe profiling system is indeed dependent on the path length covered by the laser scan. If the laser beam scans a larger area or covers a longer path, more points will be collected. This relationship between the path length of the laser scan and the size of the measurement area is a critical aspect of laser probe profiling systems. It influences the spatial resolution and coverage of the acquired data, impacting the system's ability to capture detailed surface features and variations across the scanned region. Therefore, laser probe profiling becomes particularly crucial when the size of the microstructure exceeds the measurement range of a white light interferometer.

For example, a laser probe profilometer NH-3SP (Mitaka Kouki Co., Ltd, Tokyo, Japan) was utilized to measure the geometry of machined Fresnel lenses on single-crystal germanium. This profilometer uses a semiconductor laser probe to scan the lens surface. The wavelength of the laser beam is 635 nm, and the resolution in Z-direction is 1 nm. The laser beam has an extremely small spot size, approximately 1 μm, allowing for precise measurement of narrow microgrooves without the edge-rounding effect seen in traditional stylus profiling methods. While large incident angles may occasionally introduce errors at steep rims of microgrooves, this issue can be diminished to some extent by selecting appropriate measuring conditions and processing the data using software techniques. Figure 7.15 shows the 3D topographies of the lens center and a part of the outer region. Results indicate that the Fresnel structures have been precisely fabricated both at the workpiece center and in the outer regions, without any visible defects.

Another example of laser probe profilometer measurement is given here. Figure 7.16(a) shows a photograph of a microlens array on aluminum alloy machined by slow tool servo diamond turning, which has a mirror finish. Using an optical microscope, the features of the lens surface can be observed in more detail, as shown in Figure 7.16(b). Then, the 3D topographies of the lenses were measured using a laser-probe profilometer NH-3SP, as shown in Figure 7.16(c). Cross-sectional profiles of the lenses are measured along the X-axis through the lens centers, as indicated by the red dashed lines in the 3D topographies. Form error distributions were obtained by comparing the theoretical profiles with the measured ones. The peak-to-valley (P-V) value of the form error at the lower spindle speed (15 rpm) was 0.46 μm, about 68% of that observed in the machining at the higher spindle speed (45 rpm). A lower spindle speed requires a reduced

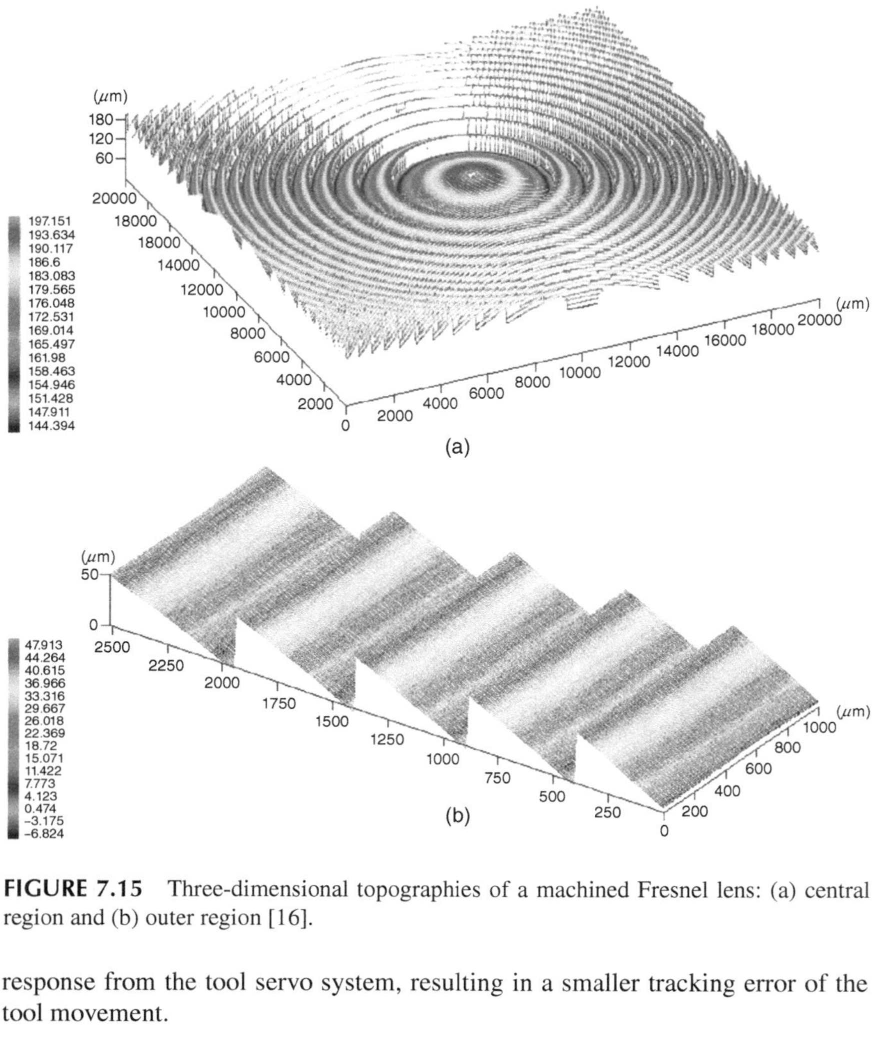


**FIGURE 7.15** Three-dimensional topographies of a machined Fresnel lens: (a) central region and (b) outer region [16].

response from the tool servo system, resulting in a smaller tracking error of the tool movement.

## 7.4 ON-MACHINE MEASUREMENT OF WORKPIECE

On-machine measurement is a process of taking measurements or collecting data directly on the machine where the machining process is occurring. This is in contrast to off-machine measurement, where measurements are performed by taking the workpiece away from the machine to a separate measurement platform. On-machine measurement can be crucial for ensuring precision, quality control, and real-time monitoring of the machining process. It can avoid a few workpiece realignment-related errors, such as mechanical deformation due to clamping/fixing forces, thermal deformation due to temperature changes, and

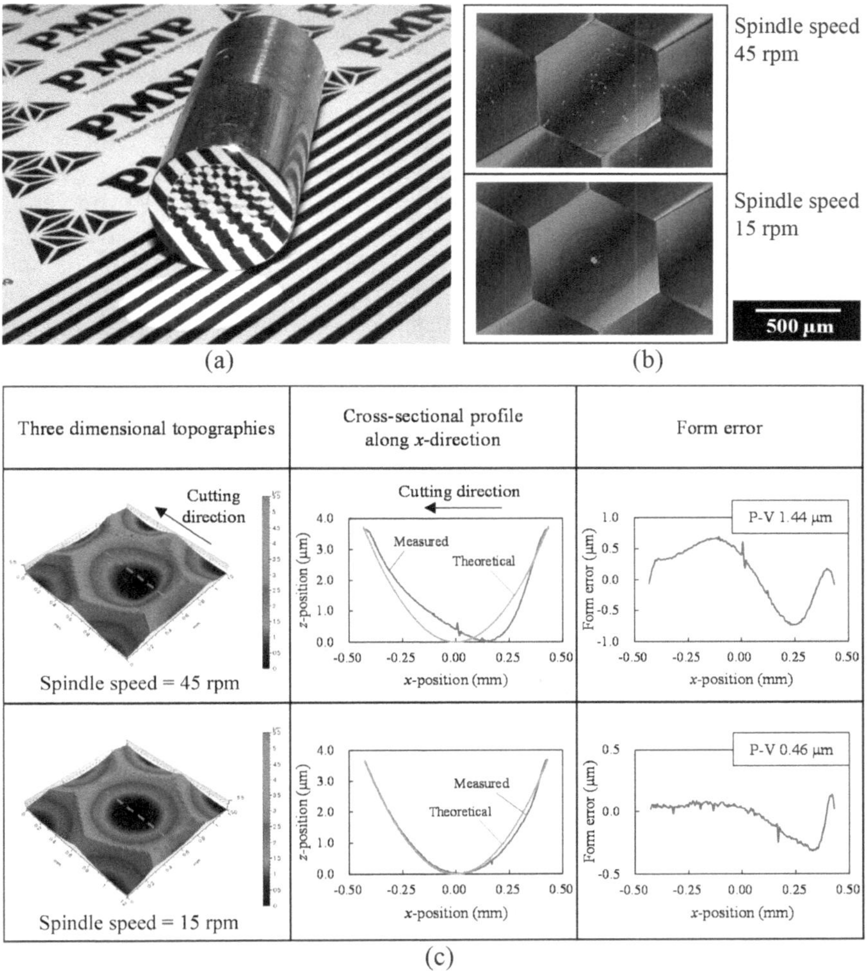


**FIGURE 7.16** (a) A hexagonal microlens array machined on aluminum alloy. (b) Differential interference contrast microscope images of the lenses. (c) 3D topographies and cross-sectional profiles of the lenses machined at different spindle speeds [17].

other environmental factor-related errors. Since on-machine inspections occur immediately after the manufacturing process, without the need to move the workpiece from the machine, the time interval from the end of manufacturing to the start of inspection can be shortened, enhancing inspection efficiency. It allows for immediate feedback and adjustments, leading to more efficient and accurate production.

A significant challenge for on-machine surface metrology is that the measurement may be affected by machine tool errors. These errors encompass spindle/

slide motion errors, thermal deformations, and vibrations. The alignment error of the sensor head on the machine tool introduces an additional source of error. Addressing and minimizing these errors in on-machine surface measurements, along with addressing calibration and traceability issues, is crucial. In addition, the physical interference of the sensor head with the machining tool must be prevented. For in-process measurement during machining, the sensor head needs to be robust enough to withstand the potentially severe machining environment, typically characterized by machining fluid, cutting chips, and machine tool vibrations. The measurement speed of the sensor head is also required to be high enough so that it can follow the machining operation. These requirements make in-process measurement more difficult than offline measurement.

Another challenge is the integration of the measuring instrument with the machine tool. Considering the limited space in the machine tool and the accessibility of the workpiece surface, the measuring instrument usually needs to be compact. The slides, spindles, positioning sensors, and numerical control (NC) controller of the machine tool itself are usually used as the constructing components of the on-machine measurement system. In the following sections, some typical on-machine measurement technologies are introduced.

### 7.4.1 On-Machine Stylus Profiling

Figure 7.17 shows a schematic of a stylus-based on-machine surface profiling system. In a stylus-scan surface profiler, usually only the height (Z-direction) sensor of the instrument is mounted, and the scanning movement relative to the workpiece surface in the XY-plane is facilitated by the X-slide and spindle of the machine tool. Consequently, it is essential to employ the position data of the slide and spindle obtained from the machine's numerical controller or directly from the output of the rotary and linear encoders of the spindle and slide to generate the topography data of the workpiece surface. Before mapping the workpiece surface, it is necessary to identify the sensor's position within the machine tool coordinate system. An artifact or a workpiece with a known shape and a known position in the machine tool coordinate system is required for this purpose. It is also necessary to establish

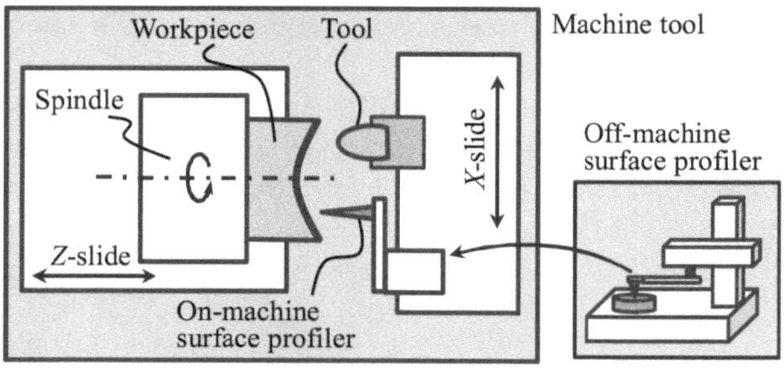


**FIGURE 7.17** Schematic of a stylus-type on-machine surface profiling system [18].

specific sampling strategies for stylus-scan measurement systems based on the machine tool configuration and the shape of the workpiece.

### 7.4.2 On-Machine Interferometer Profiling

Figure 7.18 illustrates one of the earliest on-machine applications of interferometry using a Fizeau configuration [19]. The interferometer was positioned on the machine tool to measure a diamond-turned flat workpiece. After the machining process, the tool located between the interferometer and the workpiece was moved away to enable form measurement. This setup required a considerable distance between the workpiece surface and the interferometer reference flat, which can be realized by the long coherence length of the laser source. However, the large distance significantly increased the non-common path length between the reference and measurement beams, referring to the portions of the optical paths where they traverse different routes or distances. This is a significant difference from an off-machine Fizeau interferometry system, where a nearly common-path configuration is achievable. Consequently, the results of the on-machine interferometer profiling became sensitive to environmental disturbances. To mitigate the impact of vibrations, the machine was isolated from external vibrations using an air suspension system. During measurement, the machine was completely halted and covered with a vinyl sheet to further reduce the influence of vibration and air turbulence. Under these conditions, a data frame could be captured with a stability of 1/20 wavelength (λ).

Recently, the phase-shifting technique has become a widely adopted analysis method for interferograms, enabling the calculation of surface height with high vertical resolution in almost all types of interferometers. In a phase-shifting interferometer (PSI), the reference surface is stepped by a piezoelectric (PZT) actuator by a fraction of a wavelength to record a set of data frames of interferograms (typically three or four) with precise phase shifts. A vertical resolution of better than 1/1000λ can be achieved, and the form measurement accuracy can closely approach that of the reference flat in a well-controlled off-machine environment. However, under on-machine conditions, incorrect phase shifts between the

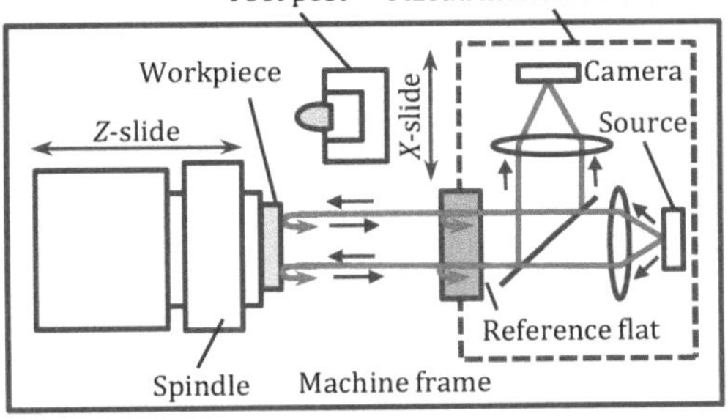


**FIGURE 7.18** Schematic of a Fizeau interferometer on a diamond turning machine [18].

recorded data frames can be induced by machine vibrations and air turbulence, leading to significant measurement errors in surface form. It is worth noting that the phase shift is sensitive to both high-frequency and low-frequency disturbances since it takes a relatively long time, typically tens of seconds, to capture the set of phase-shifted data frames. To address the instability of phase shifts between the data frames for phase-shifting analysis resulting from vibrations and air turbulence under on-machine conditions, a single-shot phase measurement method was proposed, where the set of phase-shifted data frames is captured simultaneously. Such an interferometer is often called a dynamic interferometer. Dynamic interferometers with a Twyman-Green configuration have been commercialized for on-machine surface metrology, owing to their compact structure and the robust single-shot phase measurement method for environmental disturbances. The interferometer can be directly mounted on the tool post, side by side with the cutting tool, for easy access to the workpiece without special vibration isolation and air turbulence control [18].

### 7.4.3 On-Machine Laser Probe Profiling

Compared with interferometry methods, laser probe profiling, based on laser focusing, offers better resistance to environmental disturbance and is more suitable for capturing the low-frequency topography of an optical freeform surface [20]. In laser probe profiling technology, the challenge lies in quickly identifying measurement points and effective measurement ranges while reducing measurement time and cost.

The noncontact laser focusing probe is the key component for realizing a high-precision measurement system. Capacitive and laser triangulation sensors are the two most commonly used non-contact probes for on-machine surface metrology, and they have been extensively produced by numerous sensor manufacturers. Recent advances in CCD/CMOS technology and sub-pixel resolution algorithms have enhanced the performance of laser triangulation sensors in terms of measuring range, resolution, and speed. For example, the optoNCDT series laser triangulation sensors from Micro-Epsilon Messtechnik GmbH & Co. KG, Germany, can achieve a vertical resolution of 0.1 μm over a range of 1 mm. These sensors have been developed for profiling both specular and diffuse reflection surfaces. A confocal probe is another viable choice for an on-machine optical probe. Its insensitivity to the reflectivity of the workpiece surface allows for the measurement of both specular and diffuse reflection surfaces. Keferstein et al. [21] integrated a chromatic confocal probe with a measurement range of 1 mm and a repeatability of 0.1 μm into a grinding machine. By mounting the probe onto the grinding wheel head and synchronizing the probe signal with the linear encoder signal of the slide, the measurement range of the probe could be extended to the full stroke of the slide moving the wheel head.

When integrating a laser sensor into a machine tool for on-machine probe profiling, it's essential to consider several important sources of errors, as illustrated in Figure 7.19 [18]. The most common error involves the accurate alignment and

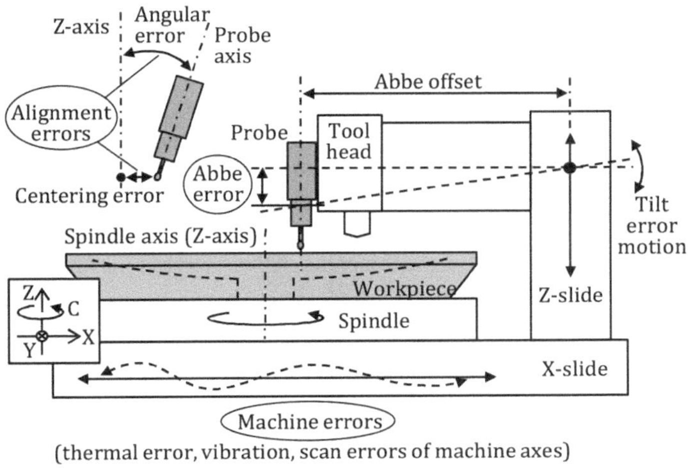


**FIGURE 7.19** Error sources in the integration of a probe into a machine tool [18].

setting of the probe on a machine tool, including centering error and angular error. Alignment errors can be identified by using the probe to detect a precision sphere with a known location on the machine tool. The angular error, that is, the tilt angle of the probe axis concerning the machine Z-axis, will introduce a cosine error, which can become a significant factor contributing to measurement uncertainty, especially over a large range. In some cases, the probe has to be mounted on the Z-slide, but the probe axis is located at a significant distance from the Z-axis; in other words, the Abbe offset is large, as shown in Figure 7.19. Under such conditions, even a small tilt error motion of the Z-slide can result in a substantial Abbe error. One way to compensate for the Abbe error is based on a measurement of the slide tilt error motion. A better solution is to select a probe with a large enough measurement range so that the use of the Z-slide for the Z-direction measurement can be avoided [22]. Additionally, in certain applications, the workpiece itself can be the primary source of errors. For example, in the in-process measurement of workpieces that generate a lot of heat, laser probe profiling faces challenges due to the elevated temperature of the workpiece. Therefore, the implementation of special thermal shielding for sensors and the use of temperature-controlled metrology frames are needed.

## 7.5 ON-MACHINE MEASUREMENT OF TOOLS

In diamond turning, the form accuracy of workpiece depends on machine tool accuracy and cutting tool accuracy. As diamond is the hardest material in the world and has significant crystallographic anisotropy, high-precision fabrication of a sharp curved edge on diamond is not easy. Currently diamond cutting tools are mostly finished by polishing with extremely fine diamond abrasives on cast ion polishing discs, as discussed in Chapter 4. At the commercially available level, the best edge sharpness and edge profile accuracy of the diamond cutting tools are both in the level of a few tens of nanometers, which is beyond the accuracy level of the ultraprecision machine tools. After machining, the tool profile error will be further increased by the tool wear. From this aspect, we may say that it is the tool profile error that finally limits the accuracy level of the diamond-cut products.

An effective method for solving this problem is to compensate for the profile error of the diamond tools in the numerical control programs used for machining. For this purpose, the tool profile must be measured precisely. To avoid additional errors occurring during removing and resetting the tools, the tool profile measurement is required to be carried out on machine. Compared to on-machine measurement of workpieces, literature on the on-machine measurement of cutting tools in ultraprecision machining is very little. The optical projection method which has been used for tool profiling in conventional metal cutting cannot be used in ultraprecision machining due to its low resolution. In recent years, several on-machine techniques have been tried to measure tool profiles.

### 7.5.1 Tool-Workpiece Transcription Method

One is the profile transcription method where a dummy workpiece made of soft ductile material, such as plastics and graphite, is plunge-cut by the tool, and then the dummy workpiece with transcribed tool marks is removed from the machine and measured by surface profiling instruments.

For example, Figure. 7.20 plots the cross-sectional profile of grooves cut on copper by the tool after cutting glass at various distances (0–50 m). Before cutting the glass, the groove cut on copper with the tool exhibits a symmetrical arc-shaped cross-sectional profile. As cutting glass gradually wears the tools, using these worn tools to cut copper transfers the worn profile of the tool to copper. The progression of tool wear can be clearly visualized through the transcription method. This method is quick and simple, but the measurement accuracy is low due to the elastic and plastic deformation of the dummy workpiece material during transcription. Especially it is difficult to use for measuring wide-angle tools which are usually required for generating micro-grooves and high-aspect ratio aspherical surfaces.

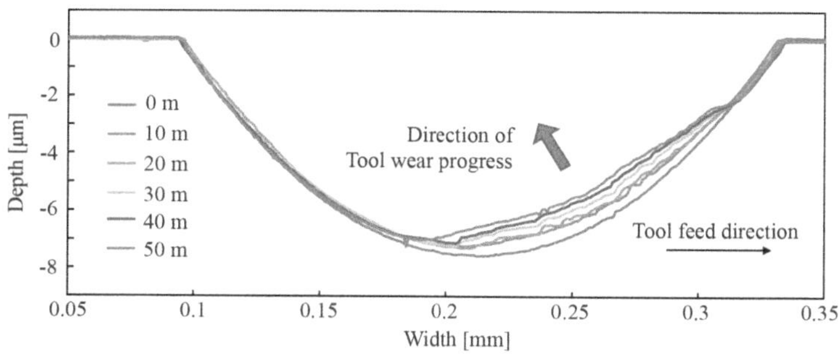


**FIGURE 7.20** Cross-sectional profile of grooves cut on copper by the tool after cutting glass at various distances.

### 7.5.2 Stylus Profiling of Tools

Stylus profiling, which has been conventionally used for workpiece measurement, has also been attempted to measure tool profiles. To meet the requirements for on-machine measurement of the edge profile of a precision cutting tool, Wen et al. [23] developed a tool edge profiler composed of a probe unit and a positioning system, as illustrated in Figure 7.21(a). The probe unit, comprising a stylus and a double-sided beam, is connected to a displacement sensor utilizing laser triangulation. This setup allows the detection of the deflection of the beam caused by contact with a target surface. For tool edge profile measurement, the probe unit is mounted on the positioning system, allowing the probe unit to move vertically and horizontally with respect to the stylus axis. The measurement of the edge profile of a cutting tool can be carried out by combining the information of the probe unit position and the deflection of the beam. With the aid of the large stroke of the positioning system and the low beam stiffness, the developed tool edge profiler can achieve the measurement of the edge profile of a precision cutting tool across a wide measuring range, using a low measuring force while maintaining reasonable stability. When a cantilever-type probe is employed to measure the steep slope of a cutting tool, the deflection of the cantilever could become unstable and unpredictable due to the asymmetrical mechanical structure of the probe. In contrast, a double-sided beam-type probe, as depicted in Figure 7.21(b), is expected to result in more stable deflection when measuring the steep slope of a tool edge.

Recently, atomic force microscopes (AFM) have also been used for on-machine measuring cutting edges of diamond tools [25]. The AFM method enables nanometer-resolution measurement within a small region, while it is difficult to measure the cross-sectional profile of the entire tool.

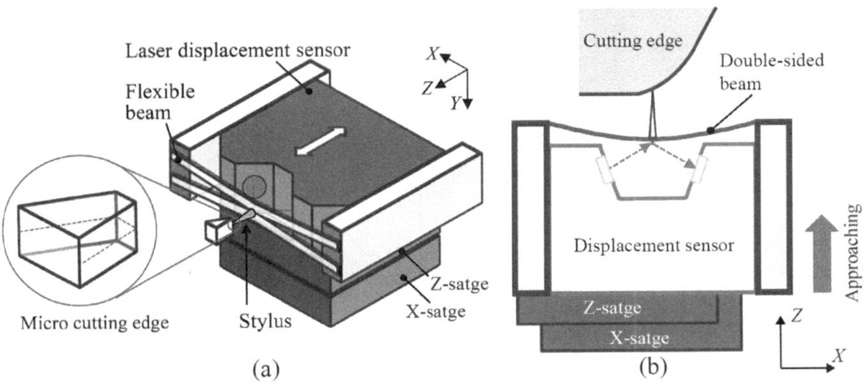


**FIGURE 7.21** Schematics of (a) a probe unit [23] and (b) deformation of the double-sided beam-type probe [24].

### 7.5.3 Noncontact Tool Measurement

Noncontact measurement systems have been used for on-machine profiling the edge profiles of diamond-cutting tools for precision micro machining. This section introduces a noncontact tool profiling system based on a compact white-light interferometer (WLI) having nanometer resolution [26]. The system enables precision measurement of tool contour accuracy for wide-angle tools in the nanometer order, and the measured data by the present system can be fed back to the numerical control system for form error compensation of the workpiece.

#### 7.5.3.1 System Configuration

Figure 7.22(a) shows a schematic of the configuration of the on-machine WLI system. The workpiece is vacuum-chucked to the Z-axis spindle of the machine. A round-nosed diamond tool is mounted on the tool post which has been fixed on the B-axis rotary table. A WLI is fixed to the Z-table of the machine and oriented parallel to the workpiece spindle. The cutting tool can be shifted between the machining position and the measurement position by the numerically controlled XZ machine tables without the need to remove the tool from the machine. The tool can also be rotated in the XZ plane by the B-axis rotary table, enabling the measurement of different orientations of a wide-angle tool.

In conventional optical projection methods, the tool profile is typically measured from the top of the tool rake face. However, the vertical resolution of the WLI is extremely high (0.01 nm) while the lateral resolution is relatively low (350 nm). Thus, profiling from the flank face can effectively make use of the high

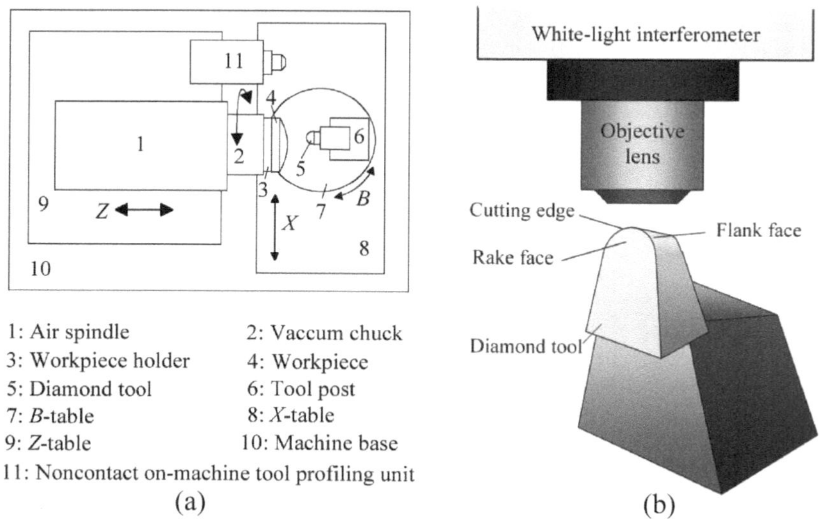


**FIGURE 7.22** (a) Configuration of the on-machine white light interferometer system (WLI). (b) Schematic of flank face profiling method.

vertical resolution. Figure 7.22(b) is a schematic of the tool-sensor position during measurement. The flank face of the tool is oriented vertically to the lens, enabling the WLI to measure the topography of the flank face near the tool apex at a resolution of 0.01 nm. After each measurement, the tool undergoes a small-angle rotation using the B-axis rotary table, facilitating an adjacent measurement. By repeating the above operation in the same way, the entire flank face of the round tool can be mapped. Subsequently, the collected data is transferred to a personal computer and processed using the developed edge-identifying program and data-stitching program, resulting in the complete profile of the tool.

### 7.5.3.2 Edge Identification

Figure 7.23(a) shows an optical image of a square region of the tool flank face with interference fringes around the tool apex. Correspondingly, Figure 7.23(b) presents the topography data obtained from a single measurement. The horizontal boundary line represents the edge of the tool, and the midpoint of the line corresponds to the tool apex. Due to depth changes at different points on the tool flank face, the figure exhibits a significant color gradation. Upon magnifying the image along the cutting edge, as shown in Figure 7.23(c), a notable dispersion of data along the cutting edge becomes apparent. In other words, there is randomness in the presence and absence of data within the bottom layers of pixels. This phenomenon may be attributed to light scattering at the edge where roundness exists. If the surface inclination angle induced by the edge roundness is too high, preventing the detection of reflected light by the objective lens, no data will be displayed at that pixel. The dispersion of data around the cutting edge complicates the precise identification of the edge location from the topography image.

Considering the pixel arrays in Figure 7.23(b) along the X- and Y-directions, respectively, the algorithm for automatic edge identification is as follows: first, the computer program examines the presence or absence of data for each pixel along

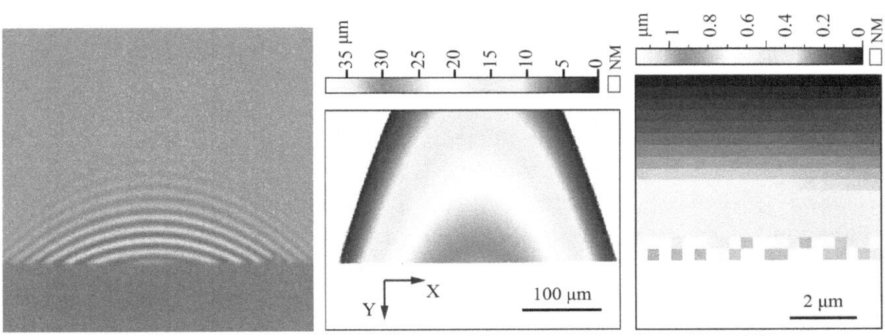


(a) (b) (c)

**FIGURE 7.23** Images of the tool flank face measured by the WLI: (a) optical image; (b) topographical image; and (c) magnified presentation of a small region along the cutting edge.

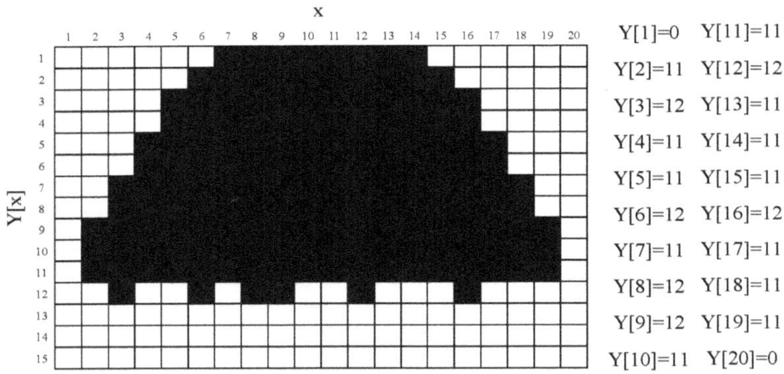


**FIGURE 7.24** Schematic presentation of edge-identification algorithm.

the Y-direction at a fixed X-coordinate. If data is present, the pixel is assigned a value of 1 and indicated in black. Conversely, in the case of data absence, the pixel is assigned a value of 0 and indicated in white. This process is repeated for all the pixel arrays along the X-direction, resulting in a map of data presence/absence, as shown in Figure 7.24. Using this map, the boundary pixels between the data presence region and the data absence region can be identified. Finally, the algorithm calculates the average value of the y-coordinates of these boundary pixels, which is denoted as the location of the cutting edge of the tool.

Strictly speaking, the tool profile should be measured precisely along the edge location identified by the above method. However, because a few data points are absent along the identified edge line, the measured profile becomes discontinuous, potentially causing errors in subsequent calculation and compensation processes. These issues can be mitigated by measuring the profile at a location slightly away (e.g., 5 μm) from the identified edge line. Since the flank face is typically a conical surface, the theoretical radius of the tool profile at any cross-section can be calculated from the nominal radius and the distance from the cross-section to the theoretical edge line.

### 7.5.3.3 Cross-Sectional Profile Extraction

Figure 7.25(a) shows the cross-sectional profile near the tool tip obtained by one measurement, where the round tool profile is clearly seen. Figure 7.25(b) is the deviation between the raw profile data and the theoretical tool profile. When the inclination angle of the flank face is within the range from −8° to 6° (the angle asymmetry on the two sides might be due to a slight discrepancy of the lens axis), the profile appears very smooth. Beyond this range, significant fluctuations occur in the profile curve because, at high incident angles, the incident light is reflected by the diamond surface and cannot be detected by the sensor in the lens. Therefore, the effective measurement range for one shot is within ±5°. Figure 7.25(c) shows the distribution of the tool profile error measured within this range. The peak-to-valley of the profile error within this range is approximately 8 nm.

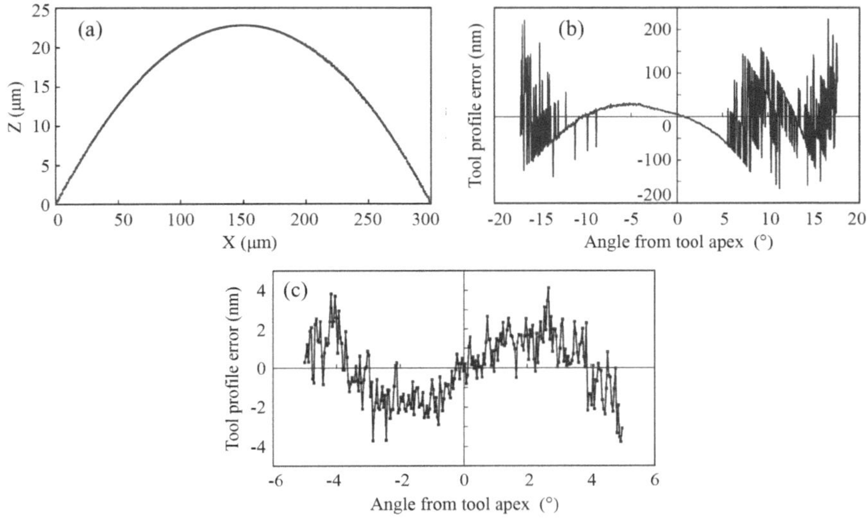


**FIGURE 7.25** (a) Raw data of a cross-sectional profile of the tool obtained by one measurement. (b) Deviation between the raw profile data and the theoretical profile of the tool. (c) Tool profile error distribution in the effective angle range.

In the figure, we can also see that extremely small variations of the surface height have also been detected by the white-light interferometer, which demonstrates the high resolution of the system.

#### 7.5.3.4 Data Stitching

In order to perform profile measurement of wide-angle diamond tools, data-stitching needs to be carried out to combine the data from neighboring measurements. The basic concept of data stitching is schematically shown in Figure 7.26(a). The first measurement is conducted near the tool apex, and the measured range is indicated by arc AB. For the second measurement, the diamond tool is rotated by an angle $\theta$, and the measurement range becomes arc CD. To perform stitching, arcs AB and CD must be overlapped for a certain length (overlapping region). The overlapping region is centered at the position where the angle from the tool apex is $\theta/2$ so that the length of the overlapping region is determined by the angle $\theta$ and the field width of the lens. From the center of the overlapping region, a small region can be extracted for data stitching. For example, as shown in Figure 7.26(b), the length of the stitching region is set to 3/5 of that of the overlapping region. The steps of data stitching can be schematically shown in Figure 7.26(c). First, a stitching region $m$ in arc AB is determined, which will be then compared with arc CD within the overlapping region $n$. Subsequently, the measured profile data of arc CD are rotated and moved to be in accordance with arc AB in their orientations within the overlapping regions. This step requires coordinate conversion of arc CD.

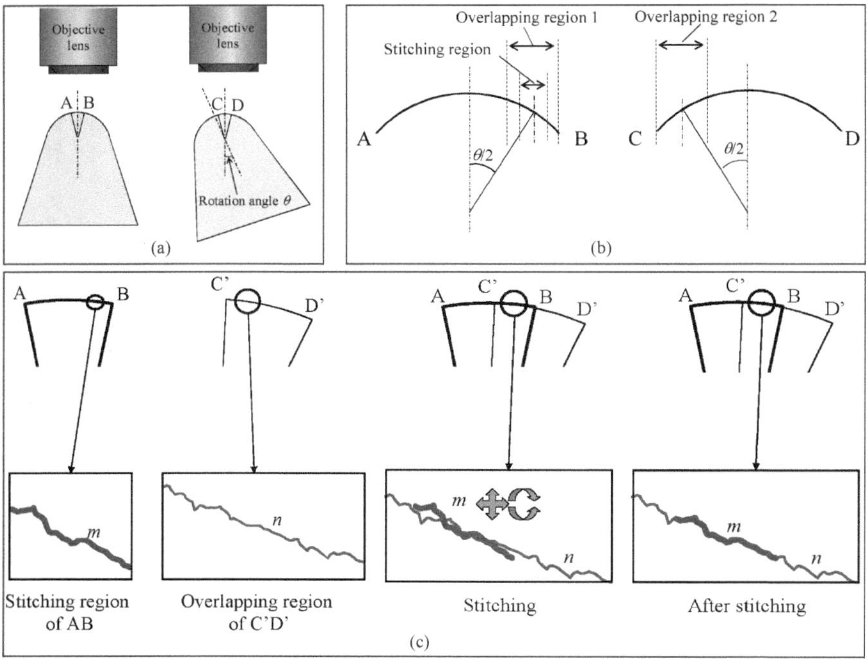


**FIGURE 7.26** Schematics of (a) wide-angle profiling through tool rotation, (b) overlapping ranges in tool profile data during data stitching, and (c) presentation of the data-stitching method.

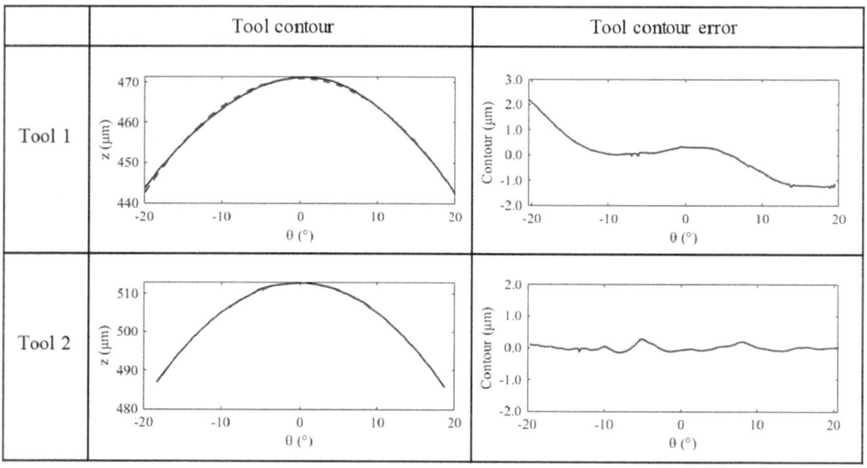


**FIGURE 7.27** Profiles of tool contour and tool contour error measured by white light interferometry [27].

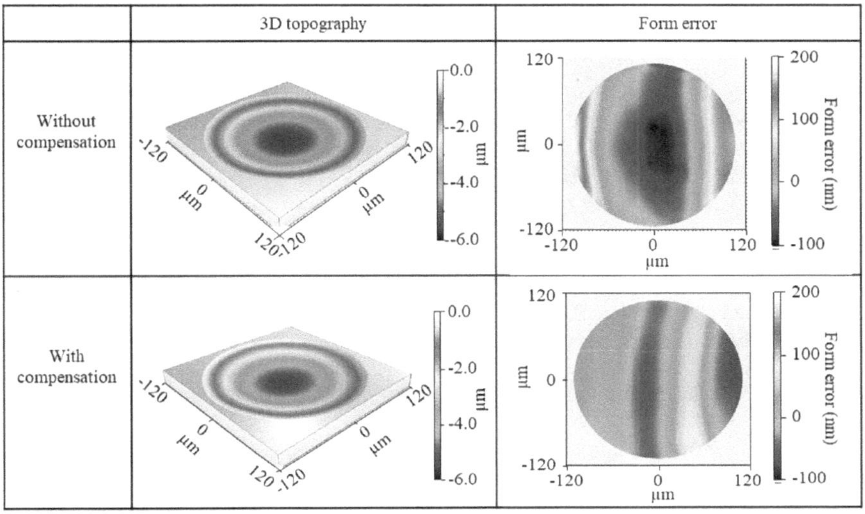


**FIGURE 7.28** 3D topographies and form errors of a microlens array machined with and without compensation of tool contour error [27].

Figure 7.27 shows the tool contours extracted from the measured raw data using the data-stitching method and the calculated contour errors for two tools (tools 1 and 2). The tool nose radius is determined to be 471.98 μm for tool 1 and 512.80 μm for tool 2. Tool 1 exhibits a contour error of 3.517 μm peak-to-valley (P-V) and 0.802 μm root mean square (RMS), while tool 2 has a smaller contour error of 0.426 μm P-V (0.086 μm RMS).

By understanding the tool profile error, it is possible to enhance the form accuracy of the structured surface to be machined through the compensation of the tool profile error in the tool path planning. Figure 7.28 shows 3D topographies and calculated form errors of a micro lens machined with and without tool contour compensation, respectively. Prior to compensation, a form error of 292 nm (P-V) is measured, whereas after compensation, the form error is reduced to 153 nm (P-V) [27].

## REFERENCES

[1] Taylor Hobson Ltd. Form Talysurf® PGI Freeform. https://www.taylor-hobson.com/products/surface-profilers/optics-pgi/pgi-freeform# (accessed Dec. 2023).

[2] Taylor Hobson Ltd. Talyrond® 595H PRO. https://www.taylor-hobson.com/products/roundness-form/high-precision-roundness/talyrond-595h-pro (accessed Dec. 2023).

[3] Ohta, T., Yan, J., Kuriyagawa, T., Kodera, S., Tomoaki, N. (2007) Prediction of subsurface damage depth of ground brittle materials by surface profiling. *International Journal of Machining and Machinability of Materials* 2 (1): 108–124.

[4] Zhang, L., Yan, J. (2022) Study on nano-graphitic carbon coating on Si mold insert for precision glass molding. *Surface and Coatings Technology* 448: 128893.
[5] Yan, J., Syoji, K., Kuriyagawa, T., Suzuki, H. (2002) Ductile regime turning at large tool feed. *Journal of Materials Processing Technology* 121 (2): 363–372.
[6] Huang, W., Yan, J. (2023) Towards understanding the mechanism of vibration-assisted cutting of monocrystalline silicon by cyclic nanoindentation. *Journal of Materials Processing Technology* 311: 117797.
[7] Kubo, K., *Laser based asphere and freeform measurement technology by UA3P, 11th Laser Metrology for Precision Measurement and Inspection in Industry* 2014, Tsukuba, Japan,
[8] Huang, W., Yan, J. (2021) Fundamental investigation of diamond cutting of micro V-shaped grooves on a polycrystalline soft-brittle material. *Journal of Manufacturing and Materials Processing* 5 (1): 17.
[9] Zhang, L., Hashimoto, T., Yan, J. (2021) Machinability exploration for high-entropy alloy FeCrCoMnNi by ultrasonic vibration-assisted diamond turning. *CIRP Annals* 70 (1): 37–40.
[10] Ohta, T., Yan, J., Yajima, S., Takahashi, Y., Horikawa, N., Kuriyagawa, T. (2007) High-efficiency machining of single-crystal germanium using large-radius diamond tools. *International Journal of Surface Science and Engineering* 1 (4): 374–392.
[11] The University of Iowa. Confocal Microscopy. https://cmrf.research.uiowa.edu/confocal-microscopy (accessed Dec. 2024).
[12] Yu, Q., Zhou, T., He, Y., Liu, P., Wang, X., Yan, J. (2021) Effects of relative tool sharpness on surface generation mechanism of precision turning of electroless nickel-phosphorus coating. *Journal of Mechanical Science and Technology* 35 (7): 3113–3121.
[13] Zhou, T., Li, Z., Guo, W., Liu, P., Zhao, B., Wang, X. (2023) A study on the material removal characteristics and damage mechanism of lapping for pressureless sintered silicon carbide (SSiC) microlens cavity. *Micromachines* 14 (6): 1162.
[14] FindLight. White light interferometry: A comprehensive guide. https://www.findlight.net/blog/white-light-interferometry-a-comprehensive-guide/ (accessed Dec. 2024).
[15] Huang, W., Yan, J. (2023) Mechanisms of tool-workpiece interaction in ultraprecision diamond turning of single-crystal SiC for curved microstructures. *International Journal of Machine Tools and Manufacture* 191: 104063.
[16] Yan, J., Maekawa, K., Tamaki, J., Kuriyagawa, T. (2005) Micro grooving on single-crystal germanium for infrared Fresnel lenses. *Journal of Micromechanics and Microengineering* 15 (10): 1925–1931.
[17] Mukaida, M., Yan, J. (2017) Fabrication of hexagonal microlens arrays on single-crystal silicon using the tool-servo driven segment turning method. *Micromachines* 8 (11): 323.
[18] Gao, W., Haitjema, H., Fang, F.Z., Leach, R.K., Cheung, C.F., Savio, E., Linares, J.M. (2019) On-machine and in-process surface metrology for precision manufacturing. *CIRP Annals* 68 (2): 843–866.
[19] Nomura, T., Yoshikawa, K., Tashiro, H., Takeuchi, K., Ozawa, N., Okazaki, Y., Suzuki, M., Kobayashi, F., Usuki, M. (1992) On-machine shape measurement of workpiece surface with Fizeau interferometer. *Precision Engineering* 14 (3): 155–159.
[20] Duan, Y., Vo, Q., Zhang, X., Wang, Y., Huang, S., Fang, F. (2019) Novel method of measuring optical freeform surface based on laser focusing probe without calibrating focus error signal. *Measurement* 148: 106961.

[21] Keferstein, C.P., Honegger, D., Thurnherr, H., Gschwend, B. (2008) Process monitoring in non-circular grinding with optical sensor. *CIRP Annals* 57 (1): 533–536.
[22] Shimizu, Y., Goto, S., Lee, J., Ito, S., Gao, W., Adachi, S., Omiya, K., Sato, H., Hisada, T., Saito, Y., Kubota, H. (2013) Fabrication of large-size SiC mirror with precision aspheric profile for artificial satellite. *Precision Engineering* 37 (3): 640–649.
[23] Wen, B., Shimizu, Y., Watanabe, Y., Matsukuma, H., Gao, W. (2020) On-machine profile measurement of a micro cutting edge by using a contact-type compact probe unit. *Precision Engineering* 65: 230–239.
[24] Wen, B., Sekine, S., Osawa, S., Shimizu, Y., Matsukuma, H., Archenti, A., Gao, W. (2021) A comparison of the probes with a cantilever beam and a double-sided beam in the tool edge profiler for on-machine measurement of a precision cutting tool. *Machines* 9 (11): 271.
[25] Jang, S., Asai, T., Shimizu, Y., Gao, W. (2012) Establishment of a measuring station on a diamond turning machine for in-process cutting edge inspection of single point diamond micro-tools. *International Journal of Nanomanufacturing* 8 (1–2): 106–122.
[26] Yan, J., Baba, H., Kunieda, Y., Yoshihara, N., Tsunemoto, K. (2007) Nano precision on-machine profiling of curved diamond cutting tools using a white-light interferometer. *International Journal of Surface Science and Engineering* 1 (4): 441–455.
[27] Nagayama, K., Yan, J. (2020) Measurement and compensation of tool contour error using white light interferometry for ultra-precision diamond turning of freeform surfaces. *International Journal of Automation Technology* 14 (4): 654–664.

# 8 Subsurface Metrology

## 8.1 INTRODUCTION

As mentioned in Chapter 1, the quality and functionality of a machined surface are determined not only by surface roughness and form accuracy but also by machining-induced subsurface structural changes of the workpiece material, namely, subsurface damage. For this reason, apart from surface metrology, subsurface metrology is attracting more and more attention from both academia and industry. This expanded scope recognizes the nuanced impact of ultraprecision machining on both the mechanical, physical, and chemical properties of materials and the functionality aspects of the machined surfaces. Subsurface metrology involves quantitatively characterizing the material microstructural changes, i.e., phase transformations, residual stresses, subsurface cracks, and other subsurface features to assess the stability of a machining process and the quality of the resulting products. It is crucial for understanding the mechanisms and optimizing the performances of ultraprecision machining techniques in various industries such as electronics, optics, and aerospace.

## 8.2 SUBSURFACE DAMAGE CLASSIFICATION

The machining-induced subsurface damage, also known as structural changes within the material beneath the machined surface, includes subsurface microcracks, dislocations, phase transformations, and residual stresses. These factors significantly impact the mechanical, optical, electrical properties, and final functionalities of the products.

### 8.2.1 MICROCRACK

Microcracks are the most common defects occurring in the machining of brittle materials, and they can be induced by various processes such as cutting, grinding, and lapping. The crack generation in the cutting of brittle materials has been introduced in Chapter 6. In this section, the crack generation mechanism of a brittle material in abrasive processes, such as grinding and lapping will be given.

Figure 8.1 illustrates the schematics of microcrack formation in the precision grinding of hard brittle materials. In grinding with a single grain, below the plastic zone, a median crack is initiated downwards into the bulk region during loading; after unloading, lateral cracks are generated along the plastic/elastic boundary and extend toward the free surface, leading to material removal via microchipping, as shown in Figure 8.1(a) [1]. The lateral microcracks are visible on the outer surface


DOI: 10.1201/9781003292388-8

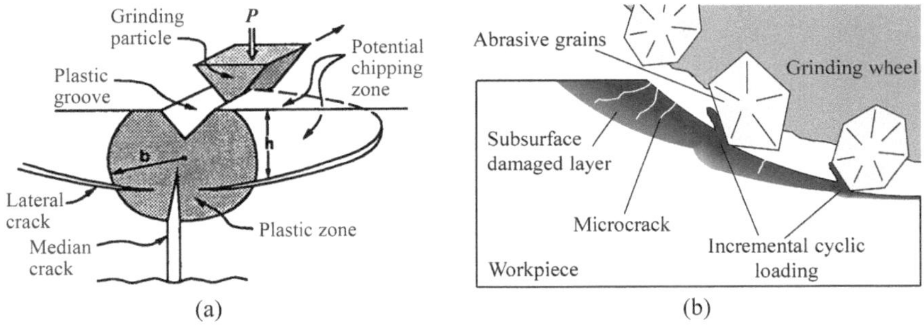


**FIGURE 8.1** Schematic of microcrack formation in grinding of hard brittle materials by (a) a single abrasive grain and (b) multiple abrasive grains.

as craters but the median cracks extend into the material and are difficult to observe or measure through surface metrology. In grinding with a cylindrical wheel, multiple grains scratch the surface intermittently, as shown in Figure 8.1(b). The incremental cyclic loading may make the crack formation/propagation process more complicated, although the cracks in the upper shoulder region will be removed by subsequent grains without affecting the final surface integrity [2].

Subsurface microcracks weaken the material, reduce its strength, and may eventually lead to failure under stress. This constitutes the most detrimental form of subsurface damage, directly determining the performance and lifetime of a component.

Subsurface microcracks can form due to a variety of reasons, including the application of excessive cutting forces, thermal gradients generated during machining, and the interaction between the cutting tool and workpiece material. High-speed machining or improper tool geometry can also contribute to the development of subsurface cracks. Different materials exhibit varying susceptibility to subsurface cracking. Hard and brittle materials are more prone to developing subsurface microcracks during machining. Factors such as material composition, microstructure, and heat treatment can influence the microcracking in materials as well. As subsurface cracks are invisible from the outside of the surface, metrology of the number, depth, and direction of the subsurface cracks is important for the quality assurance of industrial products.

### 8.2.2 Grain Refinement

For polycrystalline materials, the grain size in the near-surface layer may become finer after machining. This grain refinement phenomenon has been observed in various machining processes, such as cutting and grinding, of many materials. For example, in Chapter 6, Section 6.5.3, surface grain refinement of zirconia in diamond turning was introduced. The grains in both the chip and the subsurface region of the workpiece have been refined from 300 nm to a few tens of nanometers. The mechanism of grain refinement of zirconia has been shown in

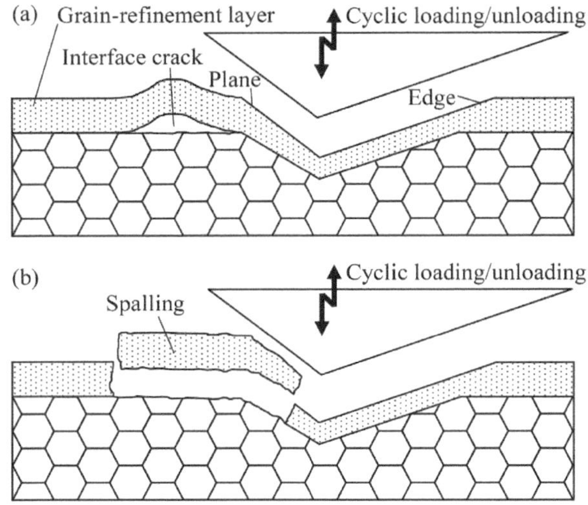


**FIGURE 8.2** Grain refinement of zirconia in ultraprecision cutting [2].

Figure 6.72. The grain refinement takes place around the shear zones in the cutting region due to the severe shear stress in these zones. The grain refinement in the primary shear zone contributes to the formation of chips with finer grains than the bulk material, while the grain refinement in the third shear zone beneath the tool edge contributes to the formation of a grain-refined surface layer on the workpiece. Such grain-refined layers may exhibit better surface properties than the bulk due to the hardness increase and compressive residual stress. However, under cyclic loading/unloading conditions, interfacial cracks may occur between the fine-grain layer and the bulk, leading to spalling of the fine-grain layer [2], as shown in Figure 8.2. Subsurface metrology is necessary to characterize the thickness and material properties of this fine-grain layer when considering industrial applications.

### 8.2.3 Phase Transformation

High-pressure phase transformation (HPPT) is a phenomenon where the extremely high pressure generated around the tool tip during cutting or other machining processes can induce a local change in the phase or crystalline structure of the material being machined. Theoretically, under high-pressure conditions, every material could undergo phase transformations, but the pressure threshold for transformation varies for each. For single-crystal silicon, the threshold is ~10 GPa. For materials with a higher threshold than the pressure induced by the machining process, phase transformation will not occur.

Figure 8.3(a) is a schematic model of subsurface damage mechanism in ductile machining of silicon [5]. As soon as the tool advances into the material (loading),

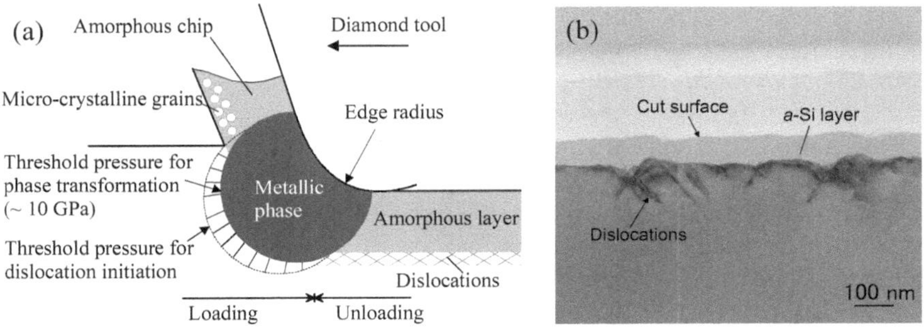


**FIGURE 8.3** (a) Schematic of the amorphization mechanism during the ductile machining of silicon and (b) amorphization and dislocations in the subsurface of single-crystal silicon [3].

a phase transformation from diamond cubic structure (Si-I) to metallic phase (Si-II) occurs in the material surrounding the tool tip. This part of the material becomes sufficiently ductile to sustain plastic flow which facilitates ductile-mode material removal. After the tool has passed (unloading), the metallic phase does not transform back into the diamond cubic structure, but rather, changes into an amorphous phase. As a result, the final subsurface damage layer contains an amorphous phase rather than the metallic phase. In the figure, as the distance from the tool tip increases, the pressure decreases and finally drops below the threshold pressure required for phase transformation (~10 GPa). However, the pressure may still be high enough to facilitate dislocation initiation in silicon if a sufficiently high shear stress exists. As a result, a dislocation layer is generated beneath the amorphous layer as the tool passes. Under a negative tool rake angle, as depth of cut increases, the shear stresses beneath the tool increase due to the downward material flow, leading to a higher dislocation density. However, the thickness of dislocation layer does not change significantly with machining conditions because the distance between the boundary of threshold pressure for phase transformation and the boundary of threshold pressure for dislocation initiation is not sensitive to cutting conditions. An example of cross-sectional TEM micrograph of a diamond-cut silicon sample is shown in Figure 8.2(b).

The phase transformation phenomena in machining processes have been explained by some researchers as a kind of thermal effect of the machining processes, i.e., the temperature rise in machining causes the phase transformation. However, the thermal effect cannot be used to explain the phase transformations taking place in indentation tests and diamond anvil tests, where no temperature rise happens. It is now popularly accepted that high pressure, namely high compressive stress, is the major reason for the phase transformations. Of course, other factors, such as shear stress and temperature rise, may assist or enhance the phase transformation to some extent.

Depending on the applications, high-pressure phase transformations induced by machining may be favorable or unfavorable. In some cases, high-pressure phase transformations can lead to improvements in material properties. For instance, during the machining of steels, the austenite-to-martensite phase transformation may occur, resulting in the material becoming harder, more wear-resistant, and exhibiting increased strength [4]. When cutting single-crystal silicon, surface amorphization occurs, which reduces the brittleness and anisotropy of the material surface, and is helpful for the formation of a smooth and uniform surface in subsequent machining steps [5]. However, in some other cases, high-pressure phase transformations may result in unexpected changes to the material properties. For example, during the machining of zirconia ceramics, which mainly consist of tetragonal phases under atmospheric temperature and pressure conditions, there will be a tetragonal-to-monoclinic phase transformation due to the high stress and heat, leading to a decrease in material strength [6]. The depth and the microstructure of the phase transformation layer are other important aspects for subsurface damage evaluation.

### 8.2.4 Dislocation

A dislocation, a crystallographic defect within the intricate lattice structure of a crystal, introduces a significant irregularity by disturbing the perfect alignment of atoms. This defect materializes as a linear displacement of atoms, deviating from their usual ordered positions within the crystal lattice. The influence of dislocations on the mechanical properties and deformation behavior of materials is significant. Their dynamic presence and mobility play a crucial role in dictating how materials respond to external forces, exerting a profound impact on fundamental characteristics such as strength, ductility, and the overall structural integrity of the material.

Dislocations typically occur easily in metals while for hard brittle materials, dislocations do not readily take place. However, under specific extreme conditions such as high pressure or high temperature, dislocations occur even for hard brittle materials. As can be seen in Figure 8.3(b), the presence of dislocations in the subsurface of single-crystal silicon has been confirmed. Understanding dislocations is essential for revealing the complex mechanisms that govern the machining processes and plastic deformation of crystalline materials. The intricate interplay between dislocations and external stressors provides valuable insights into the material's ability to undergo shape changes without catastrophic failure.

### 8.2.5 Residual Stress

Residual stress is an internal stress that remains within a material even after external forces, temperature changes, or other mechanical processes have been applied and removed. These stresses are typically locked into the material's structure and can affect its mechanical properties and behavior. The factors that affect residual stresses during machining are diverse, and include, but are not limited to, material properties, tool geometries, cutting parameters, and cooling conditions.

For instance, in cutting 2024-T351 aluminum alloy, the residual stresses in the machined surface are compressive; the maximum residual stress increases with an increase in cutting speed, while there is little change in the residual stress distribution due to a variation in the rake angle [7]. During the cutting of AISI 4340 steel, surface tensile residual stresses were detected; the residual stresses tended to be more tensile as the feed increased and cutting speed was reduced, and an increase in tool nose radius resulted in more tensile surface residual stresses. Additionally, the use of coated tools led to more tensile surface residual stresses [8].

Understanding and effectively managing the residual stresses by machining processes are paramount in the realm of manufacturing. These stresses have the potential to significantly impact the performance and longevity of a material. They can influence critical mechanical attributes such as deflection of thin-walled parts, strength, fatigue resistance, and fracture toughness. Moreover, the presence of residual stresses may contribute to the initiation and propagation of cracks, potentially leading to premature failure of the material. Measurement of the amplitudes and distributions of residual stresses is an important aspect of subsurface damage evaluation.

## 8.3 SUBSURFACE DAMAGE EVALUATION METHODS

The characterization of the above-mentioned subsurface damages is an important issue in ultraprecision machining. It provides useful information for designing, planning, and optimizing the subsequent machining steps to save total production time and cost. For example, in the manufacturing of high-precision infrared optical lenses, particularly on brittle materials such as single-crystalline germanium and silicon, grinding is usually employed as a pre-machining process to quickly generate the initial rough lens figures at a high material removal rate, followed by fine processing with diamond turning [9, 10]. However, grinding induces significant subsurface damage, especially microcracks in the workpiece materials, which need to be removed during the subsequent turning process. Therefore, evaluating the maximum depth of subsurface cracks after grinding is essential to determine the required number of tool passes in diamond turning for complete subsurface damage elimination.

However, different from surface metrology which has been popularly used in academia and industry based on a series of standards regulated by the International Organization for Standardization (ISO), the history of subsurface metrology research is still short and there are no systematic methodologies and related international standards yet. In this chapter, some methods that have been explored and proved to be useful for subsurface metrology will be introduced.

### 8.3.1 Slanted Polishing

To make samples for cross-sectional observation of subsurface cracks and measurement of their depths, slanted polishing is usually used [11]. As shown in Figure 8.4(a), a sample (machined workpiece surface) is polished with a flat polishing pad slanted at a small angle θ (approximately 1°). Diamond paste is usually selected as the polishing abrasives. Subsequently, a microscope is employed to observe the polished slant plane and measure the width ($a$) of the transition area from crack to crack-free on the plane. Then, the microcrack depth ($d$) can be calculated from the value of $a$ and $\theta$. Figure 8.4(b) presents a microscope image of a germanium surface polished using the slanted polishing method. Grinding-indued microcracks are visible on the left side, while the right side shows a crack-free area. Using this method, it is easy to obtain a large cross-section area for detailed characterization. However, this method is a destructive method, and secondary subsurface damage may be generated during the polishing step.

### 8.3.2 Small Tool Polishing

Another method for measuring subsurface crack depth is small-tool polishing. In this method, a small rotating tool made of urethane is employed to polish the machined workpiece surface, creating a small depression. The crack depth is

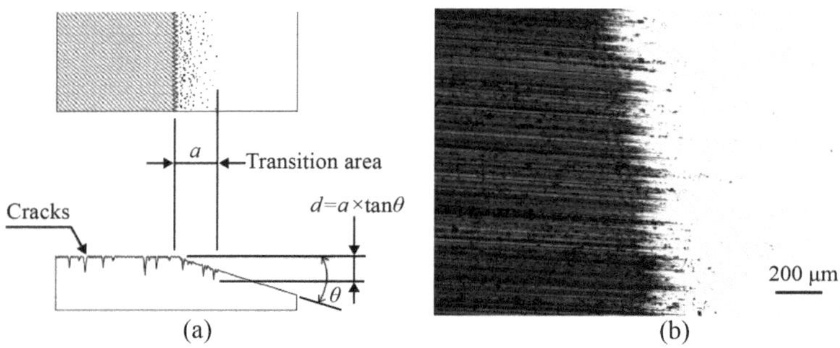


**FIGURE 8.4** (a) Schematic illustration of a sample made by slanted polishing and (b) microscope image of slanted polished germanium sample.

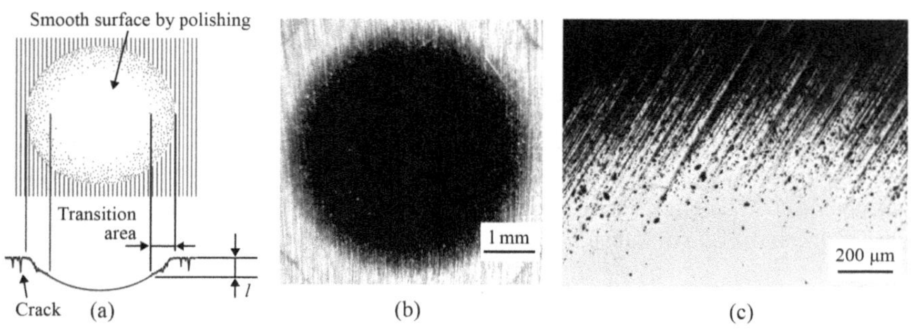


**FIGURE 8.5** (a) Schematic of a depression polished by the small tool, (b) microscope image of the depression surface polished by the small tool, and (c) close-up view of the boundary area between the ground area and crack-free area.

then evaluated by observing the cracks on the surface of the depression. The polished volume can be adjusted by varying the polishing time and pressure. Figure 8.5(a) shows a schematic of the polished surface. Similar to the slanted polishing method, the transition area between the ground area and the crack-free area can be observed using an optical microscope. The microcrack depth can be determined with a surface profilometer.

Figure 8.5(b) shows a microscope image of a shallow depression on a silicon surface ground by 20 μm diameter abrasives. The depression was generated by a small polishing tool, 6 mm in diameter, on a milling spindle. The tool was rotated at 1000 rpm, and a 100-gram polishing force was applied to the tool. Colloidal $SiO_2$ abrasives, with a diameter of 0.08 μm, were employed in the polishing slurry. The surface polished by the small polish tool has a crack-free area in the center. The smooth mirror surface region is shown as the black part in the figure. The cracks are shown as white spots in the transition area between the ground and crack-free areas. Figure 8.5(c) is an enlarged view of the transition area. In this figure, the smooth mirror surface region is shown as the white part, and cracks are shown as the black spots in the transition area. The inverse contrast against Figure 8.5(b) was caused by the difference in the sample tilting angle and the light reflection effects. The microcrack depth was determined by examining the locations where cracks occurred and then measured from the surface profile. Similar to slanted polishing, this method is a destructive method, and secondary subsurface damage may be generated during polishing.

### 8.3.3 Transmission Electron Microscope

Microscopic observation of polished cross-sectional surfaces enables the evaluation of depth of number of micro cracks. However, the magnification of microscopic observation is limited and the characterization of extremely small microcracks, dislocations, and grain refinement as well as phase transformation layers are difficult. For this purpose, transmission electron microscopy (TEM) characterization becomes necessary.

In TEM, a beam of electrons is transmitted through an ultra-thin specimen, providing detailed information about the sample's internal structure, including the arrangement of atoms and material morphology. To ensure optimal electron transmission, TEM specimens must be extremely thin, typically in the range of 50 to 100 nm. This thinness allows electrons to pass through the specimen with minimal scattering or absorption, enabling high-resolution imaging and detailed analysis of the sample's characteristics. Focused ion beam (FIB) is a typical technique used to prepare TEM specimens, which allows for precisely controlled milling of specimens at the nanoscale. This method involves using a focused beam of ions, usually gallium ions, to selectively remove material from a bulk sample.

Figure 8.6 shows an example of using FIB to prepare a TEM specimen of diamond tool edge. A single-crystal diamond tool was employed to cut silicon carbide, and SEM images of the tool before and after cutting are presented in Figure 8.6(a) and (b). Significant wear along the cutting edge is evident. To characterize

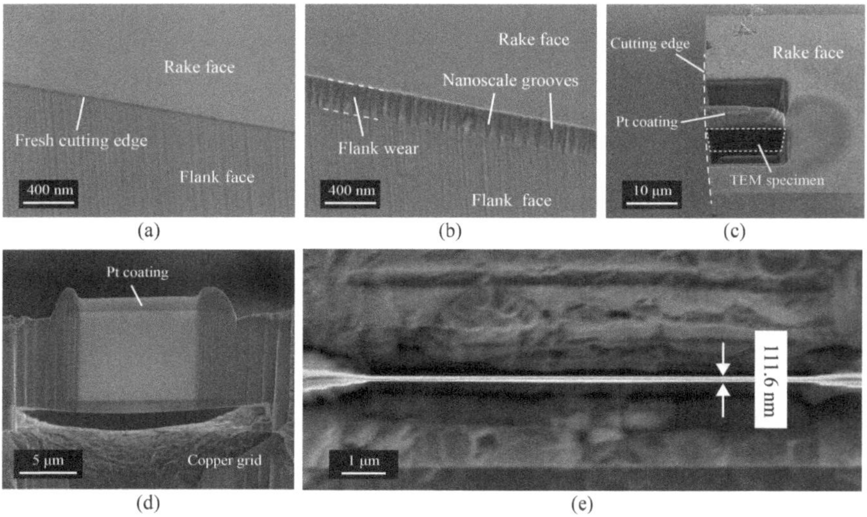


**FIGURE 8.6** SEM images of the cutting edge of the tool: (a) before, and (b) after machining SiC; (c) TEM specimen prepared by focused ion beam milling [12]; (d) a copper grid holding the TEM specimen; (e) the thickness of the TEM specimen.

the subsurface damage of the worn tool, a TEM specimen was cut out from the tool, as illustrated in Figure 8.6(c). To protect the sample from ion beam machining, platinum (Pt) was deposited on the area beneath which the TEM specimen was located while the undeposited area was removed by the ion beam. After extracting the FIB-cut specimen from the tool edge, a copper grid was used to hold it on, as shown in Figure 8.6(d). Further processing with FIB may be employed to reduce the specimen's thickness until it reaches below 100 nm for TEM observation, as shown in Figure 8.6(e).

Using the FIB sampling and TEM observation methods, cross-sectional characterization of cut surfaces of SiC was performed. Figure 8.7 shows examples of cross-sectional TEM images of the subsurface of 4H-SiC after diamond turning. Due to crystal anisotropy, three typical positions indicated in Figure 8.7(a) were characterized to examine the effect of anisotropy on subsurface damage. In the P location, cracks and dislocations were observed as subsurface damage, as shown in Figure 8.7(b). Similarly, in the S location, cracks and dislocations were observed as subsurface damage, but extended to a shallower depth compared to those in the P location, as shown in Figure 8.7(c). In the S location, cracks did not form; however, a thick subsurface damage layer, including a disordered layer and dislocations, was observed.

The FIB-TEM method for subsurface damage observation and metrology has nanoscale spatial resolution and thus can be used to identify dislocations, grain boundaries, phase transformations, and microcracks. By using selected area (electron) diffraction (SAED) equipped in TEM, the structures and orientations of

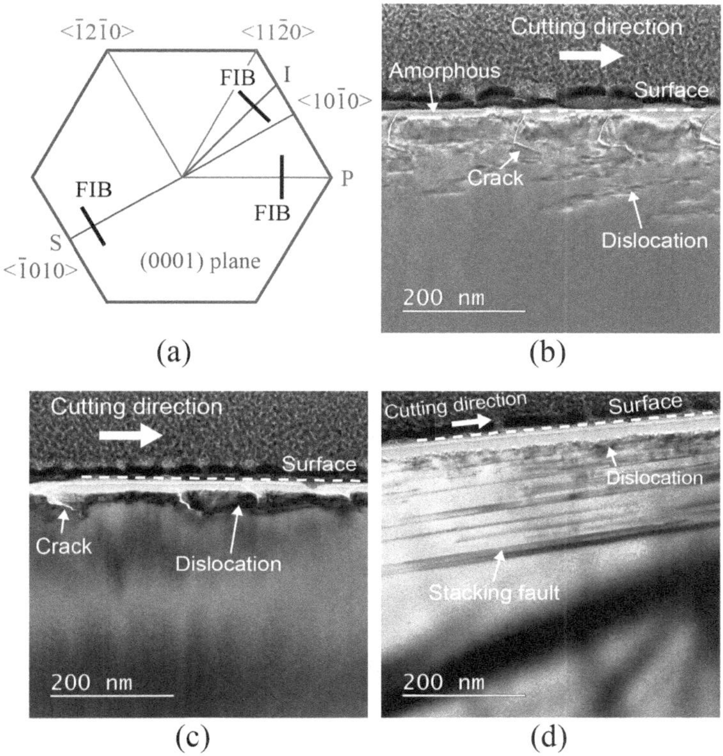


**FIGURE 8.7** (a) Schematic of FIB processing locations for TEM specimens. Cross-sectional TEM observations of the subsurface at (b) P location, (c) I location, and (d) S location in 4H-SiC [12].

crystalline lattices can be analyzed precisely, which provides a powerful tool for scientific research both for material and manufacturing researchers. However, this method is a destructive method, time-consuming, and very expensive. In some cases, slight secondary subsurface damage may be generated during sample preparation by the ion beam bombardment.

### 8.3.4 Nanoindentation Tests

Cross-sectional SEM and TEM methods provide visual information of subsurface damage and are very useful for understanding the subsurface damaging mechanism. However, these methods do not provide information of the mechanical property changes in the subsurface damage layer which is usually important for applications where mechanical contacts are involved. To solve this problem, micro/nanoindentation has been utilized to detect the existence of subsurface damage and analyze the mechanical properties of the subsurface damaged layer.

Subsurface damage resulting from machining varies across different materials. For example, in steels and alloys, it often manifests as white layers and plastic deformation layers [13]. In the cutting of high entropy alloys and ceramics, subsurface damage is characterized by grain refinement [14, 15]. When cutting hard-brittle crystals, such as Si and $Ga_2O_3$, the subsurface damage includes the formation of amorphous phase layers and dislocations [5, 16]. For soft-brittle crystals, such as $CaF_2$ and ZnSe, the subsurface damage primarily involves dislocations while no phase transformations occur [17, 18]. The analysis of load-displacement curves of nanoindentations made on material surfaces before and after machining allows us to determine how the mechanical properties of the damaged layers have changed compared to the original state and how these changes may impact subsequent machining processes or final product quality. From nanoindentation results, various mechanical properties, such as residual stress, hardness, elasticity, plasticity, and fracture toughness of the surface layer can be directly or indirectly measured.

Two typical examples are shown here to demonstrate how to analyze the load-displacement curves of micro/nanoindentation for evaluating the residual stress and hardness of a machined surface. Figure 8.8(a) presents a cross-sectional image of a turned surface of AISI 52100 steel, displaying a subsurface damage layer, i.e., a white layer, with a thickness of over 50 μm. Figure 8.8(b) presents load-displacement curves for the indents performed on the sample cross-section with different distances from the machined surface. At the near-surface region (~2 μm), the slope of the curve at initial yielding point (corresponding to indentation depth of about 100 nm) is increased, indicating that residual stress in this region is compressive; while at the depth of 51 μm, the slope of the curve at initial yielding point is decreased, indicating that residual stress turns to be tensile. It should be noted that the residual stress in the surface layer only affects the slope of the load-displacement curve at the onset of yielding, and the further plastic deformation induced by increasing the load does not provide any information about the residual stress. Also, it is not currently possible to measure residual stresses quantitatively by nanoindentation.

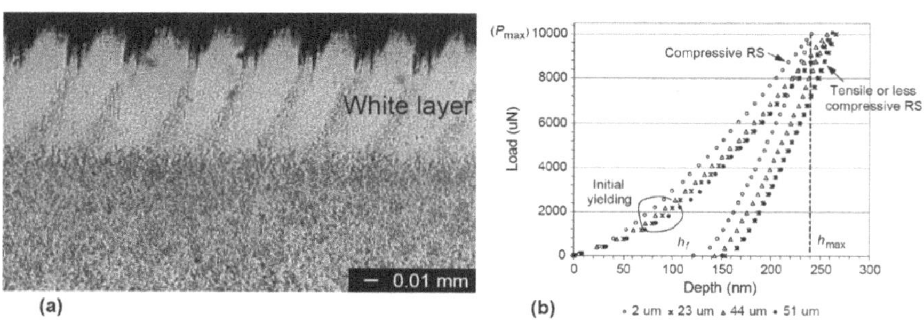


**FIGURE 8.8** (a) Cross-sectional image of a white layer in a turned surface of AISI 52100 steel and (b) the load-displacement curves for the indents performed on the sample cross-section [19].

Figure 8.9 presents load-displacement curves for nanoindentation tests on a chemo-mechanical polished silicon sample (pristine silicon) and diamond-turned silicon sample (machined silicon). It can be seen that the maximum indentation load for both samples is 0.6 mN, but a larger displacement is obtained when performing indentation on the machined silicon. This indicates the hardness of the machined surface is decreased due to the formation of subsurface damage, i.e., an amorphous silicon layer. When performing nanoindentation on pristine silicon, only elbow appears on the unloading part of the curve, as shown in Figure 8.9(a), suggesting a transformation from diamond cubic structure to amorphous. However, for the machined silicon under the same conditions, a pop-in occurs during loading, meanwhile a pop-out occurs during unloading, as shown in Figure 8.9(b). The pop-in and pop-out occurring in the indentation of machined silicon might be caused by a further phase transformation of the amorphous silicon. The hardness of pristine silicon and machined silicon under varied indentation loads are summarized in Figure 8.9(c). In general, as the indentation load increases, although the hardness of both samples is decreased due to the indentation size effect, the hardness of machined silicon gradually aligns with that of pristine silicon.

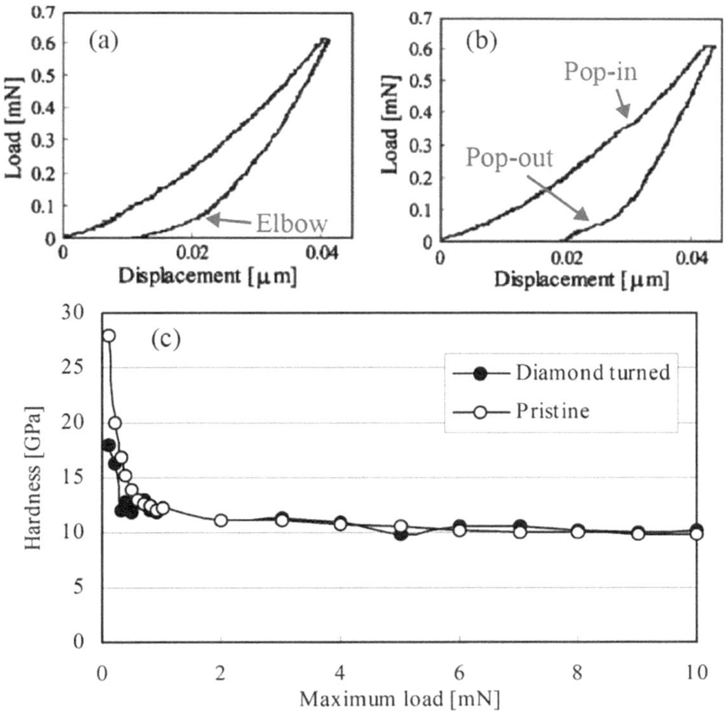


**FIGURE 8.9** Typical load-displacement curves for nanoindentation tests on (a) chemo-mechanical polished silicon wafer, and (b) diamond-turned silicon wafer. (c) Comparison of measured hardness between (a) and (b) [20].

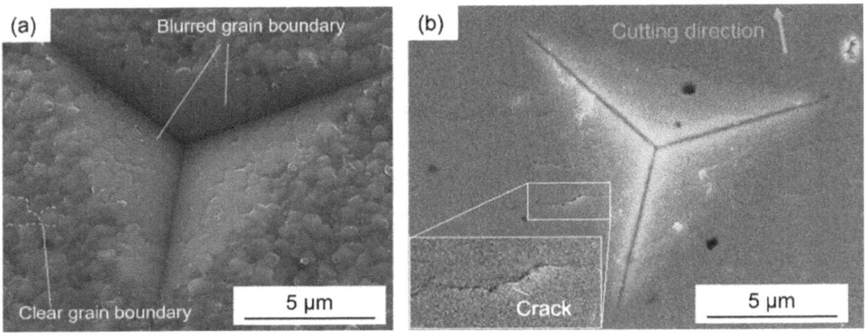


**FIGURE 8.10** SEM images of indents performed on (a) unmachined and (b) machined surface of zirconia poly crystal [2].

Therefore, by conducting a series of indentation tests on both pristine and machined samples with progressively large indentation loads, the thickness of subsurface damage layer can be estimated.

Another example is the nanoindentation on grain-refinement layer produced in diamond turning of zirconia. Figure 8.10(a) and (b) present the surface SEM morphologies of the residual indents on the unmachined and machined zirconia poly crystal, respectively. For the residual indent on the unmachined surface, the three planes of the indent are smoother than the raw surface due to the pressing of the indenter, but no brittle fractures are observed. On the raw material surface around the indent, grain boundaries are clearly seen, while the grain boundaries become blurred inside the indent. These results suggest significant plastic deformation of the crystal grains due to the pressing of the indenter. In contrast, for the residual indent on the machined surface, although the three planes of the indent are smooth without any fractures, cracks are generated around the indent. Moreover, it is found that the cracks are generated along the direction perpendicular to the cutting direction. This may imply that a residual tensile stress formed along the cutting direction in the workpiece subsurface layer during diamond turning, which consequently led to a decreased fracture toughness for this direction.

As nanoindentation is performed in an extremely small surface area, it is a nearly nondestructive method applicable to all materials. In addition, it is a fast and low-cost method without the necessity of pre-processing the samples. The measurement points can be flexibly designated and controlled by a computer program, enabling mapping measurement of hardness distribution within a certain surface area.

### 8.3.5 Chemical Etching

Chemical etching involves using a suitable etchant to open subsurface cracks so that they can be observed by microscopy. In other words, etching a machined surface can convert "subsurface" metrology into "surface" metrology. By characterizing the surface topography after different etching times, the depth of subsurface damage can be measured.

For example, Figure 8.11 presents a series of optical micrographs and corresponding surface profilometry scans captured at the same location following various etching times for fused silica surfaces ground with two types of abrasives. In the early stages of etching, a series of individual cusps formed, which are created when silica is etched from the sides and bottom of each fracture site. This allows the probe tip to detect the opened fractures, resulting in a high-frequency fluctuated surface profile. As etching time increases, individual cusps coalesce with one another resulting in the formation of successively larger cusps. Meanwhile, the number of cusps decreases, leading to a reduction in the frequency of the surface profile fluctuation.

Figure 8.12(a) plots the average cusp diameter scales with the square root of etching time. The cusp coalescence increases both the spatial period of the surface roughness and removes subsurface damage. The removal of the damaged layer stabilizes surface roughness, facilitating an accurate calculation of the depth of subsurface damage. The subsurface damage depth distributions, that is cumulative crack obscuration as a function of depth, for the two ground samples are depicted in Figure 8.12(b). The maximum subsurface damage depth was found to be ~42 μm for the two ground samples.

Etch rate is another factor that can be used to evaluate the depth of subsurface damage This is realized by identifying the inflection point in the etching rate between the damaged layer and the bulk. Figure 8.13 shows the etch rate as a function of etch depth for the InAs samples polished by different methods. In general, for both samples, the overall etch rate fitting curve exhibits a rapid downward trend, which is caused by the presence of subsurface damage in the near-surface region of the samples. As the etch depth increases, the etch rate gradually decreases and eventually fluctuates around 2 μm/min. The initial etch rate in Figure 8.13(b) is much lower than that in Figure 8.13(a), indicating that the subsurface damage of the InAs sample polished by the improved method is reduced compared to that of the conventional method. During the etching process, the subsurface damage is progressively removed, and the etch rate decreases. When the bulk material begins to be etched, the etch rate stabilizes at a constant value. At this stage, the thickness removed by chemical solutions can be roughly estimated as the depth of the subsurface damaged layer. It is noteworthy that, due to the complexity of the chemical reaction, the etch rate exhibits fluctuations rather than decreasing monotonically. The etch rate is influenced primarily by endothermic and exothermic reactions, as well as variations in the concentration of the etchant over time.

To quantitatively characterize the variation of subsurface damage along the depth direction, techniques that combine etching with scattering spectroscopy are effective. For example, etching together with Raman spectroscopy has been used to analyze the microstructural changes of the subsurface region of silicon in the depth direction [23]. In the silicon etching process, a linear fit reveals a well-established correlation between etching depth and etching time, depicted in Figure 8.14(a).

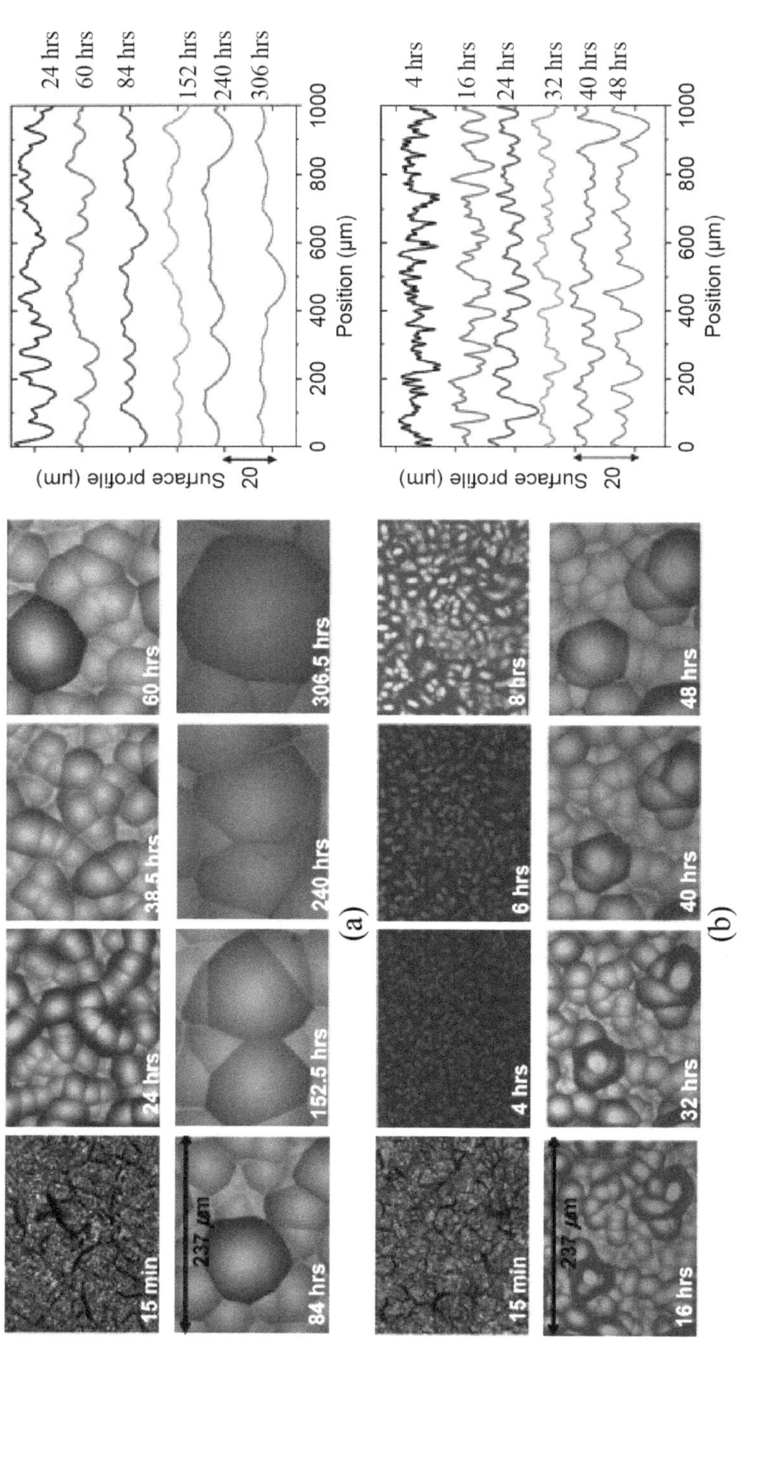


**FIGURE 8.11** Optical micrographs and surface profilometry of fused silica surfaces ground with (a) 150 grit fixed abrasives and (b) 30 μm loose abrasives, after different etch times [21].

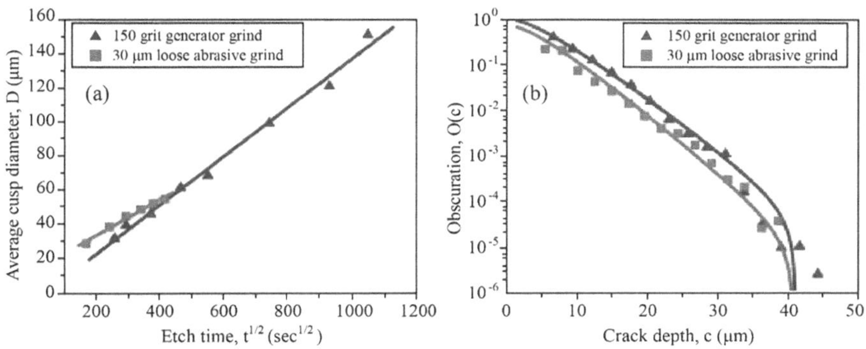


**FIGURE 8.12** (a) Measured cusp diameter versus etching time$^{1/2}$, (b) subsurface damage depth distribution in the ground fused silica surfaces [21].

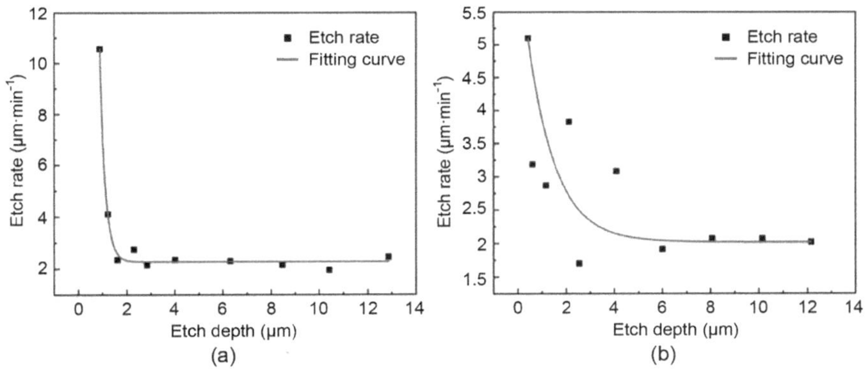


**FIGURE 8.13** Etch rate as a function of etch depth for the InAs wafers obtained through (a) conventional and (b) improved polishing methods [22].

The slope of the fitted line corresponds to an etching rate of 0.36 mm/min. Subsequently, the Raman spectra after different etching times were characterized, as shown in Figure 8.14(b). It is evident that the intensities (I) in the shoulder region (450–570 cm$^{-1}$) of the most intense band (519 cm$^{-1}$) and the second band (300 cm$^{-1}$) decrease as the etching time increases, finally become stable after 10 minutes of etching. In this case, the depth of subsurface damage is estimated to be about 3.6 μm according to Figure 8.14(a).

As another example, glow discharge optical emission spectroscopy (GDOES), a technique that combines a glow discharge (GD) with an optical emission spectrometer (OES), has been employed to rapidly determine the profiles of bulk elemental composition in solid material layers as a function of subsurface depth [24]. The plasma generated by GD is used to etch the material layer by layer, while the OES is applied for analyzing the material elements. Figure 8.15 shows the variation in the quantity of chemical components with depth from a laser-processed

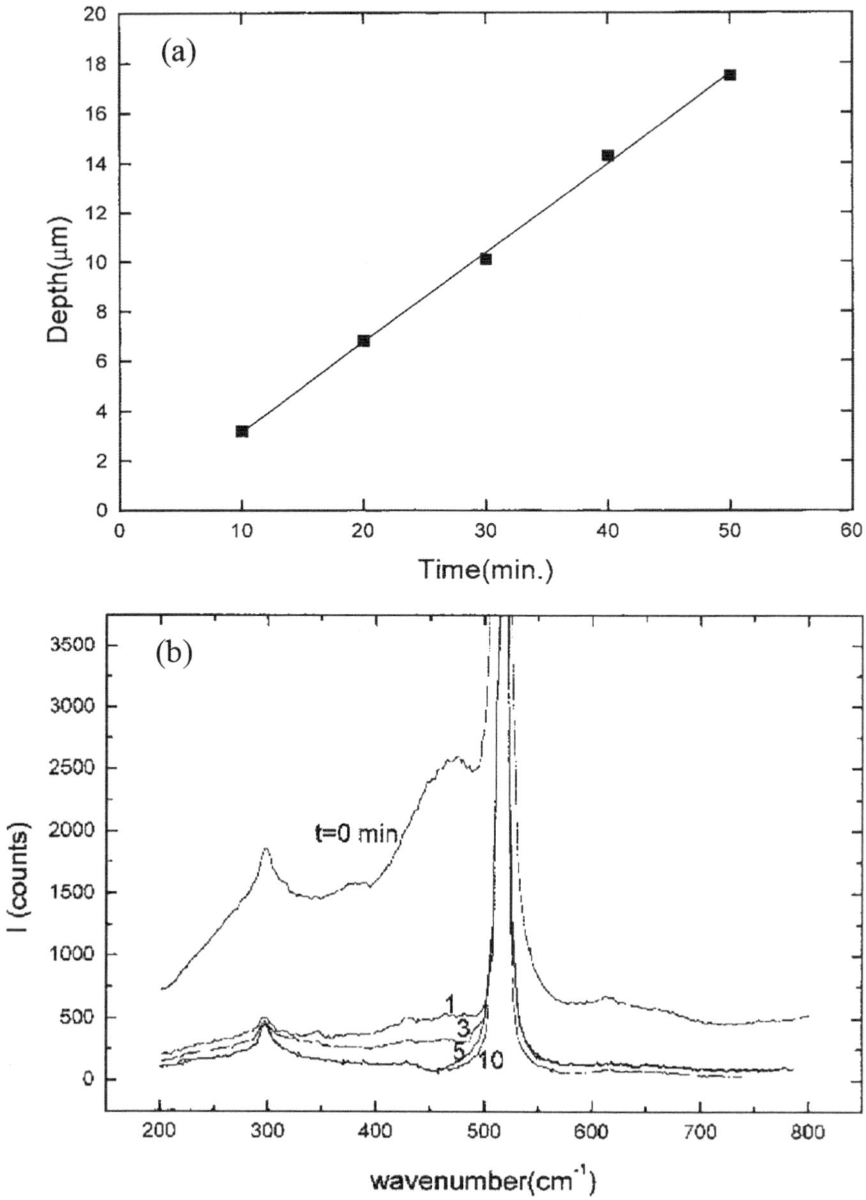


**FIGURE 8.14** Etching and scattering spectroscopy analysis of silicon: (a) etching depth as a function of etching time and (b) Raman spectra of surface after various etching times [23].

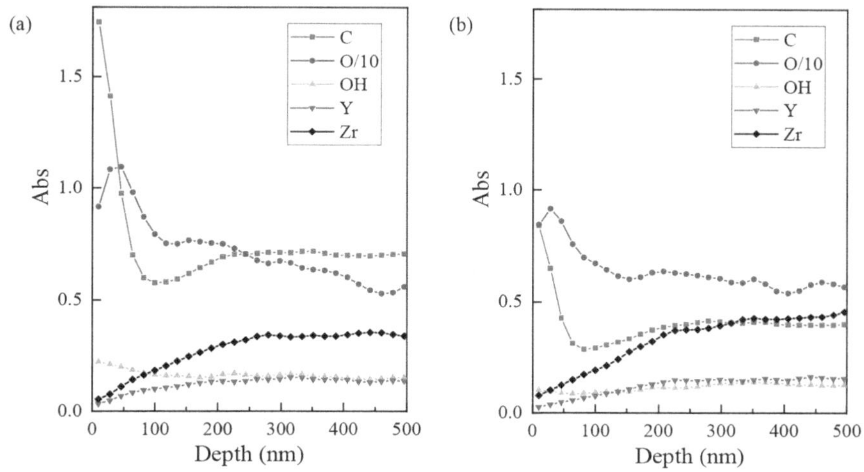


**FIGURE 8.15** GDOES analysis of yttria-stabilized zirconia: elemental distribution in subsurface (a) before and (b) after laser irradiation [24].

surface of yttria-stabilized zirconia (YSZ). After laser irradiation, a noticeable reduction in carbon (C) content on the top layer (~50 nm deep) indicates the removal of organic matter attached to the material's surface. This outcome suggests the occurrence of a surface-cleaning effect resulting from the laser treatment. In contrast, no notable alterations were observed in the levels of other components, including zirconium (Zr), yttrium (Y), and oxygen (O), indicating femtosecond laser irradiation had minimal oxidative effects on the YSZ surface.

Subsurface metrology based on chemical etching is nondestructive and can be used for evaluating a large surface area at once. However, chemical etching is not suitable for all materials. The total time for etching may be very long, even a few tens of hours. The estimation of the depth of subsurface damage is sometimes inaccurate.

### 8.3.6 X-Ray Diffraction

The aforementioned subsurface damage evaluation methods involve modification (polishing, etching, cutting, coating, indenting, etc.) of the sample surface to acquire information about subsurface damage, thus can be classified as destructive methods. For assessing subsurface damage without affecting the further use of the specimen, nondestructive methods are highly valuable.

X-ray diffraction (XRD) is a nondestructive technique for characterizing the crystallographic structure of a machined workpiece. It works by irradiating a material with incident X-rays and then measuring the intensities and scattering angles of the X-rays that leave the material, as shown in Figure 8.16. Crystals consist of regular arrays of atoms, and X-rays can be viewed as waves of

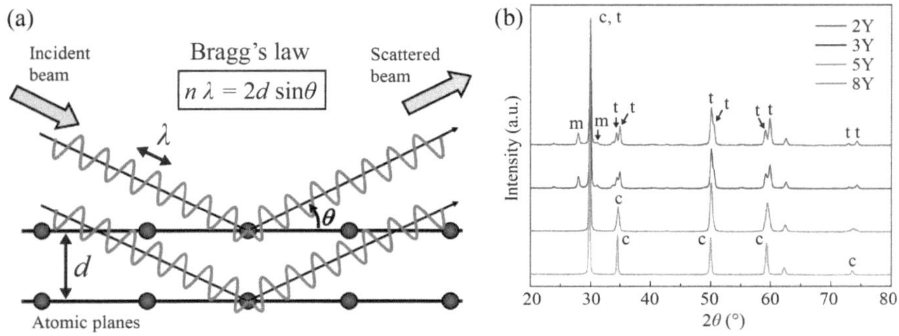


**FIGURE 8.16** (a) Principle of X-ray diffraction and (b) XRD spectra of yttria-stabilized zirconia containing different phases [25].

electromagnetic radiation. The incident X-rays are scattered by the atoms, primarily through interaction with the atoms' electrons. This scattering phenomenon is known as elastic scattering, with the electron referred to as the scatterer. A regular array of scatterers produces a regular array of spherical waves. In most directions, these waves cancel each other out through destructive interference; however, they add constructively in a few specific directions, as determined by Bragg's law. In Figure 8.16(a), $d$ is the spacing between diffracting planes, $\theta$ is the incident angle, $n$ is an integer, and $\lambda$ is the beam wavelength. The specific directions appear as spots on the diffraction pattern called reflections. Consequently, X-ray diffraction patterns result from electromagnetic waves impinging on a regular array of scatterers.

As an example, Figure 8.16(b) plots XRD spectra of yttria-stabilized zirconia (YSZ) with varying yttria dopant concentrations. In these XRD spectra, the presence of distinct peaks at various scattering angles indicates that YSZ is composed of different phases with various crystal structures, including tetragonal (t), monoclinic (m), and cubic (c) phases, and these phases vary with changes in the yttria dopant concentration. By analyzing the XRD spectra, machining-induced subsurface phase transformations in zirconia can be easily detected without the necessity of modifying the workpiece.

Moreover, X-ray diffraction can determine elastic strains and residual stress in a workpiece by measuring interplanar atomic spacing ($d$) in machined surfaces when subjected to applied or internal stress, as illustrated in Figure 8.17. Based on this principle, there have been serval studies of characterizing residual stress with XRD [26]. XRD is also applicable to grain size analysis for powder materials and is effective for characterizing the grain refinement layers in polycrystal material machining.

It should be pointed out that the penetration depth of X-ray into materials is highly dependent on the incident angle $\theta$. For thin film analysis and machining-induced subsurface damage layer analysis, the incident angle should be set to an extremely small angle (<1°) at a very high angular resolution (0.01°). The small-angle XRD

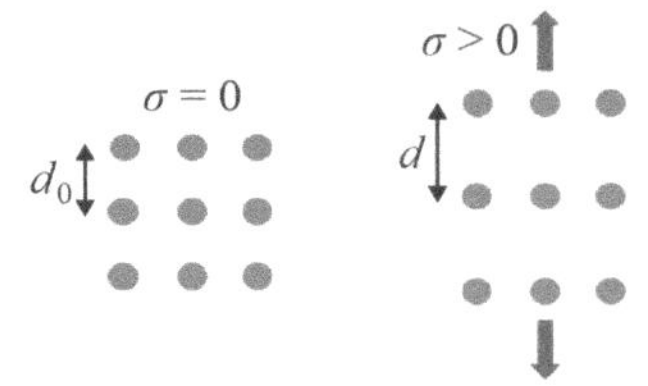


**FIGURE 8.17** Schematic of the principle of XRD measurement for residual stress.

enables an analysis depth from a few tens of nanometers to a few tens of micrometers, depending on the material to be analyzed. For a subsurface damage layer thinner than the minimum analysis depth, it will be difficult to discriminate the scattering from the subsurface damage layer and that from the bulk region.

## 8.4 LASER MICRO-RAMAN SPECTROSCOPY

The depth and structure of subsurface damage undergo significant changes with machining conditions, including the depth of cut, tool geometry, and crystal orientations of the workpiece. Predicting the relationship between subsurface damage depth and machining conditions remains challenging. Therefore, there is an increasing demand for rapid and cost-effective nondestructive measurement of subsurface damage depth, enabling a comprehensive evaluation of subsurface damage across a large number of instances. Laser micro-Raman spectroscopy is recognized as a powerful nondestructive characterization technique for subsurface damage layers in machining of various semiconductor and insulator materials.

### 8.4.1 Principle of Raman Scattering

The Raman effect is based on the inelastic scattering of laser [27]. Figure 8.18 schematically shows the principle of laser scattering. Scattering occurs when a beam of monochromatic light passes through a crystal, and the scattering involves Rayleigh scattering and Raman scattering. The Rayleigh scattering is caused by the elastic collision between the incident photons with the phonons in materials, thus generating light with the same vibration frequency as the incident ray. On the other hand, the Raman scattering is induced by the inelastic collision between the incident photons with the phonons in materials, hence produces light with different vibration frequencies. In the Raman scattering, the scattered light with vibration frequency $\nu_1 - \nu_R$ is termed stroke scattering, and those $\nu_1 + \nu_R$ is called anti-strike scattering, where $\nu_R$ ($\nu_R > 0$) is the Raman shift. The Raman shift is strongly influenced by microscopic structural changes, impurity, and residual strains, which lead to changes in phonon frequencies, broadening of Raman peaks, and breakdown of Raman selection rules.

For example, laser micro-Raman spectroscopy has been widely used to detect the structural change and residual stress of silicon in scratching, dicing, lapping,

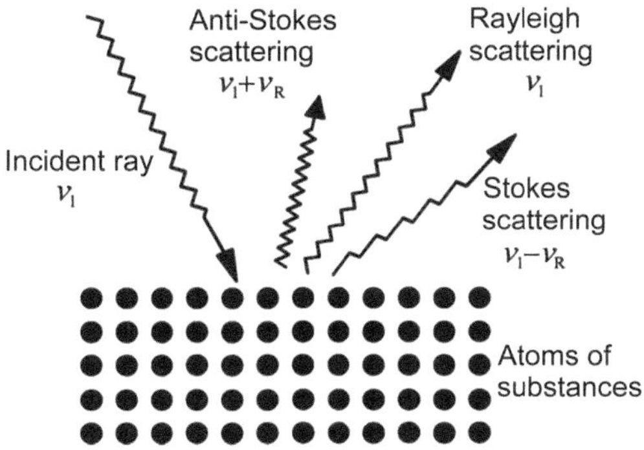


**FIGURE 8.18** Schematic of laser scattering.

grinding, and diamond-turning processes [28, 29]. The spectra of different silicon phases exhibit distinct differences. For bulk crystalline silicon (c-Si), the triple degenerate optical phonons display in the first-order Raman spectrum a sharp peak at the Raman shift of 521 $cm^{-1}$, and for amorphous silicon (a-Si), the first-order Raman spectrum reflects the phonon density of states and presents an optical band peak at 470 $cm^{-1}$ [30].

### 8.4.2 Raman Spectroscopy of Subsurface Damage

Figure 8.19 shows three-dimensional topographies of the surfaces diamond-turned at various undeformed chip thicknesses. At an undeformed chip thickness of 20 nm, as shown in Figure 8.19(a), the machined surface is extremely smooth, showing no evidence of damage across the entire surface. The surface roughness was on the level of a few nanometers, which indicates that the surface has been produced through a completely ductile regime material removal. Figure 8.19(b) shows the surface machined at an undeformed chip thickness of 200 nm. The surface consists of regular microgrooves formed by periodical cross-tool feeds. Besides the saw-toothed tool feed profiles, a few extremely small microcracks, approximately 1 μm in size, can be observed. This kind of surface topography indicates that the machining mode has gone beyond the crack initiation threshold and transitioned to a brittle regime. When the undeformed chip thickness is further increased to 800 nm, as shown in Figure 8.19(c), the resulting surface is far from smooth and is pitted with microfractures, the sizes of which range on the order of 1–10 μm. This kind of surface indicates that brittle fracture has been predominant in the material removal process.

To investigate subsurface damage in diamond-turned silicon, surfaces machined at undeformed chip thicknesses of 20, 50, 200, and 800 nm were characterized using a laser micro-Raman system. These Raman spectra were recorded over a range of 200–600 $cm^{-1}$ to cover the characteristic peaks associated with crystalline and amorphous silicon, as shown in Figure 8.20. In the Raman spectrum of

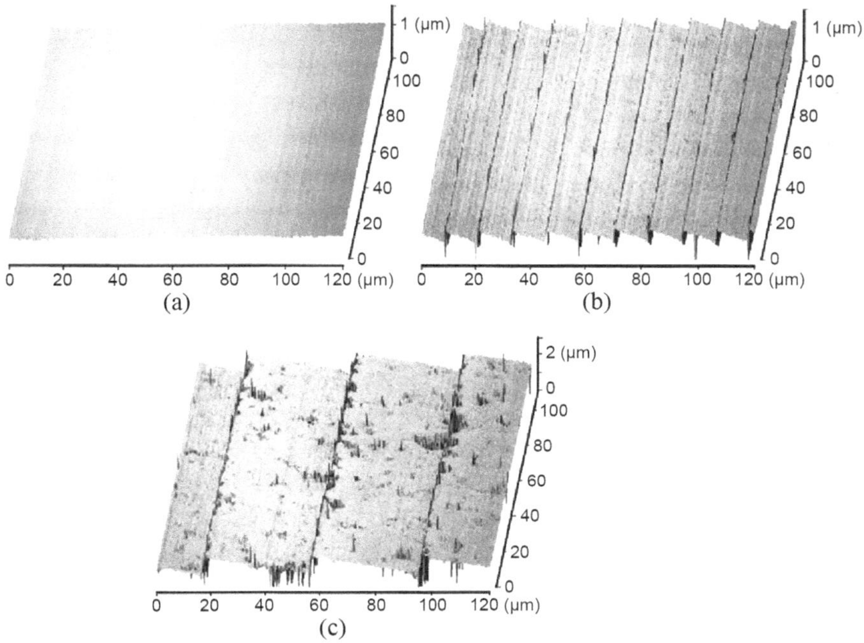


**FIGURE 8.19** Scanning laser microscopic topography of the silicon surfaces machined at an undeformed chip thickness of (a) 20 nm, (b) 200 nm, and (c) 800 nm [28].

the surface machined at an undeformed chip thickness of 20 nm, as shown in Figure 8.20 (a), there is a characteristic Raman peak of c-Si at 521 $cm^{-1}$, whereas the intensity at other frequencies is negligibly low. This indicates that the subsurface layer is mainly crystalline. As the undeformed chip thickness increased to 50 nm, as shown in Figure 8.20(b), the intensity of the Raman peak of c-Si at 521 $cm^{-1}$ clearly became lower than that in Figure 8.20(a). Additionally, a significant broadband peak centered at 470 $cm^{-1}$ appears, indicating that the subsurface layer has been partially transformed into an amorphous state. Figure 8.20(c) shows the Raman spectrum of the surface machined at an undeformed chip thickness of 200 nm. The broadband peak at 470 $cm^{-1}$ becomes more significant, whereas no peak can be observed at 521 $cm^{-1}$. The absence of a feature at 521 $cm^{-1}$ indicates the lack of crystalline silicon and suggests that the subsurface layer within the laser penetration depth has been entirely transformed into an amorphous state. However, as the undeformed chip thickness was increased to more than 300 nm, the measurement results of Raman spectra began to exhibit distinct differences depending on the location of the test points, specifically in the nonfractured or fractured areas. For instance, Figure 8.20(d) shows the Raman spectrum of the nonfractured areas on the surface machined with an undeformed chip thickness of 800 nm. Both a small peak at 521 $cm^{-1}$ and a broadband peak at 470 $cm^{-1}$ can be observed. However, when the laser spot was positioned within the micro-fractures, only

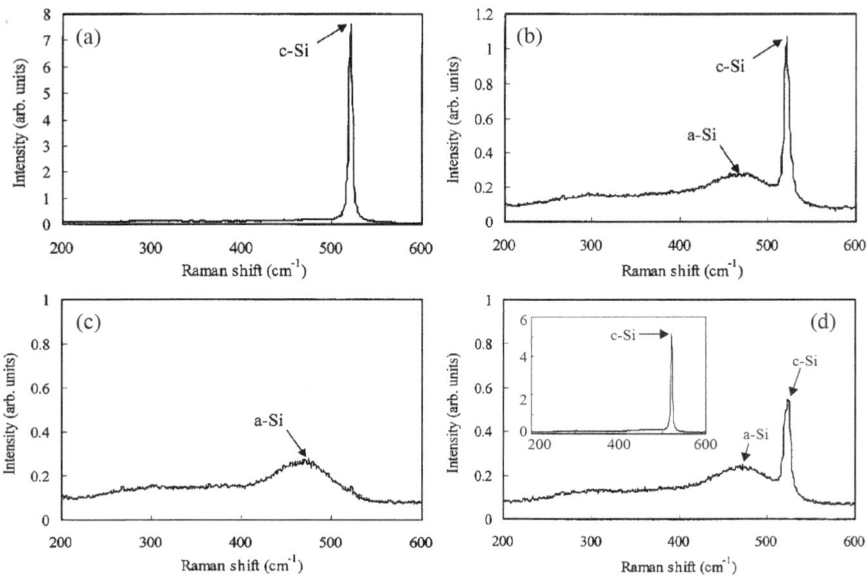


**FIGURE 8.20** Laser micro-Raman spectra of silicon surfaces machined at an undeformed chip thickness of (a) 20 nm, (b) 50 nm, (c) 200 nm, and (d) 800 nm [28].

peaks at 521 $cm^{-1}$ appeared, with no peak observed at 470 $cm^{-1}$, as shown in the inset of Figure 8.20(d). This indicates the absence of the amorphous phase.

### 8.4.3 Quantitative Evaluation of Subsurface Damage Depth

The Raman intensity can be utilized to quantitatively assess the depth of subsurface damage (amorphous layer) in the silicon [31]. This is achieved by comparing the intensities of Raman scattering from the amorphous region with that from the crystalline region, denoted as the Raman intensity ratio ($r$). The ratio $r$ can be expressed as follows:

$$r = \frac{I_a}{I_c} \tag{8.1}$$

where $I_a$ is the total Raman intensity of the amorphous silicon, and $I_c$ is the total Raman intensity of the crystalline silicon. It can be seen that the bigger the value of $r$, the thicker the amorphous layer. The two total Raman intensities, $I_a$ and $I_c$, mean the integrations of the amorphous and crystalline Raman peaks with respect to the Raman shift. Compared to the maximum peak value, the total Raman intensity can precisely represent the laser scattering at different vibration frequencies.

On most occasions, the Raman spectra of a machined surface contain two typical peaks: a sharp crystalline peak at 521 $cm^{-1}$ and a broadband amorphous peak around 470 $cm^{-1}$. These two peaks overlap each other, as schematically shown in Figure 8.21(a). Due to this overlap, it is difficult to directly obtain the integrations of Raman intensities for individual peaks from the raw Raman spectra. Therefore, peak separation processing becomes necessary.

Curve fitting is an effective method for separating overlapping peaks. Gaussian distribution and Lorentzian distribution are two common functions used for fitting light spectra in the optical domain. Generally, Gaussian distribution is suitable for fitting broadband curves, and Lorentzian distribution is better for fitting sharp peaks. In this work, we used Gaussian distribution for fitting amorphous peaks and Lorentzian distribution for fitting crystalline peaks, respectively. The functions of Gaussian distributions $f_{Ga}$ and Lorentzian distribution $f_{Lc}$ with respect to Raman shift $v$ can be expressed as follows:

$$f_{Ga}(v) = h_a \cdot \exp\left\{-\frac{(v - v_a)^2}{b_a^2}\right\} \tag{8.2}$$

$$f_{Lc}(v) = \frac{h_c}{\left\{1 + \frac{(v - v_c)^2}{b_c^2}\right\}} \tag{8.3}$$

where $v_a$ and $v_c$ are the Raman shifts, $h_a$ and $h_c$ are the heights, $b_a$ and $b_c$ are the half-power bandwidths of the amorphous and crystalline peaks, respectively. Therefore, the Raman intensity ratio $r$ in Eq. (8.1) can be obtained by:

$$r = \frac{\int f_{Ga}(v)dv}{\int f_{Lc}(v)dv} = \frac{h_a b_a \sqrt{\pi}}{h_c b_c \pi} = \frac{1}{\sqrt{\pi}} \frac{h_a b_a}{h_c b_c} \tag{8.4}$$

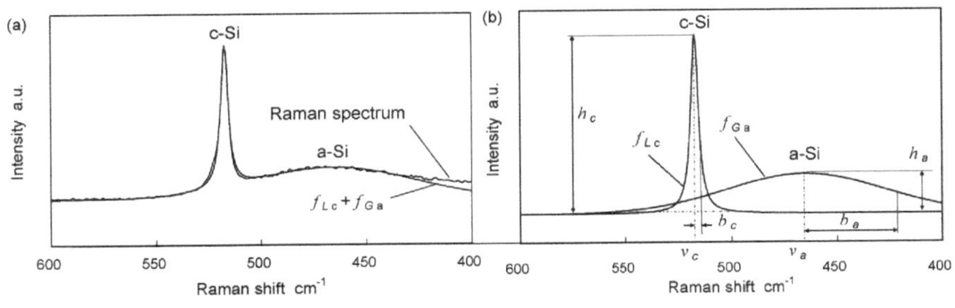


**FIGURE 8.21** Curve fitting method for separating Raman peaks: (a) comparison of experimental Raman spectra and fitted curve; (b) separated Raman peaks of amorphous and crystalline silicon [31].

To perform curve fitting of the Raman peaks automatically, a computer program based on the least squares method was developed in Ref. [31]. Using this program, the synthetic curve $f_{Ga} + f_{Lc}$, namely the sum of the amorphous peak and the crystalline peak, can be fitted to coincide with the experimental Raman spectra in a few tens of seconds. It should be pointed out that the baseline of Raman spectra, where the theoretical Raman intensity is zero, sometimes goes up and down due to the background noise or drift of the Raman system. To avoid the drifts and noises, we at first offset the baseline of the Raman spectra to zero before curve fitting.

An example of curve fitting result is given in Figure 8.21(a) together with the experimental Raman spectra. It can be seen that the fitted curve $f_{Ga} + f_{Lc}$ is generally consistent with the experimental Raman spectra and the fitting error is ignorable. Then the curves $f_{Ga}$ and $f_{Lc}$ are separated and the parameters $h_a$, $h_c$, $b_a$, and $b_c$ can be simply read from the individual curves of $f_{Ga}$ and $f_{Lc}$, as shown in Figure 8.21(b). Finally, the Raman intensity ratio $r$ can be calculated easily according to Eq. (8.4).

As an example, the relationship between the Raman intensity ratio and the thickness of the amorphous layer was examined experimentally. Figure 8.22(a) is the Raman spectrum of a surface plunge-cut at a depth of cut of 60 nm. There are two characteristic peaks: a sharp peak at 521 cm$^{-1}$ and a broadband peak centered at 470 cm$^{-1}$. This is a typical Raman spectrum for a diamond-machined silicon surface, indicating that the near-surface layer has been transformed into the amorphous phase. In this case, the amorphous layer is relatively thin so the Raman peak of crystalline is significant compared to the amorphous peak. Figure 8.22(b) is the Raman spectrum of the surface plunge-cut at a depth of cut of 120 nm. The broadband peak at 470 cm$^{-1}$ has grown significantly compared to that in Figure 8.22(a), whereas the peak at 521 cm$^{-1}$ becomes very small. The lack of features at 521 cm$^{-1}$ is indicative of the increase in the thickness of the amorphous layer. In other words, the Raman-scattered light from the deep bulk crystalline region becomes very weak.

Raman intensity ratios were calculated using the aforementioned method and plotted against the depth of amorphous layers measured by TEM, as shown in Figure 8.23. Since the Raman intensity ratio varied over a broad range, a logarithmic scale was used for the vertical axis in the graph. In the figure, despite the limited number of data points, a clear correlation can be identified between the Raman intensity ratio and the depth of the amorphous layer. The correlation is approximately linear within the experimental range. Therefore, by utilizing this linear relationship, it is possible to quantitatively measure the depth of subsurface amorphous layers.

It must be pointed out that the Raman intensity ratio is meaningful only when both the crystalline peak and the amorphous peak simultaneously appear in the Raman spectra. If the amorphous layer is too thin to generate detectable Raman scattering around 470 cm$^{-1}$ or too thick to allow the Raman-scattered light from the bulk crystalline region (521 cm$^{-1}$) to be detected from the surface, there will be only one visible peak in the Raman spectra; thus, the Raman intensity ratio cannot be calculated. In other words, there exist an upper sensing limit and a lower sensing limit in the depth range of the amorphous layer.

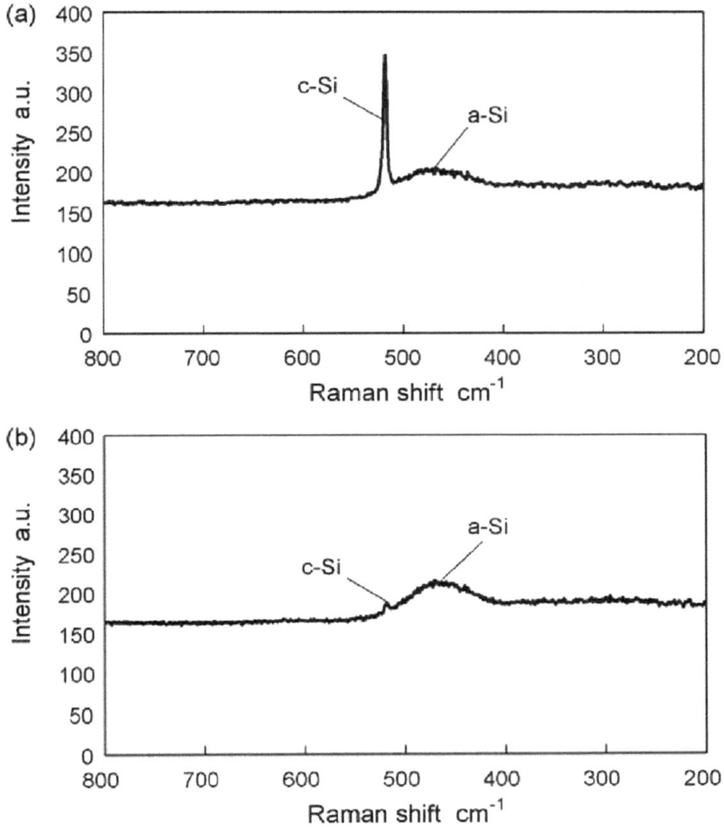


**FIGURE 8.22** Laser micro-Raman spectra of surface regions machined at various depths of cut: (a) 60 nm and (b) 120 nm [31].

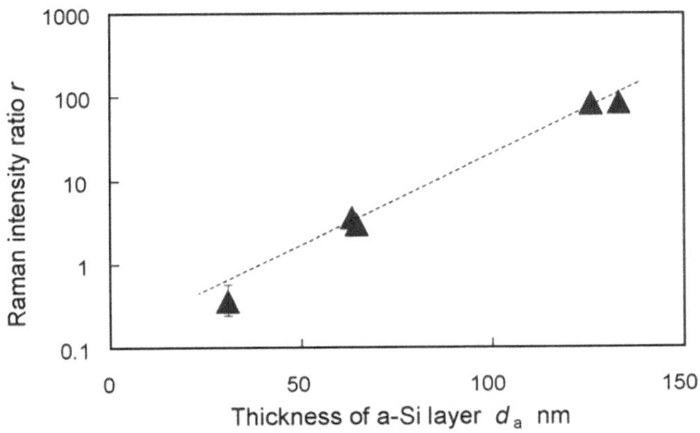


**FIGURE 8.23** Plot of Raman intensity ratio versus amorphous layer thickness [31].

To determine the sensing limits of amorphous layer depth, Raman test points were systematically moved along the plunge-cut microgroove. It was observed that the upper and lower sensing limits were approximately 10 nm and 150 nm, respectively, assuming that the thickness of the amorphous layer is proportional to the depth of cut in plunge cutting. Theoretical analysis revealed that the relationship between the Raman intensity ratio and the amorphous layer thickness approached linearity within the range from 20 to 140 nm, and beyond this range, nonlinearity occurred. Fortunately, the range of 20 to 140 nm covers the majority of machining-induced amorphous layer depths reported in the literature [32–37]. Therefore, measurements based on the linear correlation are applicable for surfaces produced by various precision machining processes, including single-point machining and multipoint abrasive machining. The following equation can quite well describe the relationship between the Raman intensity ratio ($r$) and the thickness of the amorphous layer ($d_a$).

$$d_a = 33.3 \times \ln\left\{\frac{8.84 + 15r}{8.84 + 0.167r}\right\} \tag{8.5}$$

Using Eq. (8.5), the thickness of the amorphous layer can be simply calculated from the Raman intensity ratio.

Laser micro-Raman spectroscopy can be used for quantitative measurement of subsurface damage depth of silicon. This measurement method is very quick, inexpensive, and completely nondestructive. Other advantages of laser Raman measurement include that it can be done at room temperature, in the air, and without contacts. Furthermore, the measurement of subsurface damage depth can be combined with the residual stress measurement by using a single Raman system to detect the shift of the Raman peaks. If a compressive stress is exerted on the crystal lattice, the lattice constant decreases and the Si peak shifts toward higher wavenumbers, and vice versa.

### 8.4.4 Raman Mapping Measurement

Raman spectroscopy can also be used for on-machine measurement and mapping of specific surface areas. For example, Figure 8.24 analyzes the subsurface microstructure of a diamond-turned micro lens dimple on a silicon substrate before and after laser recovery. The laser recovery process, proposed by Yan et al, is a post-machining process for improving the surface integrity of the machined surface by laser melting and recrystallization of the subsurface damage layers. The details of the mechanisms, laser specifications, system constructions, and industrial applications of this technology can be found in references [38–41].

According to the micrograph in Figure 8.24(a), no microfractures were observed on the diamond-turned surface. However, the color of the dimple's surface appeared uneven, indicating the presence of surface unevenness. After the recovery with laser irradiation, the surface exhibited a uniformly bright color, indicating the surface smoothing effect of laser recovery.

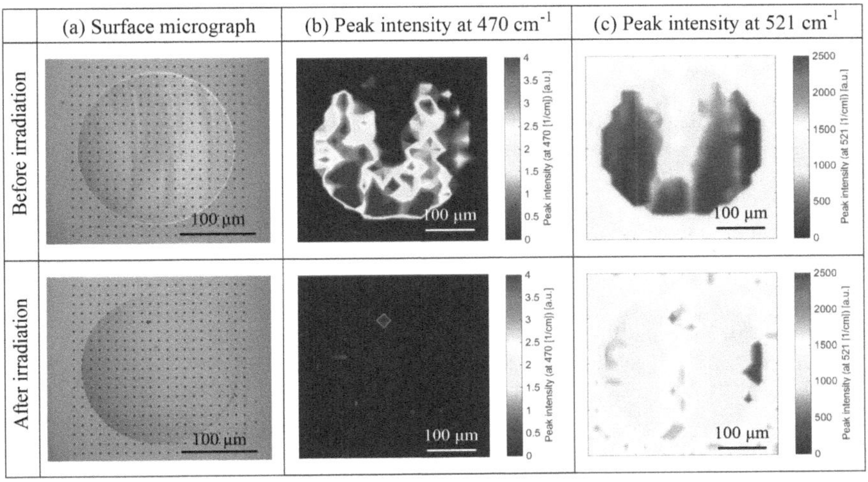


**FIGURE 8.24** Silicon dimple surface machined by diamond turning: (a) surface micrograph with Raman mapping points and Raman mapping images of (b) 470 cm$^{-1}$ (a-Si) and (c) 521 cm$^{-1}$ (c-Si) of the dimples.

Figure 8.24(a) also indicates the Raman mapping points; at each point, a Raman spectrum will be measured. The corresponding Raman mapping images of the dimple before and after laser irradiation at 470 cm$^{-1}$ and 521 cm$^{-1}$ are presented in Figure 8.24(b) and (c), respectively. In Figure 8.24(b), high-intensity amorphous Raman peaks were observed on the dimple surface after diamond turning, especially in the area where the tool cut into the material, contrasting with the peaks found on its surrounding unmachined surface. Meanwhile, in Figure 8.24(c), a low-intensity peak region of single-crystal silicon on the dimple surface is distributed in the same area as those with high-intensity amorphous phase. After laser irradiation, in the machined dimple, the amorphous peak intensity decreased significantly, while the single-crystal silicon peak intensity increased. This indicates that the amorphous silicon in the subsurface induced by diamond cutting was completely recovered to a single-crystal structure by laser irradiation, while preserving the shape and improving the surface quality.

## REFERENCES

[1] Malkin, S., Hwang, T.W. (1996) Grinding mechanisms for ceramics. *CIRP Annals* 45 (2): 569–580.

[2] Kosai, K., Yan, J. (2020) Effects of cyclic loading on subsurface microstructural changes of zirconia polycrystals in nanoscale mechanical processing. *International Journal of Machine Tools and Manufacture* 159: 103626.

[3] Yan, J., Asami, T., Harada, H., Kuriyagawa, T. (2009) Fundamental investigation of subsurface damage in single crystalline silicon caused by diamond machining. *Precision Engineering* 33 (4): 378–386.

[4] Fu, C.H., Sealy, M.P., Guo, Y.B., Wei, X.T. (2014) Austenite–martensite phase transformation of biomedical Nitinol by ball burnishing. *Journal of Materials Processing Technology* 214 (12): 3122–3130.

[5] Yan, J., Asami, T., Harada, H., Kuriyagawa, T. (2012) Crystallographic effect on subsurface damage formation in silicon microcutting. *CIRP Annals* 61 (1): 131–134.

[6] Gupta, T.K., Lange, F.F., Bechtold, J.H. (1978) Effect of stress-induced phase transformation on the properties of polycrystalline zirconia containing metastable tetragonal phase. *Journal of Materials Science* 13 (7): 1464–1470.

[7] Jeelani, S., Biswas, S., Natarajan, R. (1986) Effect of cutting speed and tool rake angle on residual stress distribution in machining 2024-T351 aluminium alloy — unlubricated conditions. *Journal of Materials Science* 21 (8): 2705–2710.

[8] García Navas, V., Gonzalo, O., Bengoetxea, I. (2012) Effect of cutting parameters in the surface residual stresses generated by turning in AISI 4340 steel. *International Journal of Machine Tools and Manufacture* 61: 48–57.

[9] Yan, J., Maekawa, K., Tamaki, J., Kuriyagawa, T. (2005) Micro grooving on single-crystal germanium for infrared Fresnel lenses. *Journal of Micromechanics and Microengineering* 15 (10): 1925–1931.

[10] Militão Dib, M.H., Otoboni, J.A., Jasinevicius, R.G. (2022) Mapping ductile-to-fragile transition and the effect of tool nose radius in diamond turning of single-crystal silicon. *The International Journal of Advanced Manufacturing Technology* 120 (1): 843–867.

[11] Ohta, T., Yan, J., Kuriyagawa, T., Kodera, S., Tomoaki, N. (2007) Prediction of subsurface damage depth of ground brittle materials by surface profiling. *International Journal of Machining and Machinability of Materials* 2 (1): 108–124.

[12] Huang, W., Yan, J. (2023) Mechanisms of tool-workpiece interaction in ultraprecision diamond turning of single-crystal SiC for curved microstructures. *International Journal of Machine Tools and Manufacture* 191: 104063.

[13] Chen, L., Tai, B.L., Chaudhari, R.G., Song, X., Shih, A.J. (2017) Machined surface temperature in hard turning. *International Journal of Machine Tools and Manufacture* 121: 10–21.

[14] Zhang, L., Hashimoto, T., Yan, J. (2021) Machinability exploration for high-entropy alloy FeCrCoMnNi by ultrasonic vibration-assisted diamond turning. *CIRP Annals* 70 (1): 37–40.

[15] Yan, J., Okuuchi, T. (2019) Chip morphology and surface integrity in ultraprecision cutting of yttria-stabilized tetragonal zirconia polycrystal. *CIRP Annals* 68 (1): 53–56.

[16] Gao, S., Wu, Y., Kang, R., Huang, H. (2018) Nanogrinding induced surface and deformation mechanism of single crystal β-Ga2O3. *Materials Science in Semiconductor Processing* 79: 165–170.

[17] Kakinuma, Y., Azami, S., Tanabe, T. (2015) Evaluation of subsurface damage caused by ultra-precision turning in fabrication of CaF2 optical micro resonator. *CIRP Annals* 64 (1): 117–120.

[18] Huang, W., Yan, J. (2021) Chip-free surface patterning of toxic brittle polycrystalline materials through micro/nanoscale burnishing. *International Journal of Machine Tools and Manufacture* 162: 103688.

[19] Warren, A.W., Guo, Y.B., Weaver, M.L. (2006) The influence of machining induced residual stress and phase transformation on the measurement of subsurface mechanical behavior using nanoindentation. *Surface and Coatings Technology* 200 (11): 3459–3467.

[20] Yan, J., Takahashi, H., Tamaki, J.I., Gai, X., Harada, H., Patten, J. (2005) Nanoindentation tests on diamond-machined silicon wafers. *Applied Physics Letters* 86 (18): 181913.

[21] Wong, L., Suratwala, T., Feit, M.D., Miller, P.E., Steele, R. (2009) The effect of HF/NH4F etching on the morphology of surface fractures on fused silica. *Journal of Non-Crystalline Solids* 355 (13): 797–810.
[22] Feng, Y., Shen, G., Zhao, Y., Liu, J., Yang, J., Xie, H., He, J. (2023) Evaluation of polishing induced subsurface damage in InAs substrates by an acid solution etching method. *Materials Science in Semiconductor Processing* 167: 107770.
[23] Chen, L., Zhang, X., Zhang, T., Lin, H.Y., Lee, S. (2000) Micro-Raman Spectral Analysis of the Subsurface Damage Layer in Machined Silicon Wafers. *Journal of Materials Research* 15 (7): 1441–1444.
[24] Yamamuro, Y., Shimoyama, T., Yan, J. (2023) Generation of nanopore structures in yttria-stabilized zirconia by femtosecond pulsed laser irradiation. *Journal of Materials Research and Technology* 23: 1155–1176.
[25] Yamamuro, Y., Shimoyama, T., Nagata, H., Yan, J. (2024) Effect of dopant concentration on femtosecond pulsed laser irradiation of yttria-stabilized zirconia for generating nanopores. *Journal of Alloys and Compounds* 980: 173596.
[26] Lodh, A., Thool, K., Samajdar, I. (2022) X-ray Diffraction for the Determination of Residual Stress of Crystalline Material: An Overview. *Transactions of the Indian Institute of Metals* 75 (4): 983–995.
[27] Harris, D.C., Bertolucci, M.D. (1989) *Symmetry and spectroscopy: an introduction to vibrational and electronic spectroscopy*, Dover Publications.
[28] Yan, J. (2004) Laser micro-Raman spectroscopy of single-point diamond machined silicon substrates. *Journal of Applied Physics* 95 (4): 2094–2101.
[29] Gogotsi, Y.G., Domnich, V., Dub, S.N., Kailer, A., Nickel, K.G. (2000) Cyclic Nanoindentation and Raman Microspectroscopy Study of Phase Transformations in Semiconductors. *Journal of Materials Research* 15 (4): 871–879.
[30] Carles, R., Zwick, A. (1993) Multiple-order Raman scattering in crystalline and amorphous silicon. *Physical Review B* 48 (9): 6024–6032.
[31] Yan, J., Asami, T., Kuriyagawa, T. (2008) Nondestructive measurement of machining-induced amorphous layers in single-crystal silicon by laser micro-Raman spectroscopy. *Precision Engineering* 32 (3): 186–195.
[32] Yan, J., Asami, T., Kuriyagawa, T. (2007) Response of machining-damaged single-crystalline silicon wafers to nanosecond pulsed laser irradiation. *Semiconductor Science and Technology* 22 (4): 392–395.
[33] Shibata, T., Ono, A., Kurihara, K., Makino, E., Ikeda, M. (1994) Cross-section transmission electron microscope observations of diamond-turned single-crystal Si surfaces. *Applied Physics Letters* 65 (20): 2553–2555.
[34] Jeynes, C., Puttick, K.E., Whitmore, L.C., Gärtner, K., Gee, A.E., Millen, D.K., Webb, R.P., Peel, R.M.A., Sealy, B.J. (1996) Laterally resolved crystalline damage in single-point-diamond-turned silicon. *Nuclear Instruments and Methods in Physics Research Section B: Beam Interactions with Materials and Atoms* 118 (1): 431–436.
[35] Puttick, K.E., Whitmore, L.C., Chao, C.L., Gee, A.E. (1994) Transmission electron microscopy of nanomachined silicon crystals. *Philosophical Magazine A* 69 (1): 91–103.
[36] Zarudi, I., Zhang, L.C. (1998) Effect of ultraprecision grinding on the microstructural change in silicon monocrystals. *Journal of Materials Processing Technology* 84 (1): 149–158.
[37] Yan, J., Takahashi, H., Tamaki, J.I., Gai, X., Kuriyagawa, T. (2005) Transmission electron microscopic observation of nanoindentations made on ductile-machined silicon wafers. *Applied Physics Letters* 87 (21): 211901.
[38] Yan, J., Sakai, S., Isogai, H., Izunome, K. (2009) Recovery of microstructure and surface topography of grinding-damaged silicon wafers by nanosecond-pulsed laser irradiation. *Semiconductor Science and Technology* 24 (10): 105018.

[39] Yan, J., Kobayashi, F. (2013) Laser recovery of machining damage under curved silicon surface. *CIRP Annals - Manufacturing Technology* 62 (1): 199–202.
[40] Niitsu, K., Tayama, Y., Kato, T., Maehara, H., Yan, J. (2018) Characterization of recrystallized depth and dopant distribution in laser recovery of grinding damage in single-crystal silicon. *Materials Science in Semiconductor Processing* 82: 54–61.
[41] Niitsu, K., Tayama, Y., Kato, T., Yan, J. (2019) Laser recovery of grinding-induced subsurface damage in the edge and notch of a single-crystal silicon wafer. *Surface Topography: Metrology and Properties* 7 (1): 015013.

# Index

Pages in *italics* refer to figures and pages in **bold** refer to tables.


3D imaging, 232

## A

abrasive grain, 16, 36–39, 97, 224, 253
abrasive machining, 16, **26**, 34–37, 208, 278
absolute encoder, 71
active vibration suppression, 76–77
additive manufacturing, **2**, 21, 46
advanced manufacturing, 1
air mount, 61, 117–118
alignment error, 43, 238, 241
alloy, 5, 60, *85*, 86, *87*, 88, 105, 130, 134, 262
  aluminum, 4, 15, 109, *110*, 113, *114*, 115, *116*, 117, 124, 235, *237*, 257
  copper, 115, *116*, 120, 122
  high entropy, 83, 135, 230, 262
  magnesium, 134–135
  nickel, 125
  titanium, 88, 134
aluminum, **3**, 4, 9, *13*, 15–16, 71, 82, 89, 105, 107, 109, *110*, 113–117, 120, 122–125, 128, 134–135, 168, 204–205, 235, *237*, 257
amorphization, 180, 207, *255*, 256
amorphous phase, 169, 175, 207, 255, 262, 274, 279
angle of incidence, 50
application
  biomedical and biochemical, 18
  electronic, 15
  optical, 1, 20, 26, 34, 51
artificial joints, 18–19
aspheric surface, 4, 28
aspherical lens, **3**, 4–7, 10, 28, 35, 52, 157, *168*, 169, 172–173, 177, 195, *203*, 204, 227
atomic force microscope (AFM), 96, *117*, *126*, 168, 176–177, 196, 202, 225–227, 232, 261–264
atomic-scale removal, 50
atomic-scale, 41
axis drives, 65

## B

backlash, 64, 66, 70
bearing, 26, *62*, **63**, 66–67, 75, 86, 196
  air, 15, 59, 64–65, 113, 227
  hydrostatic, 27, 61, 63–65
  magnetic, 65
  pneumatic, 64–65
  rolling, 61, 63
  sliding, 61–63
binder material, 86, 210
biomaterial, 18, 213
bonding strength, 86, 210–211
brass, 110, 120, 122–123, 131
brittle materials, 6, 8, 16, 28, 32, 35, 37, 45, 80, 90, 93, 99, 114, 151–152, 155–156, 159–161, 167, 186, 252–253, 256–257
brittle mode, 152–153, 155, *168*, 206
brittle-ductile transition, 97
brittleness, 53, 99, 140, 158, 165, 171, 185, 256
broadband peak, 207, 273, 276
brushless motor, 66
brushless, *66*
build-up edge, 113, 134
bulk material, 100, 254, 265
burr formation, 11, 32, 34, 55, 89, 118–119, *129–130*

## C

calcium fluoride ($CaF_2$), 83, *110*, *115*, 151, 165, 173–177, 262
camera, 1, **3**, 5, 10, 135, 140, 195
carbon diffusion, 98, 128
cast iron, 36, 60, 64, 86, 97
catalytic reaction, 86, 132
cellulose nanofiber-reinforced plastics, 143
cemented carbide, 5, 88, 130, 210
ceramics, 18, 38, 45–46, 60, 64, 86, *87*, 151, 204–205, 213, 256, 262
chatter, 101, 117–118
chemical
  action, 41
  effect, 39, 117
  reaction, 40, 97, 117, 132
  solution, 39
chemical etching, 54, 264, 269
chip adhesion, 124, 126, 130, 139
chip formation, 34, 93, 106–107, 114, 121–123, 126, 135, *152*, 162–163, 207
chip morphology, 115, 122, 214
chucking force, 73
chucking, 71–73
clamping, 54, 71–73, 236
cleavage fracture, 153, 166, 174

cleavage plane, 165, **167**, **170**, 173–174, *175*, **178**, 179, 183
coated tool, 124, 257
color-aberration compensation, 173
complicated surface, 21, 227
composite, 18, 44, 53, 60, 88, 129, 204
compound semiconductor, 177
conductive material, 44, 225
confocal laser scanning microscope, 231–232
contact force, 226
contact measurement, 223, 228, 235
contact-type, 225
continuous cutting, 81, 141, *142–143*, 145
contour error, *248*, 249
cooling effect, 141
cooling, 4, 51–52, 73, 75, 80, 90, 117, 124, 126, 139, 141–143, 256
coordinate conversion, 247
copper, **3**, 4–5, 8–9, 11, 15, 36, 41–42, 50, *55*, *74*, 82, 96, 98, 105, 107, 115, *116–117*, 120–122, 128, 135, 168, 242, *260*
crack generation, 155, 161, 216, 252
crack-free surface, 93, 153, 156, 161, 184, 196
critical chip thickness, 107, *158*, 160, 164, 184, 197–198
cross-sectional profile, *10*, 197, 201, 246
cryogenic coolant, 73
crystal anisotropy, 260
crystalline orientation, 122
crystallographic anisotropy, 165, 241
curve fitting, 275–276
curved surface, 31, 47, 90, 226
cusp, 265, *267*
cutting edge angle, 28–29, 92, 112–113, 157, 176
cutting fluid, 73–75, 118, 120, 125, 128, 135, 212
cutting speed, 32, 34, 81, 105, 109–110, 113, *121*, 124, *131–132*, 134–135, 138–141, 143–*144*, 145–146, 152, 169, *213*, 257
cyclic loading, 253–254

## D

damage depth, 265, *267*, 271, 274, 278
damping characteristics, 60–62
data stitching (data-stitching), 245, 247, *248*
defect density, 153
deformed layer, 161, 185, *186*
deliquescent material, 190
dental implant, 213
depth of cut, 28, 31, 34–35, 106–107, 109, *110*, 111–113, *114*, 129–130, 134–135, 139–140, *141*, 152, 157–159, 176, 196, 201, 255, 271, 276, 278
deviatoric stress, 166
diamond anvil test, 255
diamond cubic structure, 255, 263
  BcBN, 86–87, *88*
  PcBN, 86–87, *88*, **89**, 92
diamond tool, 80–101, 108, 115, 128–146, 157–211, 244, 259
  round-nosed, 28, 90, 112–113, 156, 158, 244
  straight-nosed, 90–92, 112–113, 157, 169, 173
diamond turning, 107–146, 156–126
diamond-coated, 130
dielectric fluid, 41
diffraction optical elements, 8
diffraction pattern, 180, 270
diffractive optical element, 50, 52, 195
digital cameras, 5, 10, 195
dilatational stress, 215
dislocation, 17, 29, 166–167, 169–170, 180, 183, 185, *186*, 188, 194, 207, 252, 256, 259–260, 262
dislocation density, 170, 255
dislocation mobility, 169, 207
dislodgement, 205, 207, *209*, 211–215
disordered layer, 194, 260
dopant concentration, 270
downward flow, 161
dry condition, 135, 175, *176*, 190
dry cutting, 118, 124, *125*, 126, *127*, 128, 176
ductile cutting, 156, 169, 184, 201, 207, *209*
ductile machining, 28–30, 83, 92, 151, 155–216
ductile material, 18, 23, 93, 105, 114, 118, 151, 154, 157, 168, 242
ductile response, 28, 90, 155, 172, 178, 182, 200
ductile-regime machining, 156
dwell time, 38, 50
dynamic interferometer, 40

## E

edge geometry, 159
edge identification, 245
edge location, 245
edge profile measurement, 243
edge radius, 32, 82, 96, 101, 106–111, 114–115, 128–130, 155, 159, 161–163, 166, 216, 230, 232
edge roundness, 106, 165–166, 245
elastic emission machining, 40–41
elastic recovery, 107, 110, 116, 143, 163, 165, 178, 183, 188
electric field-assisted machining, 132
electrical discharge machining, 41, 53
electrical machining, 41
electro chemical machining, **26**, 42
electrochemical reaction, 44
electroless plating, 52

electrolytic, 43, 54
electromagnetic force, 14, 65
electron backscatter diffraction (EBSD), 186, *187*, 188
electron beam, 9, **26**, 47–48, 228–229
electron beam polishing, 47
elliptical vibration cutting, 212, *213*
energy scaling, 154
energy-dispersive X-ray (EDS), 229–230
environmental control, 52, 68, 75
environmentally friendly, 45, 73–74
equivalent stress, 163
error compensation, 244
etch rate, 265, *267*
etching process, 49, 265
etching time, 265, 267, *268*
exit burr, 129, *130*
extruding, 162

## F

fan-shaped pattern, 181
feedback control, 26, 65, 67–68, 77
femtosecond laser, 45, 100, 269
ferrous metal, 26, 28, 82, 84, 130
  alloy steel, 120, 130
  carbon steel, 105, 109, 120, 130, 134
fine-grain layer, 254
finite element method (FEM), 155, 162–163
fixed abrasive, 34, *266*
flammability, 135
flank wear, 84, 86–88, 115, 122, 128, 137, 163, 192, 198, 230–231
flattening, 37
flood lubrication, 73
flow-type chip, 182
fly cutting, 2, 15, 26, 29, 34, 80, 121, 191
fracture toughness, 16, 18, 81, 83–85, 105, 151–152, 173, **193**, 197, 215, 257, 262, 264
freeform optics, 12–14, 223–227
Fresnel structure, 7, 28, 90
friction coefficient, 61–62, **63**, 90, 108–109, 111, 117, 134
functional surface, 34
functionality, 1, **2**, 22, 130, 252
fused quartz, 40
future perspectives, 20

## G

gallium arsenide (GaAs), 182–184, *185–186*
gas environment, 131–132
Gaussian distribution, 275
germanium, 6–8, 28, *30*, *110*, 151, 160, 164–165, 170–173, 179, 188, 206, 224, 235, 257–258
glass
  BK7, 203
  borosilicate, 151, 195–197
  fluoride, 203–204
  soda lime, 195, 197–198, *199*, 203
glass lens, 5, 35, 52, 195, 204
glass molding, 5, *6*, 10, **26**, 32, 51–52, 195, 210
glow discharge optical emission spectroscopy (GDOES), 267, *269*
gold, 9, 17, *18*, 105, 120, 128, *129–130*, 229
grain boundaries, 116, *117*, 186, 205, 208, 210, 215, 229, 260, 264
grain boundary step, 116, 121, 126, 188–189
grain refinement, 214, 216, 253–254, 259, 262, 270
grain size, 21, 84–86, **88**, 186, *187*, 214–215, 253, 270
granite, 27, 60
graphitization, 86, 98, 131–132
grating scale, 68–70
grinding wheel, 35–36
grinding, 34–38, 46–47, 53–54, 224, 252–253, 257
grooving, 30, 34, 100, 118–120, 141, 158, 184, 198
guideways, 61

## H

hard brittle material, 6, 8, 35, 45, 151–216, 252, 256
hardness distribution, 264
HDPE, 140, *141*
heat treatment, 3, 53, 121–122, 126, 253
high pressure, 63, 82, 86, 97, 169, 175, 207, 216, 254–256
high-aspect ratio, 242
high-density dislocation, 180, 185
high-efficiency, 86, 157
high-power device, 192
high-pressure phase, 166, 175, 207, 254–256
high-resolution imaging, 259
hybrid processes, 53
hydrodynamic, 40–41
hydrostatic pressure, 93, 155, 157, 159–160, 166, 169, 184, 208

## I

imprinting, **26**, 50
indentation test, 155, 161, 169, 172, 200, 255, 261, 263–264
indium phosphide (InP), 16, 177, **178**, 179–180, *181–182*

infrared material, 170
infrared optical material, 167
in-process measurement, 222, 238, 241
interfacial condition, 117
interferometer profiling, 239
intermittent cutting, 80, 141, *142*, *143*
internal stress, 52, 73, 256, 270
inverse pole figure (IPF), 186–188, *190*
ion beam, **26**, 49–50, 99–100, 259–261
ion beam figuring, 49–50
iron core, 67
ironless linear motor, 67

## K

KDP, 159, 164–165, 189–191

## L

lamellae, 115
lamellar structure, 214
lapping, 36–39
laser
  machining, 2, **26**, 45–46, 100
  polishing, 45–47
  processing, 45
  recovery, 46–48, 278
laser diode, 177
laser heating, 197
laser interferometer, 68, 227
laser irradiation, 197, *198*, 269, 278–279
laser micro-Raman spectroscopy, 271, 278
laser probe profiling, 235, 240–241
laser probe profilometer, 235
laser scale, 68–69
laser triangulation sensor, 240
lateral microcrack, 252
lateral resolution, 232, 244
lathe, 26, 59, 71–72
lattice structure, 193, 256
lens
  aspherical, 3–5, 10, 168–177
  Fresnel, 3, 7–8, 28–30, 52, 54–55, 90, 236
lens substrate, 173
light guiding plate, 3, 30
light-sensitive sensors, 177
linear motor, 27, 66–67
load capacity, 61–62, **63**, 64
load–displacement curve, 262–263
loose abrasive, 36–38
Lorentzian distribution, 275
LSI substrate, 17–18, 129
lubricating, 63, 73, 107

## M

machinability, 6, 8, 28, 42, 105, 120, 125–126, 130, 138, 140, 143, 151, 165, 177, 182–183, 186, 194–195, 201, 204–205, 210
machine structures, 59–60
machine tool, 13, 27, 33, 59–77
machining mechanism, 23, 105, 159, 195, 207
magnesium, 124, 134–135
magnetic disk, 15–16
magnetic field, 47, 65–66, 134–135, 228
maintenance, 2, 21, 62, **63**, 77, 157, 195
manufacturing
  methods, 9, 15, 204
  process, 1, 22, 25, 177, 222
  technology, 1, 20
manufacturing cost, 62
material adhesion, *120*, 124, 135, 137
material composition, 204, 253
material heterogeneity, 115–116
material property, 107, 116, 170, 177, 204
material removal mechanism, 37–38, 42, 161–162, 207
material spalling, 158
measurement platform, 236
measurement range, 225–232, 235, 240–241, 246–247, 256
measurement speed, 238
mechanical clamping, 71–72
median crack, 27, 252–253
melting, 48, 105, 124, 135, 138–139, 143, 198, 278
Merchant's model, 106
metal material, 9, 25, 50, 75, 120, 128, 153
metallic phase, 155, 255
micro groove, 8, 32, 100, 199
micro lens array, 10–12, 31, 47, 52
microcrack depth, 258–259
microfluidics, 19–20, 31, 196
microfracture, 129, 138–140, 151–153, 157, 159–160, 167, 172–176, 200, 214–215, 272, 278
micro-groove, 199
micropit, 205–206, 210
microprism, 1
microstructural change, 22, 206, 252, 265
microstructured surfaces, 11, 31
milling
  end, 32–33, 92–93
  face, 31
  side, 31
minimum chip thickness, *107*, 109, 113–114, 134
minimum quantity lubrication (MQL), 73, *74*

mirror
laser reflecting, 3, 28
off-axis, 3–4
parabolic, 4
planar, 4, 68
polygon, 2, 4, 30, 124
spherical, 3
mirror finishing, 38
misorientation angle, 186, *187*, 188
mobile phone, 5, 195
Moiré fringe, 70
mold insert, 35
molding
die, 5–6
glass, **26**, 32, 51–52, 195, 210
injection, 5, 9–10, 20, **26**, 28, 32, 50–51
plastic, 5
molding temperature, 10, 52
molecular dynamics (MD), 107, 162
monoclinic, 215–216, 256
multiphoton absorption, 45

N

nanoindentation, 169, 172, 200, 208, 225, *227*
nanoindentation tests, 172, 200, 261, 263
nanoparticle, *74*, 75, 193, 211
nanostructure, 45
near-dry machining, 73
nickel, 5, 8, 15, 52, 60, 82, 105, 120, 125
nickel plating, **3**, 50, 126, 128
pure nickel, 125–126, 128
noncontact measurement, 228, 235, 244
noncontact tool measurement, 244
noncontact type, 225
non-contact, 45, 50, 64, 240
nondestructive method, 264, 269
nonferrous metals, 28, 82, 84
nonlinearity, 278
numerical control program, 242

O

offline measurement, 222, 238
oil film, 64
on-machine measurement, 241, 278
optical components, 1, 21, 38, 50–51, 185
optical element, 1, 8, 12, 28, 35, 41, 50, 52, 186, 195
optical glass, 40, 83, 197, 200, 203–204
optical microscopy, 232
optical probe, 240
optical projection, 242, 244
optoelectronic communication, 177
orthogonal cutting, 106, 110, 119
oxygen shielding, 132

P

pad, 37–39
passive vibration suppression, 76
penetration depth, 229, 270, 273
permanent magnet, 66–67
phase transformation, 17, 22, 155, 165–166, 169, 172, 175, 207, 215–216, 252, 254–256, 259–260, 262–263, 270
phase-shifting, 239–240
photolithography, 9, 20, 54, *55*, 138, 173
photoresist, 17–18, 54, *129*, *138–140*
pileup, 185, *186*, 201
plane strain condition, 157
planning, 33–34
plasma channel, 42
plastic deformation, 21, 105, 107, 113, 152–153, 155, 161, 163, 165, 169, 178, 183, 208, 214, 242, 256, 262, 264
plastic flow, 154, 169, 200, 207, 226, 255
plastic-elastic boundary, 161
ploughing, 36, 38
plowing, 36, 110, 113–114, 162
plunge-cutting, 158, 201
PMMA, 5, *110*, 140, *141*, 142–143
polishing fluid, 40
polishing slurry, 259
polishing, 37–41
chemomechanical, 38
float, 40
mechanical, 38
polycrystalline, 53, 116, 151, 186, 229, 253
polycrystalline diamond (PCD), 11–12, 33, 53
polygon mirror, 2–3
polymers, 18, 20, 28, 138
polymetric material, 137
polytype, 192, **193**
pop-in, 263
pop-out, 263
positioning accuracy, 67
positioning, 14, 35, 43, 59, 64, 67–70, 238, 243
pre-existing defect, 159, 210
preferential slip system, 179
pressure threshold, 254
pressure, 9, 17–18, 36, 38, 40–41, 51, 63–64, 68–69, 71, 73, 75, 80, 84, 86, 97, 109, 114, 129, 133, *197*, 204, 207, 216, 254–256, 259
air, 27
hydrostatic, 93, 155, 157, 159–160, *166*, 169, 194, 207
oil, 63
pressurized air, 64
primary cracking direction, 179
probe, 223, 225, 227, *228*, 235, 240–241, 243, 265

productivity, 2, 14, 21, 80, 89
profile extraction, 246
pulse duration, 45

## Q

quantitative evaluation, 274

## R

radial crack, 178, 201, 226
rake angle, 32, 93, 110–111, 114, 140, 152, 155, 158–162, 184, 190–191, 205–209, 216, *231*, 255
Raman intensity ratio, 274–278
Raman mapping, 278–279
Raman scattering, 271, 274, 276
Raman spectra, 207, *208*, 267, *268*, 272–276, *277*
Raman spectrum, 206–208, 272–273, 276, 279
rare metal, 10, 128–129
reaction–bonded silicon carbide (RB–SiC), 151, 204–212
reactive ion etching, 54
real-time control, 77
real-time monitoring, 236
recrystallization, 47, 121, 278
reflecting mirror, 4, 28, 41, 121
relative tool sharpness, 161, *162*
relief angle, 93, 111, 114–115, 159, 164–165, 231
residual stress, 17, 21–22, 43, 51, 60, 121, 125, 161, 252, 254, 256–257, 262, 270–271, 278
residual tensile stress, 161, 264
resolidified layer, 43–44, 197
resolution, 5, 12, 20, 27, 68–70, 101, 113, 195, 222, 229–232, 235, 239–240, 242–245, 259–260
retro-reflective film, 15, 34
ribbon chip, 128
rotary cutter, 31
rotary encoder, 70–71
rotary table, 2, 70, 211, 244–245
rotating tool, 29, 101, 258
rotation accuracy, 65
rotator, 64–65, 70, 101
round-nosed tool, 90–92, 111–113, 156–159, 208
rubbing, 89, 107, 114–115, 162
runout error, 64

## S

sapphire, 40
scanning electron microscope (SEM), 11, 18, 39, 43–44, 46, 55, 74, 82–83, 96, 99–100, *115*, 119–127, 173–179, 203–206, 225–232, 259–261, 264
scanning probe microscope (SPM), 225
scanning tunneling microscope (STM), 225
scratch speed, 201
secondary electron (SE), 228
semiconductor material, 16, 155, 167, 170, 177, 192
semiconductor wafer, 16–17
semiconductor, 16–17, 35, 121
service life, 6, 19, 62, **63**
servo motor, 65–67
shape flexibility, 21
shaping, 33–34
shear angle, 114–115
shearing, *94*, **95**, 107, 110, 162
side burr, *119*, 129, *130*
silicon
  amorphous silicon (a–Si), 169, 263, *272*, 274, 279
  crystalline silicon (c–Si), 272–274, *275*
  single-crystal silicon, *13*, *17*, 83, 153, 157, *158*, 160, 165–168, 171, 192, **193**, 254–256, 279
silicon carbide (SiC), **3**, 5, 11, *12*, 16, 45, 52–53, 60, *115*, 151, 192, **193**, 194–195, 204–212, 233–234, 260
silicon dioxide, 39, 197, 204
single crystal material, 151–216
single-crystal silicon carbide (SiC), 53, 192–193
single-point diamond turning (SPDT), 16, 26, 35
sintered diamond, 53, 84
size effect, 109, *153*, 263
slanted polishing, 257–259
slides, 18, 26, 33, 59, 61, 113, 238
slip band, 180
slipping system, 165
sloped surface, 226
slurry, 39–41
small tool polishing, 258
softening point, 9, 137, 143
solid lubricant, 211
spark, 41–43
spatial resolution, 230–231, 235, 260
specific energy, 109, *110–111*
speed capacity, 62, **63**
spindles, 26, 59, 61, 65, 80, 113, 238
sputtering, 49, 99
stator, 64, 66, 70
steep slope, 243
steep surface, 227
stepper lens, 173
stiffness, 4, 27, 35, 60–62, 64–65, 111, 204–205, 213
straight-nosed tool, 90–92, 112–113, 126, 157, 169, 173
strain energy, 154
stress concentration, 19, 75, 152, 155, 158, 166
stress state transition, 155
stylus profiling, 223, 235, 238, 243

stylus profilometry, 223, 226
subsurface
  damage, 17, 23, 29, 32, 159–160, 163, 169, 174, 190, 193–195, 252–265, *267*, 269–272, 274, 278
  defect, 41–42, 169
  metrology, 22–23, 252, 254, 257, 269
  microcrack, 29, 42–43, 208, 252–253, 257–260, 272
  structural change, 169, 252, 265, 271
surface amorphization, 256
surface defect, 41–42, 46, 130, 136–137, 143, 145, 169
surface energy, 127, 154
surface integrity, 21, 90, 121, 151, 165, 253, 278
surface metrology, 21–23, 222, 237, 240, 252–254, 257, 269
surface morphology, *142*, 180–181, *185*, 201, *203*, 205
surface quality, 1, 9, 17, 19–21, 27–28, 32, 35, 42, 44, 52–54, 86, 90, 97, 105, 110, 114, 121, 126, 130, 138, 151, 164–165, 199, 213, 235, 279
surface roughness, 4–6, 11–12, 15–16, 18, 21–23, 25, 28, 31–32, 34, 36, 48, 54, 84, 86–87, 89, 94–95, *98–99*, 105, 111–116, 118, 121–122, 125–126, 129–131, 143, *145*, 161, 164–165, 168, 172–173, 177, 182, *184*, 188–190, 199–200, 204, 224–225, 230, 233, 252, 265, 272
surface structure, 14, 31, 47, 52, 99
surface tension, 46
surface topography, *14*, 90, 99, 118, *125*, 188, *203*, 232, 265, 272
sustainability, 22
system configuration, 244

## T

tearing, 140, 143, 145–146
TEM, 169–170, 174–175, 180, 186, 214–215, 255, 259–261, 276
temperature, 5, 10, 42, 50–52, 59, 64, 68, 73, 75–76, 80
  distribution, 75, 90
  fluctuation, 76
  rise, 64, 117, 141, 153, 164, 176, 255
  stability, 75
tensile stress, 105, 139–140, 146, 152, 155, 160–161, 173, 179, 215, 264
tetragonal, 173, 189, 192, 215–216, 256, 270
thermal conductivity, 75, 86, **88–89**, 105, 134, 143, 170, 175–176, **193**, **205**, 213
thermal control, 75
thermal deformation, 73, 75, 117, 236, 238
thermal effect, 45, 176, 197, 201, 255
thermal erosion, 41
thermal expansion coefficient, 64, 173, 175–176, **205**
thermal expansion, 60–62, 64, 75, 86, 105, 117, 143, 173, 175–176, 195, **205**
thermal plasticity, 140
thermal stability, 60, 62, 86, **88**, 205
thermal-chemical reaction, 204
threshold depth, 159
titanium, 88–89, 134, 210
tool apex, 245, 247
tool electrode, 41, 53
tool feed rate, 28, 125, 128–129, 137–144, 156, 158, 206, 214
tool feed, 28–29, 31–32, 84, 111–113, 125, 128–129, *136–137*, 139–143, *144*, 156–158, 175–176, 184, 205–206, 214, 225, 272
tool mark, 12, 136, 199, 242
tool metrology, 222
tool profile error, 242, 246, *247*, 249
tool vibration, 83, 117, 213, 238
tool wear, 28–29, 73, 75, 83, 88, 94, 97, 111, 114–115
tool-swinging cutting, 193, *211*, 212
tool-workpiece interface, 211
tool-workpiece transcription, 242
topographic map, 225
topographical imaging, *245*
transmission electron microscope, 259
truing/dressing, 36
tungsten carbide (WC), **3**, 11, 33, 52, 81, 86–89, 92, 96, 101, 151, 195, 210–213
twin boundary, 188–189

## U

ultrahigh accuracy 3D profilometer (UA3P), 226–227, *228*
ultraprecision cutting, **3**, 26, 82, 89, 96, 151, 188
ultraprecision machine tool, 9, 27–28, 59, *60*, 62, 64, 66, 68–69, 75, *76*, 242
ultraprecision machining
  definition, 25
  methods, 25–55
ultraprecision metrology, 22, 96
ultrasonic impact, 137
undeformed chip thickness, 28–32, 91–92, 107, 126–128, 146

## V

vacuum, 9, 49, 52
  chuck, 27, 71, *72*, 73, 121, 192, 228
  pressure, 71, 73
  pump, 71
vertical resolution, 232, 239–240, 244–245
vibration isolation, 26, 35, 59, 240

vibration isolator, 76
vibration, 59, 61–63, 65, 76–77, 113, 115
vibration-assisted, *43*, 193
viscoplasticity, 201
viscous flow, 201–202
volumetric shrinkage, 51

## W

wafer planarization, 38
wavelength conversion, 189
white light interferometer, 54, 188, 233–235, *244*, *248*
wide bandgap semiconductor, 192
wide-angle tool, 242, 244, 247, *248*

## X

X-ray, **3**, 229, 269–270

## Y

Yttria-stabilized zirconia, 269–270

## Z

zinc selenide (ZnSe), **3**, *115*, 151, 165, 185–186, *187*, 229, 262
zinc-blende lattice, 180
zirconia, 18, 45, *46*, 60, 213, 253, *254*, 256, 264, 269–270

FSC
www.fsc.org
MIX
Papier | Fördert gute Waldnutzung
FSC® C083411

Kolibri 360 GmbH
Ferdinand-Jühlke-Straße 7
99095 Erfurt, Deutschland
produktsicherheit@kolibri360.de